Optics of Cosmic Dust

NATO Science Series

A Series presenting the results of scientific meetings supported under the NATO Science Programme.

The Series is published by IOS Press, Amsterdam, and Kluwer Academic Publishers in conjunction with the NATO Scientific Affairs Division

Sub-Series

I. Life and Behavioural Sciences	IOS Press
II. Mathematics, Physics and Chemistry	Kluwer Academic Publishers
III. Computer and Systems Science	IOS Press
IV. Earth and Environmental Sciences	Kluwer Academic Publishers
V. Science and Technology Policy	IOS Press

The NATO Science Series continues the series of books published formerly as the NATO ASI Series.

The NATO Science Programme offers support for collaboration in civil science between scientists of countries of the Euro-Atlantic Partnership Council. The types of scientific meeting generally supported are "Advanced Study Institutes" and "Advanced Research Workshops", although other types of meeting are supported from time to time. The NATO Science Series collects together the results of these meetings. The meetings are co-organized bij scientists from NATO countries and scientists from NATO's Partner countries – countries of the CIS and Central and Eastern Europe.

Advanced Study Institutes are high-level tutorial courses offering in-depth study of latest advances in a field.
Advanced Research Workshops are expert meetings aimed at critical assessment of a field, and identification of directions for future action.

As a consequence of the restructuring of the NATO Science Programme in 1999, the NATO Science Series has been re-organised and there are currently Five Sub-series as noted above. Please consult the following web sites for information on previous volumes published in the Series, as well as details of earlier Sub-series.

http://www.nato.int/science
http://www.wkap.nl
http://www.iospress.nl
http://www.wtv-books.de/nato-pco.htm

Series II: Mathematics, Physics and Chemistry – Vol. 79

Optics of Cosmic Dust

edited by

Gorden Videen
Army Research Laboratory,
Adelphi, Maryland, U.S.A.

and

Miroslav Kocifaj
Astronomical Institute,
Slovak Academy of Sciences,
Bratislava, Slovak Republic

Springer-Science+Business Media, B.V.

Proceedings of the NATO Advanced Research Workshop on
Optics of Cosmic Dust
Bratislava, Slovak Republic
16–19 November 2001

A C.I.P. Catalogue record for this book is available from the Library of Congress.

ISBN 978-1-4020-0820-7 ISBN 978-94-010-0628-6 (eBook)
DOI 10.1007/978-94-010-0628-6

Dedication

This workshop was held in the shadow of the terrible events of September 11, 2001. At the time, feelings and fears were especially strong, and these were exacerbated by another tragic event: American Airlines flight 587 crashed in Rockaway NY, on November 12. Because of these events, it was not clear who would be able to attend or even whether the workshop would be cancelled altogether. The events were certainly on the minds of all the members of the workshop, and were a cloud hanging over us, not knowing when the next attack would come, and if they might personally affect some of the attendees. Although unpleasant, it is important to experience such times as it forces us to remember and appreciate what we have, the preciousness of life and the bonds we have made across the world. We are a diverse group of researchers representing various cultures and nationalities, whose only tenuous link is our study of various aspects of dust, but in reality, there is very little to distinguish us. We dedicate this book to both the victims and the heroes, to their memory, and to our memories.

TABLE OF CONTENTS

Characterization Methodologies

Backscatter Polarization

Dynamics

Preface

When light and other electromagnetic radiation interacts with dust or other particles, it is scattered in all directions. These scattering interactions are relevant in diverse fields of science: dust affects communications, specifically in battlefield environments; dust particles affect the output of aerosol monitors used to detect both pollutants and chemical and biological warfare agents; and dust particles are an important astrophysical component making up a large part of the dark matter in the universe, and the electromagnetic forces exerted on these particles may play an important role in bringing these particles together to form larger astronomical bodies. On 16-19 November 2001, a NATO Advanced Research Workshop was held in Bratislava, Slovakia on the "Optics of Cosmic Dust." The aim was to focus researchers' attentions, not only on the primary astrophysical needs, but also to draw researchers within the astrophysical community to the needs of some other fields for which their tools are applicable.

The beginning of the workshop focused on the chemical and morphological properties of dust particles, both cosmological and terrestrial and some of the techniques used to gain the information. For instance it is well known that the shape of a particle affects how it scatters light, but it is almost completely unknown what the effects of specific shape irregularities are. For instance, Figure 1 shows three very different scattering systems. The cluster of *B. subtilis* is used as a simulant for anthrax. The cluster of spheres is a system that can be more easily characterized and modeled. Dust particles are a common atmospheric and astronomic constituent. Although these systems have fundamentally different shapes and we expect that certain properties of their scattering signals are fundamentally different, we are currently unable to differentiate the particles from their scattering signals. Part of the problem is that theories and computational codes to calculate the scatter from any non-symmetric particle system have only been developed in the last decade and are extremely computationally expensive. Since the field is very much in its infancy, the necessary initial studies have yet to be conducted.

To illustrate some of our current inadequacies, Figure 2 shows a spatial distribution of the intensity of light scattered from one of the particle systems shown in Figure 1. Although we can reproduce certain features of the scattering using theoretical models, there is currently little that can be said about the particle system from the scattered light. For example, we are currently unable even to determine what type of system in Figure 1 the scatter of Figure 2 might belong.

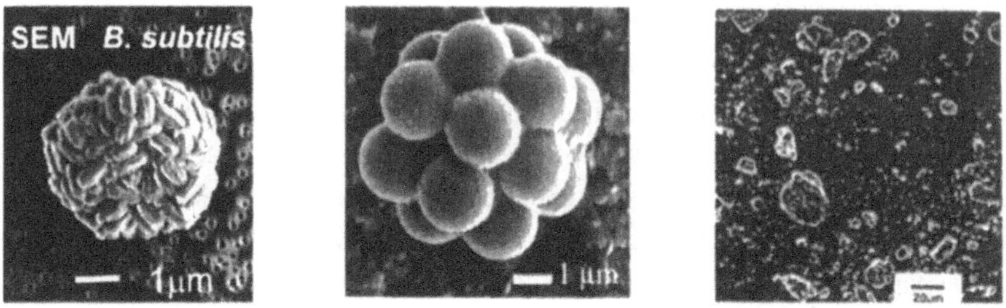

Figure 1: Three irregularly shaped particle systems of interest: a cluster of *B. subtilis* spores that is an anthrax simulant (left); a cluster of spheres (middle); and dust particles (right).

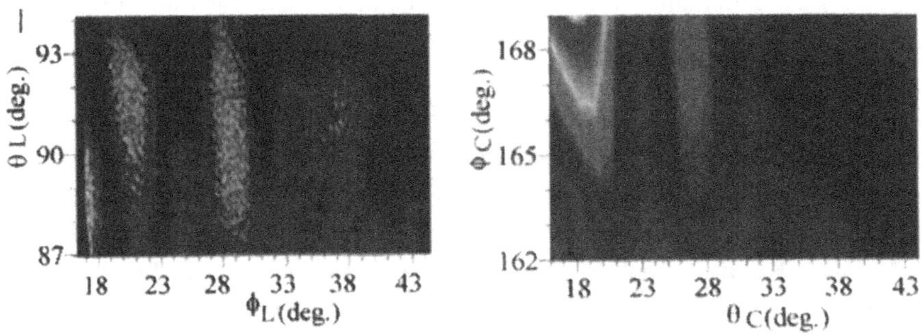

Figure 2: Experimental (left) and Theoretical (right) intensities of the light scattered by a non-spherical particle system similar to what is shown in Figure 1. Differences result because the exact orientation and morphology of the system are unknown. If you guessed that the scatter came from a cluster of spheres, you would be correct.

One focus of the NATO ARW is to explore what computational techniques are now available to address our needs. However, in order to model the scatter from particle systems, it is necessary to know what types of particles we may be dealing. The first portion of the workshop and this book deals with this topic. Although we have only discussed a few simple particle types relevant to bio-detection, many other particle systems were discussed at the workshop, especially chemically heterogeneous systems and fractal-like systems, both of which are commonly present in the interstellar medium and especially relevant for pollution monitoring.

Ideally, we would like to characterize a particle system by its scattered light and by its spectrum. In practice, things are much more difficult, in part because of noise and incomplete data sets, and also because the problem is ill-posed. At present we hope to be able to partially characterize particle systems; i.e., we are focusing on methods to find properties that will allow us to characterize a particle system within a specified degree of confidence. Because the field is in its infancy we are exploring methodologies that may allow us to accomplish this. For instance, the structure of the features in Figure 2 provides information on the size and asymmetry of the particle system. These particular features may be used to identify particles of interest.

Particular properties and features of particle systems may provide us with clues to particle characterization. One property of light is its polarization. Interaction with dust and other irregular particles tends to mix or degrade the polarization state of the light. A transmitted signal, or image, can be degraded by the particles it interacts with before being detected. This degradation also tends to affect this polarization, so in some cases it is possible to improve a signal or image by analyzing the polarization properties of the light. One topic of which there was enormous interest at the NATO ARW was the negative polarization of light scattered in the near-backward direction. Four chapters in this book address this effect directly through astronomical observations, experiments, theory, and modeling. Special attention was devoted to orbital evolution of cosmic dust particles under the action of electromagnetic radiation. Two chapters in this book deal with this topic: the first one discusses the Poynting-Robertson effect, which is commonly used in orbital simulations, the second one concentrates on more general equations of motion for arbitrarily shaped dust particle.

Details of these and other advances are to be found in this volume.

Scientific Organizing Committee

Gorden Videen	Miroslav Kocifaj	Bo Gustafson
Alfons Hoekstra	Jozef Klačka	Karri Muinonen

Local Organizing Committee

Miroslav Kocifaj	L'ubomír Turňa	Jozef Klačka
Andrea Chovancová	Štefan Gajdoš	

Primary support for the NATO ARW was provided by the NATO Science Committee. In addition, this material is based upon work supported by the European Research Office of the US Army under Contract No. N68171-01-M-5712. Any opinions, findings and conclusions or recommendations expressed in this material are those of the author(s) and do not necessarily reflect the views of the European Research Office of the US Army. In fact, since science continually evolves they may not even express those of the author(s) anymore. Additional support was provided by the Astronomical Institute FMPI, Comenius University. The meeting took place at the Hotel Družba, and we are thankful of the management and staff for their hospitality. We are especially thankful of our Scientific and Local Organizing Committees and especially of Jozef Klačka. We are especially grateful to our families for their patience, understanding and support.

May 2002

IN THE KITCHEN OF DUST MODELING

N.V. VOSHCHINNIKOV
Sobolev Astronomical Institute, St. Petersburg University,
Universitetskii prosp., 28, 198504 St. Petersburg, Russia,
e-mail: nvv@astro.spbu.ru

Abstract. Dust grains have been detected in various astronomical objects. Interpretation of observations of dusty objects includes three components: 1) determination of the materials which can exist in the solid phase and the measurements or acquisition of their optical constants; 2) selection of the light scattering theory in order to convert the optical constants into the optical properties of particles; and 3) the proper choice of the object model which includes, in particular, the correct treatment of the radiative transfer effects. The current state of the components of dust modeling and the reliability of information obtained on the cosmic dust from transmitted, scattered and emitted radiation are discussed.

1. Introduction

1.1. OBSERVATIONS

Dust grains have been detected in almost all astronomical objects from the local environment of the Earth to very distant galaxies and quasars. The interaction of radiation with grains includes two main processes: dust grains *scatter* and *absorb* radiation. The scattered radiation has the same wavelength as the incident one and can propagate in any direction. The radiation absorbed by a grain is transformed into its thermal energy and the particle *emits* at wavelengths usually longer than the absorbed radiation. Both processes contribute to *extinction* when the radiation from celestial bodies is attenuated by the foreground dust in the line of sight, i.e.

Extinction = scattering + absorption.

In general, it is possible to investigate the processes of extinction, scattering and emission of radiation by cosmic dust. Extinction is observed

G. Videen and M. Kocifaj (eds.), Optics of Cosmic Dust, 1–36.

when the light is scattered at the scattering angle $\Theta = 0°$ (forward-transmitted radiation). Corresponding observational phenomena are interstellar extinction and polarization. In the case when the scattering dominates, an observer sees the radiation scattered at different angles from $\Theta = 0°$ to $\Theta = 180°$. The scattered radiation comes from the comets, zodiacal light, reflection nebulae, circumstellar shells, galaxies. Dust emission occurs in the H II regions, circumstellar shells, interstellar clouds, galaxies, etc.

1.2. MODELING

Interpretation of observations of dusty objects can be divided into three steps. By analogy with cooking, the "kitchen of dust modeling" would outline three primary factors to prepare this dish:

1. laying-in *provision*
   The primary task is to find elements which can be converted into the solid species in the circumstellar/interstellar conditions and to determine the resulting *materials.* The next task is to find or to measure *optical constants* (refractive indices) of the materials under consideration.
2. choice of *equipment*
   Selection of *light scattering theory* ("equipment") is an essential aspect in dust modeling. The chosen method must provide the possibility of reproducing the most significant features of the observational phenomenon and to work rather fast in order to give results in a reasonable time.
3. *cooking*
   This most important part of the procedure is related to the skill of the cook (modeler) and includes not only a selection of provision and equipment but also a proper choice of the method of cooking — *object modeling.*

Lastly, one needs to taste the prepared dish, i.e. to compare the model with observations. The latter are performed with a limited accuracy which imposes a corresponding limitation on the claims of the model. From another point of view, a very complicated and detailed model with many parameters is ambiguous in principle. Further complicating the model, one should not forget about the principle of optical equivalence introduced by George Gabriel Stokes 150 years ago:

> *It is impossible to distinguish two beams which are the sum of non-coherent simple waves if they have the same Stokes parameters.*

So, a judicious restriction on the detailed elaboration of different components in dust modeling should be found.

2. Abundances and optical constants

2.1. ELEMENT ABUNDANCES AND DEPLETIONS

Ultraviolet (UV) and optical absorption-line studies have shown that the interstellar (gas-phase) abundances of many elements are lower than *cosmic (reference) abundances*. The rest of the elements is assumed to be locked in solid particles. The depletion of an element X is defined by

$$D_{\rm X} = \left[\frac{\rm X}{\rm H}\right]_{\rm g} \bigg/ \left[\frac{\rm X}{\rm H}\right]_{\rm cosmic}, \tag{1}$$

where [X/H] is an abundance of element X relative to that of hydrogen. Here X (or $N_{\rm X}$) and H (or $N_{\rm H} = N_{\rm HI} + 2N_{\rm H_2}$) are the column densities of an element X and hydrogen in a given direction. The abundances by number are usually expressed as the number of X atoms per 10^6 hydrogen nuclei (particles per million, ppm, hereafter).

For a long time, the reference abundances were assumed to be equal to solar abundances. However, Snow and Witt [49] found that many species in cluster and field B stars, and in young F and G stars were significantly underabundant (by a factor of 1.5–2.0) relative to the Sun. The old (solar) and new (stellar) abundances for the five main elements forming cosmic dust grains are as follows: C (398ppm/214ppm), O (851ppm/457ppm), Mg (38ppm/25ppm), Si (35.5ppm/18.6ppm) and Fe (46.8ppm/27ppm). New abundances limited the number of atoms incorporated into dust particles. For example, the dust-phase abundances in the line of sight to the star ζ Oph are 79 ppm for C, 126 ppm for O, 23 ppm for Mg, 17 ppm for Si and 27 ppm for Fe. The most critical situation occurs for carbon which is the main component of many dust models. The inability to explain the observed interstellar extinction using the amount of carbon available in the solid phase resulted in so called "*carbon crisis*" which has not been resolved up to now.

2.2. REFRACTIVE INDICES AND THEIR MIXING

The complex refractive indices (m) or dielectric functions (ε) of solids are called *optical constants*. The refractive index is written in the form $m = n(1+\varkappa i)$ or $m = n + ki$, where $k = n\varkappa \geq 0$. The sign of the imaginary part of the refractive index is opposite to that of the time-dependent multiplier in the presentation of fields. Note that in the book of van de Hulst [52] the refractive index is chosen as $m = n - ki$ whereas in the book of Bohren and Huffman [3] as $m = n + ki$.

The physical sense of n and k becomes clear if one considers the solution to the wave equations in an absorbing medium which is performed in the Cartesian coordinate system xyz. For an electric field propagating, for example, in the z-direction, we have

$$\vec{E} = \vec{E}_0 \exp\left(-\frac{\omega}{c}kz\right) \exp\left[-i\omega\left(t - \frac{nz}{c}\right)\right], \tag{2}$$

where $\omega = 2\pi\nu = 2\pi c/\lambda$ is the circular frequency, c the speed of light, t time. As seen from Eq. (2), the imaginary part k (often named the *extinction coefficient or index*) characterizes damping or absorption of the wave. The real part n (the *refraction index*) determines the phase velocity of the wave in the medium, $v_{\rm phase} = c/n$.

The real and imaginary parts of the optical constants are not independent and may be calculated one from another using the Kramers–Kronig relations. When applied to n and k, the relations are

$$n(\omega) = 1 + \frac{2}{\pi}\mathcal{P}\int_0^\infty \frac{\Omega k(\Omega)}{\Omega^2 - \omega^2}\,\mathrm{d}\Omega, \tag{3}$$

$$k(\omega) = -\frac{2\omega}{\pi}\mathcal{P}\int_0^\infty \frac{n(\Omega)}{\Omega^2 - \omega^2}\,\mathrm{d}\Omega, \tag{4}$$

where $\mathcal{P}$ denotes the principal part of the integral. Equations (3) and (4) allow one to make some conclusions on the behavior of the optical constants in different wavelength ranges. In particular, it is impossible to have a material with $k = 0$ at all wavelengths because in this case the radiation does not interact with the material ($n = 1$ everywhere).

The optical constants for amorphous carbon plotted in Fig. 1 allow us to understand the relation between the real and imaginary parts of the refractive indices. In particular, the limiting and asymptotic values of n and k are clearly seen. Note that the absorption features appear as loops in the $k - n$ diagrams. In the last few years, special measurements of the optical constants for cosmic dust analogues have been made. These data are being collected in the database of optical constants for astronomy (Jena–Petersburg Database of Optical Constants, JPDOC; see [20], [25] for details).

The conditions in which cosmic dust grains originate, grow and evolve should lead to the formation of heterogeneous particles with complicated structure. The problem of electromagnetic scattering by such composite particles is so difficult that a practical, real-time solution is currently unfathomable, in particular keeping in mind the unknown real structure of grains. Therefore, the thought of obtaining the optical properties of heterogeneous particles using homogeneous particles with *effective* dielectric functions $\varepsilon_{\rm eff}$ found from a mixing rule (generally called an Effective Medium Theory; EMT) is an extremely attractive proposition.

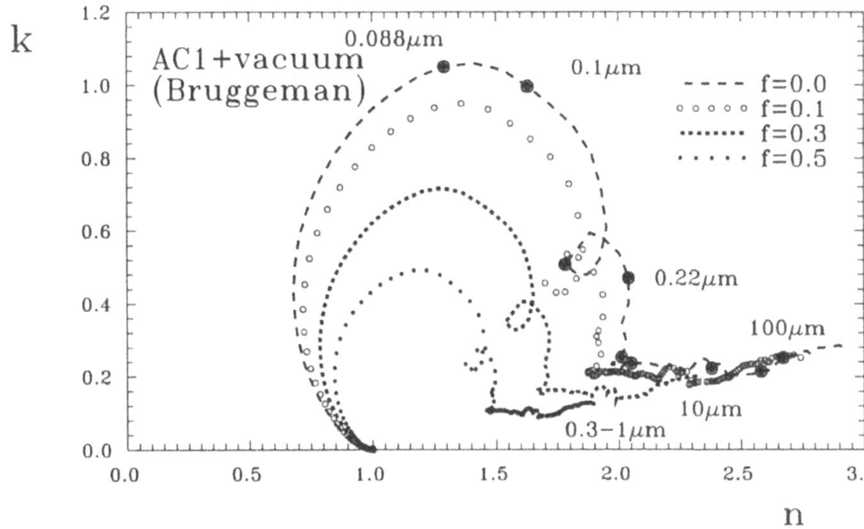

Figure 1. Refractive indices of amorphous carbon calculated with the Bruggeman mixing rule for a different fraction of vacuum (f=0–0.5). The values of wavelengths in μm are indicated.

Many different mixing rules exist (see [8] and [47] for a review). They are rediscovered from time to time and sometimes can be obtained one from another. The most popular EMTs are the classical mixing rules of (Maxwell) Garnett and Bruggeman. In the case of spherical inclusions and a two-component medium the effective dielectric constant may be calculated easily from the dielectric permittivities ε_1, ε_2 and volume fractions f, $1-f$ of the components. The Garnett rule assumes that one material is a matrix (host material) in which the other material is embedded. It is written in the following form:

$$\varepsilon_{\rm eff} = \varepsilon_2 \left[1 + \frac{3f \frac{\varepsilon_1 - \varepsilon_2}{\varepsilon_1 + 2\varepsilon_2}}{1 - f \frac{\varepsilon_1 - \varepsilon_2}{\varepsilon_1 + 2\varepsilon_2}} \right]. \tag{5}$$

When the roles of the inclusion and the host material are reversed, the inverse Garnett rule is obtained

$$\varepsilon_{\rm eff} = \varepsilon_1 \left[1 + \frac{3(1-f) \frac{\varepsilon_2 - \varepsilon_1}{\varepsilon_2 + 2\varepsilon_1}}{1 - (1-f) \frac{\varepsilon_2 - \varepsilon_1}{\varepsilon_2 + 2\varepsilon_1}} \right]. \tag{6}$$

The Bruggeman rule is symmetric with respect to the interchange of materials

$$f \frac{\varepsilon_1 - \varepsilon_{\rm eff}}{\varepsilon_1 + 2\varepsilon_{\rm eff}} + (1-f) \frac{\varepsilon_2 - \varepsilon_{\rm eff}}{\varepsilon_2 + 2\varepsilon_{\rm eff}} = 0. \tag{7}$$

Figure 1 shows variations of the refractive indices of amorphous carbon calculated with the Bruggeman mixing rule for a different fraction of vacuum.

A special question is the range of applicability of the EMTs. A general conclusion made from calculations (see [53], [67]) is that an EMT agrees well with the exact theory if the inclusions are Rayleigh and their volume fraction f is below $\sim$ 40–60%. In the case of non-Rayleigh inclusions, apparently, f should not exceed $\sim$10%. Kolokolova and Gustafson [30] performed a comprehensive study of the possibility of applying nine mixing rules comparing calculations for organic spheres with silicate inclusions using microwave analog experiments. They recommend the use of EMTs if the volume fraction of inclusions is not more than 10%.

3. Light scattering theories

When the optical constants are chosen, they can be converted into *optical properties* of particles: various cross-sections, scattering matrix, etc. using a light scattering theory. Theories of light scattering by particles may make it possible to calculate the following:
1) intensity and polarization of radiation scattered in any direction;
2) energy absorbed or emitted by a grain and to find its temperature;
3) emission spectra of dusty objects;
4) radiation pressure force on dust grains which frequently determines their motion.

Theoretical approaches to solve the light scattering problem are divided into *exact methods* and *approximations* by their dependence on the following:
1) the size parameter $x = 2\pi r/\lambda$, where r is the typical size of a particle (e.g., the radius of the equi-volume sphere) and λ the wavelength of incident radiation in the surrounding medium;
2) the module of the difference between the refractive index and unity $|m - 1|$;
3) the phase shift $\rho = 2x|m - 1|$.
Approximations may be applicable if at least two of these quantities are much smaller or much larger than unity [52]. In particular, approximate approaches allow one to estimate the wavelength dependence of extinction cross-sections easily. The latter determine, for example, interstellar extinction $A(\lambda)$ (see Sect. 5.1). Multi-color observations show that $A(\lambda) \propto \lambda^{-1}$ in the visible part of the spectrum and no approximation predicts such a wavelength dependence. Therefore, astronomers in general are doomed to work with exact methods. Spherical grains do not explain the interstellar polarization and another feature required of the theory is the ability to treat non-spherical particles with sizes close to or larger than the radiation wavelength.

At present, only a few methods satisfy astronomical demands and three of them are widely used in astrophysical modeling. These are the separation of variables method for spheroids, the T-matrix method for axially symmetric particles, and some modifications of the method of momentum (and first of all the discrete dipole approximation, DDA).

The current state of methods and techniques of calculating light scattering by non-spherical particles are described in special issues of *Journal of Quantitative Spectroscopy and Radiative Transfer* [23], [32], [37], [54], review papers [27], [68] and the collective monograph [36]. A detailed description of different approximations can be found in [27], [29] and [40].

4. Objects' models

In the modeling of dusty objects one needs to select not only optical constants and a light scattering approach, but also an appropriate model of the object. The model of a dusty object includes an appropriate choice of the spatial distribution of scatterers (dust grains) and illuminating sources and correct treatment of radiative-transfer effects.

Radiative transfer methods are rather conservative and making changes or modifications usually require tremendous efforts. As a result, the *radiative transfer code available determines the final result of modeling,* but different radiative transfer codes may give different results. During its long history, radiative transfer theory passed through several stages: analytical, semi-analytical and numerical. The problems which can be solved analytically are very simple ones like the fluxes from a sphere found in the Eddington approximation assuming grey opacity and the spherical phase function. Modern observational techniques give the spectral energy distributions, images and polarization maps of very complicated objects like fragmented molecular cloud cores, circumbinary and circumstellar disks. These data cannot be modeled without very complicated radiative transfer programs, and consideration of polarization usually requires application of Monte Carlo methodologies.

Some characteristics of radiative transfer programs created during the last 25 years and their applications to the interpretation of observed data on dusty cosmic objects are collected in Table 7 in [58]. The major part of the work mentions that these are based on two methods:
1) iterative schemes to solve the moment equations of the radiative transfer equation and
2) Monte Carlo simulation.
Standard applications include the following: circumstellar shells and envelopes around early (pre-main-sequence) and late-type stars and young stellar objects, reflection nebulae, interstellar clouds and globules, diffuse galactic light and in recent years galaxies and active galactic nuclei.

5. Interstellar extinction and polarization

5.1. OBSERVATIONS

Observational analysis of interstellar extinction and polarization is twofold: "in depth" and "in breadth". The first involves examination of the *wavelength dependence* and gives information about the properties of interstellar grains. The second includes the study of the distribution of dust matter and relates to work on galactic structure.

The wavelength dependence of *extinction* $A(\lambda)$ is rather well established in the range from near-infrared (IR) to far-UV. It has often been represented using simple analytical formulae. The most recent analytical fitting of the extinction curves was obtained in [13]. Observations in the far-UV show that the growth of the interstellar extinction continues almost up to the Lyman limit [42]. At wavelengths shorter than 912 Å, the extinction is dominated by photo-ionization of atoms, not by the scattering and absorption by dust. In the extreme UV (λ 100–912 Å), almost all radiation from distant objects is "consumed" by neutral hydrogen and helium in the close vicinity of the Sun. In the X-range, photo-ionization of other abundant atoms (C, N, O, etc.) becomes important. The resulting wavelength dependence of extinction in the diffuse interstellar medium is plotted in Fig. 2. It shows extinction cross-sections $\sigma_{\rm d}(\lambda)$ related to the column density of H-atoms, which can be found from the following expression (see [58] for details):

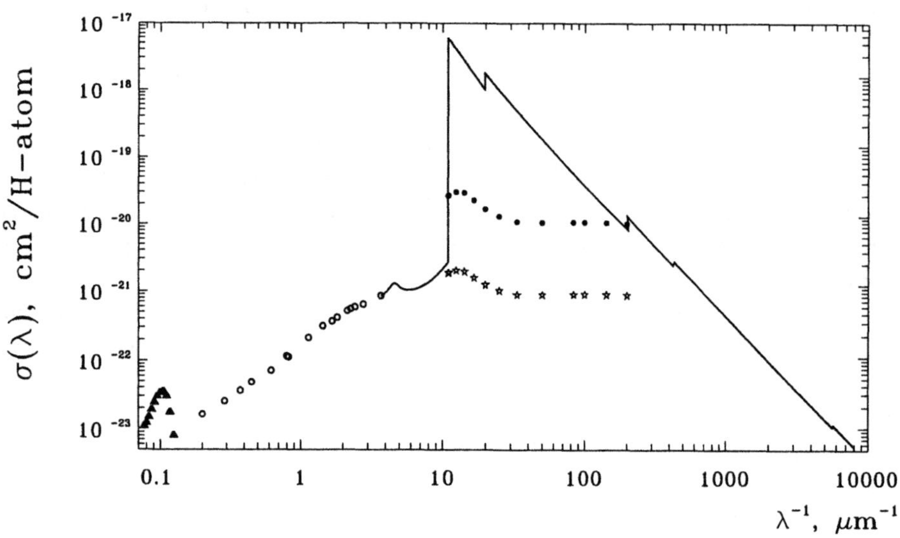

Figure 2. Extinction in the diffuse interstellar medium. The sources of the data are given in Table 9 in [58].

$$\sigma_{\rm d}(\lambda) = 4.18 \times 10^{-22}\,\frac{A(\lambda)}{A_{\rm V}}. \tag{8}$$

The coefficient in Eq. (8) was calculated for the ratio of the total extinction to the selective one $R_{\rm V} = A_{\rm V}/E({\rm B-V}) = 3.1$ and the gas to dust ratio $N({\rm H})/E({\rm B-V}) = 6.83 \times 10^{21}$ atoms cm^{-2} mag^{-1}.

The phenomenon of *interstellar linear polarization* is connected to the effect of the linear dichroism of the interstellar medium which arises because of the presence of aligned non-spherical grains. Such particles produce different extinctions of light depending on the orientation of the electric vector of incident radiation relative to the particle axis. The wavelength dependence of polarization $P(\lambda)$ now is known in the spectral range $\lambda\,0.12$–$12\,\mu$m. The polarization degree usually has a maximum in visible and declines in the IR and UV (Fig. 3). As a rule, the dependence $P(\lambda)$ is described by an empirical formula suggested by Serkowski [44]

$$P(\lambda)/P_{\rm max} = \exp[-K\ln^2(\lambda_{\rm max}/\lambda)]. \tag{9}$$

Initially, the *Serkowski's curve* had only two parameters: the maximum degree of polarization $P_{\rm max}$ and the wavelength corresponding to it $\lambda_{\rm max}$. The coefficient K was chosen by Serkowski [44] to be equal to 1.15.

An example of the behavior of polarization is shown in Fig. 3. The solid curves fit the ground-based data using only Serkowski's curve (Eq. (9)) and the relation $K = 1.86 \cdot \lambda_{\rm max} - 0.10$ found in [65]. The interstellar polarization in the direction of HD 7252 ($\lambda_{\rm max} = 0.52\,\mu$m) displays a clear excess over the extrapolated curve ("super-Serkowski behavior"). Note that polarization features have been discovered in only a few directions in the Galaxy, although the bump near λ2175 Å is a common attribute of all extinction curves. In order to represent the wavelength dependence of polarization from IR to UV, Martin *et al.* [33] suggested a five-parameter interpolation formula consisting of two terms describing UV and visual-IR polarization. The result of fitting is shown by the dashed curve in Fig. 3.

Note also that historically the direction of starlight polarization is associated with the orientation of the plane-of-the-sky component of the interstellar magnetic field, $B_\perp$. The data of polarimetric surveys together with other observations like Zeeman splitting of the H I (21 cm) line are used to study the magnetic field structure at different scales.

5.2. INTERPRETATION

The intensity of radiation after passing through a dust cloud $I(\lambda)$ is equal to

$$I(\lambda) = I_\star(\lambda)e^{-\tau_{\rm ext}(\lambda)}, \tag{10}$$

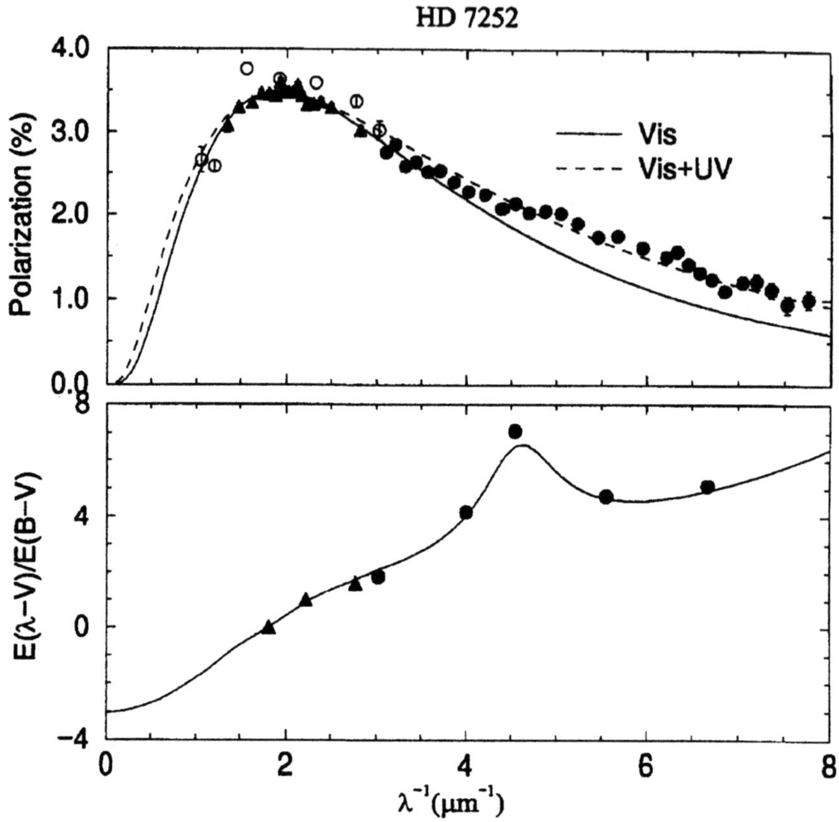

Figure 3. Polarization and extinction curves in the direction of the star HD 7252 (after [63]).

where $I_\star(\lambda)$ is the source (star) intensity and $\tau(\lambda)$ the optical thickness along the line of sight. The interstellar extinction is

$$A(\lambda) = -2.5 \log \frac{I(\lambda)}{I_\star(\lambda)} \approx 1.086 \tau_{\rm ext}(\lambda). \tag{11}$$

For spherical particles of radius $r_{\rm s}$, we have

$$A(\lambda) = 1.086\, \pi r_{\rm s}^2\, Q_{\rm ext}(m, r_{\rm s}, \lambda)\, N_{\rm d} = 1.086\, \pi r_{\rm s}^2\, Q_{\rm ext}(m, r_{\rm s}, \lambda)\, n_{\rm d}\, D, \tag{12}$$

where $N_{\rm d}$ and $n_{\rm d}$ are the column and number densities of dust grains, correspondingly, and D is the distance to the star. From Eq. (12) it follows that the wavelength dependence of interstellar extinction is determined

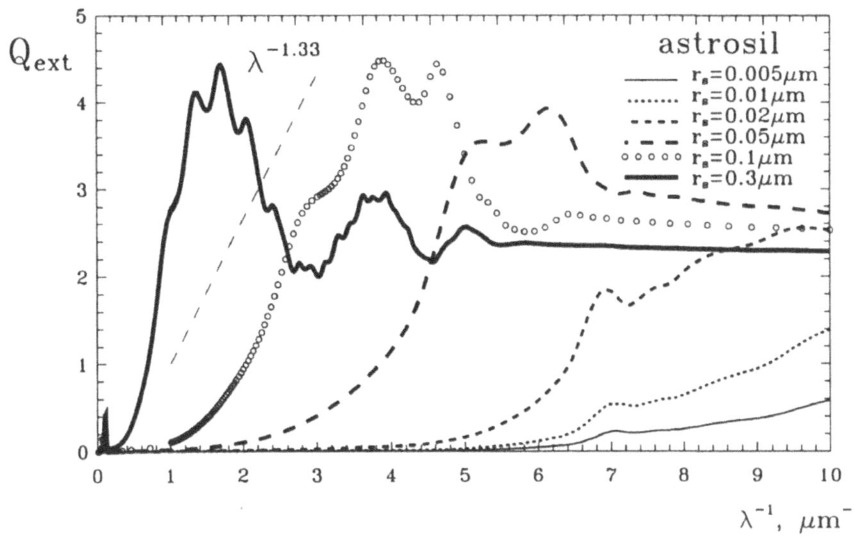

Figure 4. **Wavelength dependence of the extinction efficiency factors for homogeneous spherical particles of different sizes consisting of astronomical silicate. The dashed segment shows the approximate wavelength dependence of the mean galactic extinction curve at optical wavelengths (after [58]).**

completely by the wavelength dependence of the extinction efficiencies $Q_{\rm ext}$. Such a dependence is plotted in Fig. 4 for particles of astronomical silicate.

Dust grains are considered to have some size distribution. Very often a grain size distribution like that suggested in [34] ("MRN") $n(r_{\rm s}) \propto r_{\rm s}^{-3.5}$ is used. In this case, the extinction is proportional to

$$A(\lambda) \propto \int Q_{\rm ext}(\lambda)\, r_{\rm s}^{-1.5} {\rm d}r_{\rm s}. \tag{13}$$

Figure 5 allows us to estimate the contribution of particles of different sizes to the extinction for a given wavelength (see Eq. (13)) if the distribution is like MRN.

From Fig. 4 the influence of the size and chemical composition of particles on the extinction for a given wavelength can be examined. In all cases, the rate of growth increases until we approach the first maximum of $Q_{\rm ext}(m, x)$. As follows from Fig. 4, spheres of astrosil with $r_{\rm s} \approx 0.1\,\mu$m can produce the dependence $A(\lambda)$ resembling the observed one. The same is possible for spheres of amorphous carbon of smaller radius (see discussion in [58]). So, from the wavelength dependence of extinction one can deter-

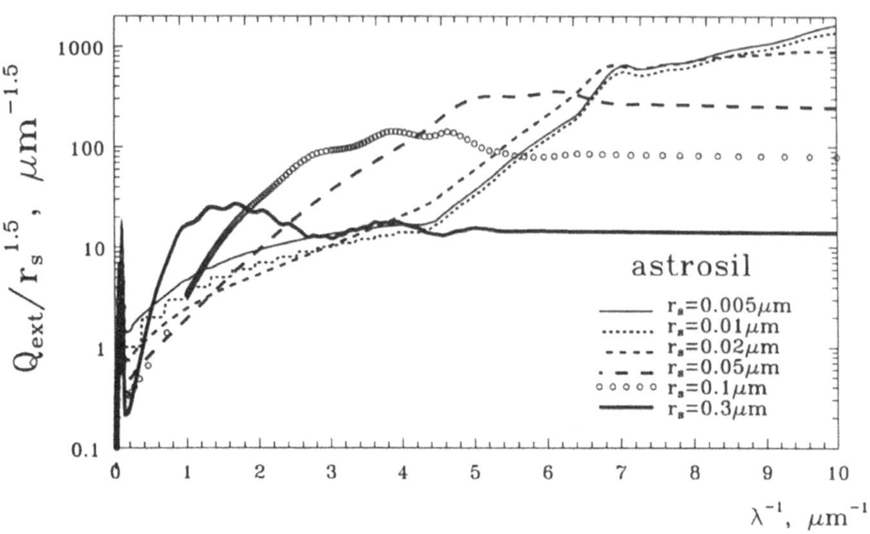

Figure 5. Wavelength dependence of the integrand in Eq. (13) for homogeneous spherical particles of different sizes consisting of astronomical silicate.

mine only the product of the typical particle size on refractive index but not the size or chemical composition of dust grains separately. It is also possible to show that particles of different structure (for example, with mantle or voids) as well as of different shape may represent the dependence of $A(\lambda)$ rather well. Thus, neither chemical composition, nor structure and shape of dust particles can be uniquely deduced from the wavelength dependence of the interstellar extinction.

Along with the wavelength dependence, it is important to reproduce the *absolute value of extinction* using the dust-phase abundances found for a given direction $[X/H]_d$. These abundances can be expressed via
1) observed quantities: interstellar extinction A_V and hydrogen column density $N(H)$;
2) model parameters: mass of constituents in a grain, the relative part of the element X in the constituent i, density of grain material and relative volume of the constituent in a particle V_i/V; and
3) a calculated quantity: the ratio of the extinction cross-section to the particle volume C_{ext}/V.
The last ratio must be maximal in order to produce the largest observed extinction and simultaneously to save the material. However, at the moment it is rather difficult to explain the carbon crisis using particles of different structure and shape (see discussion in [58]). For example, in order to ex-

plain the absolute extinction of the star ζ Oph (HD 149757; $A_V = 0^m\!.94$) using homogeneous spherical particles, the minimum abundance of carbon must be 320 ppm in the case of amorphous carbon and 267 ppm in the case of graphite (the dust-phase value is 79 ppm, see Sect. 2.1). If the particles are of astrosil ($MgFeSiO_4$), 52.5 ppm of Fe, Mg and Si and 211 ppm of O are required. All these abundances are larger than those estimated from observations of dust-phase abundances. Possibly, the way to resolve the problem is a thorough analysis of cosmic abundances which probably are not the same in different galactic regions.

To model the *interstellar polarization* one needs to calculate the forward-transmitted radiation for an ensemble of non-spherical aligned dust grains. This procedure consists of two steps: 1) computing the extinction cross-sections for two polarization modes, and 2) averaging the cross-sections for given particle size and orientation distributions.

Let non-polarized stellar radiation passes through a dusty cloud with a homogeneous magnetic field. The angle between the line of sight and the magnetic field is Ω ($0° \leq \Omega \leq 90°$). As follows from observations and theoretical considerations [10], the magnetic field determines the direction of alignment of dust grains. The linear polarization produced by a rotating spheroidal particle is

$$\begin{aligned} P(\lambda) = \int_{\varphi,\omega,\beta} \frac{1}{2}\left[C_{\rm ext}^{\rm TM}(m, r_V, \lambda, a/b, \alpha) - C_{\rm ext}^{\rm TE}(...)\right] \times \\ \times \hat{f}[\xi(r_V, \chi'', n_{\rm H}, B, T_d, T_g), \beta] \cos 2\psi \, {\rm d}[\varphi, \omega, \beta] \cdot N_d \cdot 100\%, \end{aligned} \quad (14)$$

where $C_{\rm ext}^{\rm TM,\,TE}$ are the extinction cross-sections for two polarization modes, $N_{\rm d}$ the column density of dust grains. $\hat{f}(\xi, \beta)$ is the function of orientation, depending on the alignment parameter ξ and the precession angle β.

Of special interest are the variations of the polarization in cold dark clouds and star-forming regions. It was found that in several dark clouds (L 1755, L 1506, B 216–217 [2]) the increase of polarization with growing extinction stops beginning at some value of A_V. This fact may be related with the following:
1) optics of dust ($C_{\rm ext}^{\rm TM} \simeq C_{\rm ext}^{\rm TE}$ in Eq. (14));
2) physics of dust (no grain alignment);
3) object's properties (line of sight is parallel to the direction of alignment, $\Omega \simeq 0°$ or two or more clouds are present on the line of sight which leads to the cancellation of polarization).

The proximity of cross-sections $C_{\rm ext}^{\rm TM}$ and $C_{\rm ext}^{\rm TE}$ may indicate that the shape of grains is more spherical in darker parts of clouds. In a like manner, the reduction of polarization occurs if the size of non-spherical particles becomes larger than the radiation wavelength (see Fig. 6). This Figure

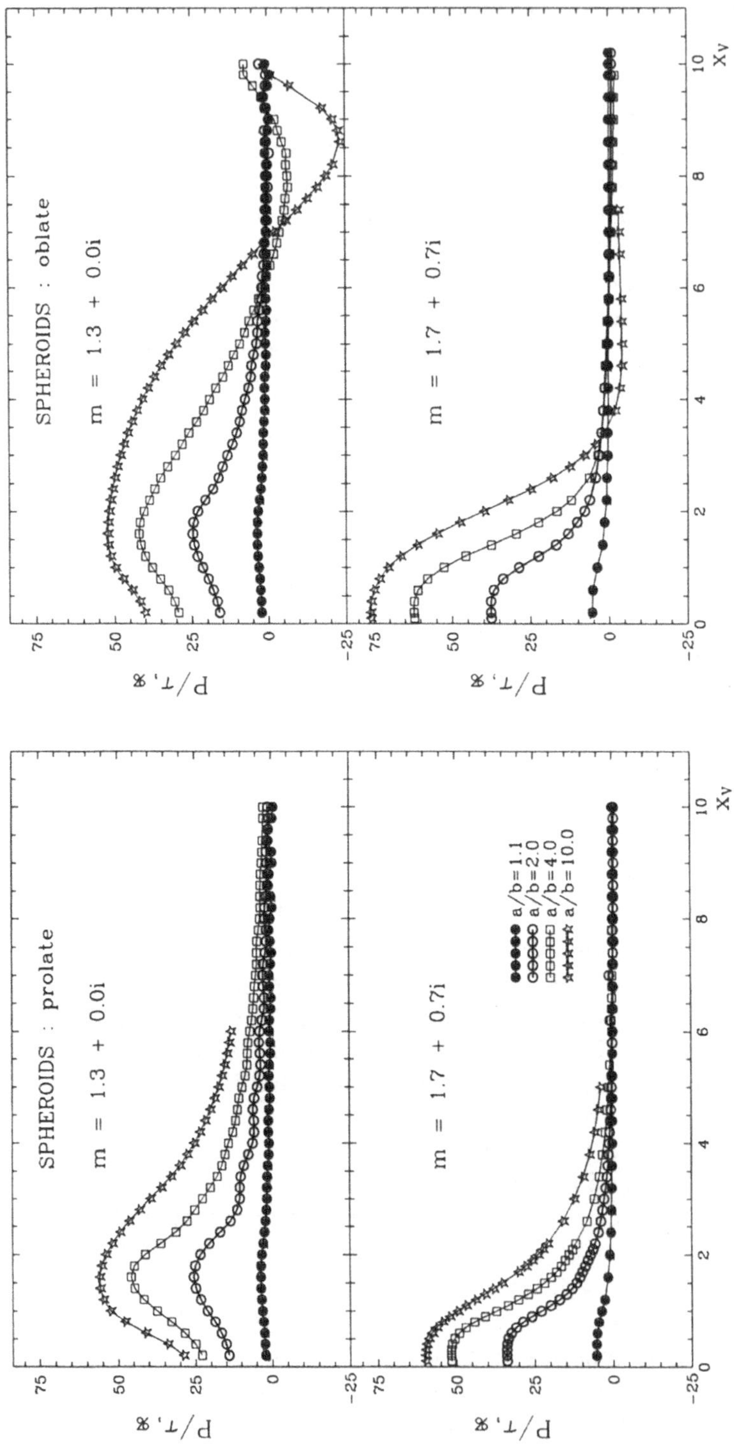

Figure 6. Polarization efficiency as a function of x_V for prolate and oblate spheroids with $m = 1.3 + 0.0i$ and $1.7 + 0.7i$, picket fence orientation, $\alpha = 90°$. The size/shape effects are illustrated (after [62]).

illustrates the behavior of the polarization efficiency P/τ (the upper/lower sign corresponds to prolate/oblate spheroids):

$$\frac{P(\lambda)}{\tau(\lambda)} = \pm\frac{C_{\rm pol}}{C_{\rm ext}} = \pm\frac{C_{\rm ext}^{\rm TM} - C_{\rm ext}^{\rm TE}}{C_{\rm ext}^{\rm TM} + C_{\rm ext}^{\rm TE}} \cdot 100\% = \pm\frac{Q_{\rm ext}^{\rm TM} - Q_{\rm ext}^{\rm TE}}{Q_{\rm ext}^{\rm TM} + Q_{\rm ext}^{\rm TE}} \cdot 100\% \qquad (15)$$

for non-absorbing and absorbing spheroids for the case when a maximum polarization is expected (non-rotating particles, radiation propagates perpendicular to the symmetry/rotation axis of the spheroid, $\alpha = 90°$). It is seen that relatively large particles produce *no polarization* independently of their shape. For absorbing particles, non-zero polarizations occur at smaller x_V values[1] than for non-absorbing ones. However, the position at which the ratio P/τ reaches a maximum is rather stable in every panel of Fig. 6 and is independent of a/b.

These results no longer remain valid when the radiation is incident obliquely as is demonstrated in Fig. 7. The angular dependence of the extinction and linear polarization factors in Eq. (15) differs: if α decreases, the position of the maximum for $Q_{\rm ext}$ shifts to smaller values of x_V while that for $Q_{\rm pol}$ shifts to larger x_V (Fig. 7, upper panels). As a result, the maximum polarization efficiency for prolate spheroids is reached at smaller x_V in the case of tilted orientation (Fig. 7, lower left panel), and the picture is reversed for oblate particles (see [58]).

Thus, for particles larger than the radiation wavelength, the linear polarization is expected to be rather small. This does not allow one to distinguish between the particle properties like refractive index, shape, orientation. Even in the case of the most preferable orientation, large particles (possibly available in dark clouds) do not polarize the transmitted radiation significantly.

6. Scattered radiation

6.1. OBSERVATIONS

If a dusty object radiates at a given wavelength, *scattered light* is always present in its radiation. The amount may be negligible, but it exists. This is due to the fact that the particle albedo can never be equal to zero.

The necessary condition for the appearance of scattered radiation is the existence of a source of illumination. The spectrum of the scattered light generally resembles that of the source. However, the color of the reflection nebulae may be bluer or redder than that of the illuminating stars depending on the properties (primarily the size) of the dust grains and the

[1] $x_V = 2\pi r_V/\lambda$ where r_V is the radius of a sphere whose volume is equal to that of the spheroid, λ the wavelength of incident radiation.

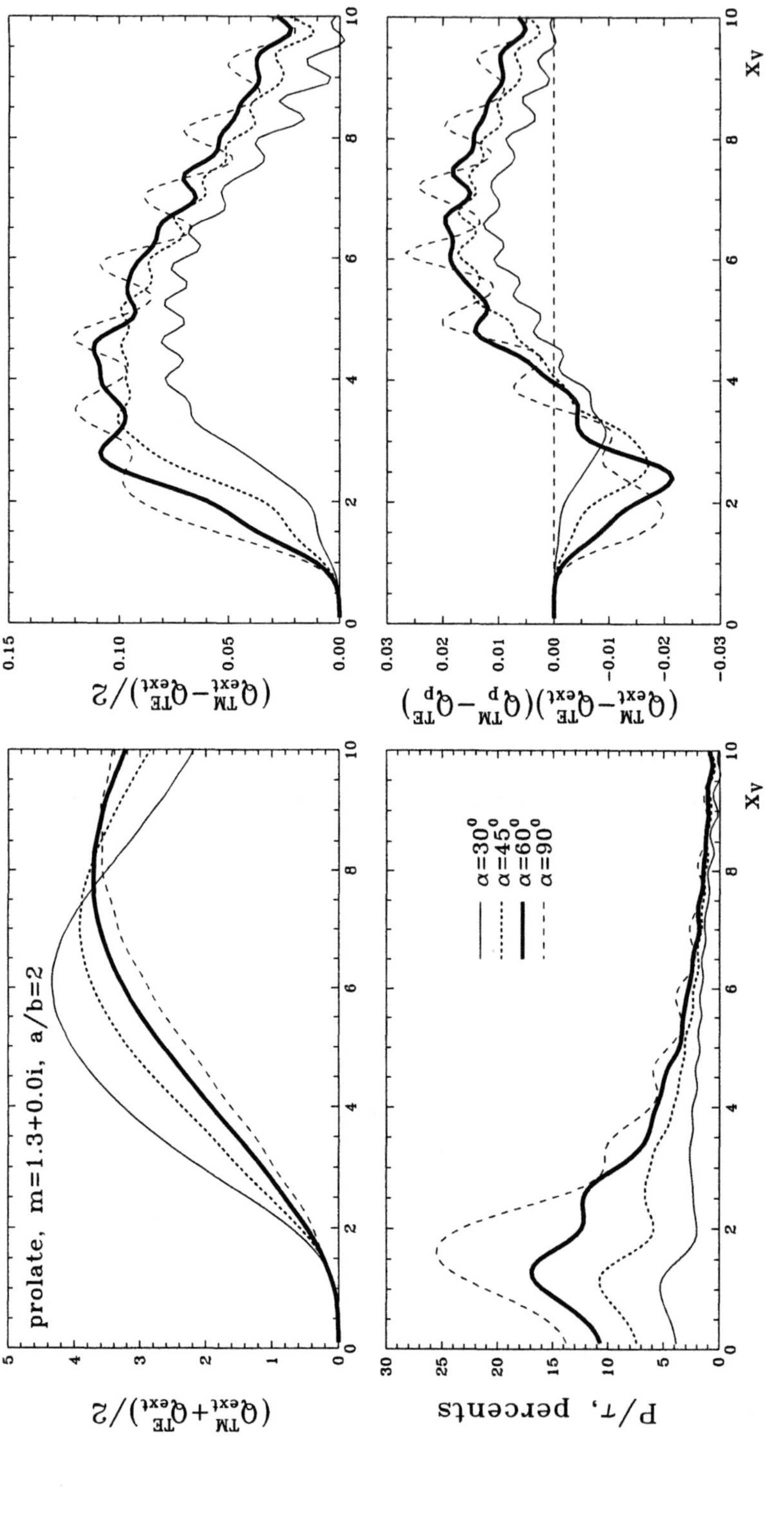

Figure 7. Extinction and linear polarization factors, polarization efficiency and circular polarization factors as a function of x_V for prolate spheroids with $m = 1.3 + 0.0i$ and $a/b = 2$, picket fence orientation. The effect of variations of particle orientation is illustrated (after [58]).

scattering geometry (see [55] for discussion). The latter (i.e., the mutual position of an illuminating source, scattering volume and an observer) also strongly influences the observed *polarization*, which is the usual attribute of the scattered radiation. The degree of linear polarization in reflection nebulae may reach 50% and more. The typical intrinsic polarization in circumstellar shells is several percent, but may sometimes reach 20–30%. The centro-symmetric polarization pattern in nebulae is used to search for the positions of illuminating stars.

However, simple cases, like single scattering from one source, are not typical in astronomy. More often, complex or mixed cases are observed: attenuation of a part of the scattered volume by a foreground dust cloud, multiple scattering effects, etc. This leads to the appearance of so-called polarization null points on the polarization maps where a reversal of the polarization occurs, change of the sign of the polarization with wavelength, non-centrosymmetric structure of the polarization maps, variations of the positional angle of polarization with wavelength, or very high degrees of circular polarization (see [66] for references and discussion). Some of these features are seen in Fig. 8 where the polarimetric map of the circumstellar shell around the carbon star IRC +10 216 is shown.

6.2. INTERPRETATION

The process of light scattering is described in terms of the particle *albedo* and *scattering matrix*. But it is enough to calculate the first element of the scattering matrix F_{11} (*phase function*) only if the polarization is not considered. Frequently, the phase function is used in the simplified form suggested by Henyey and Greenstein [21] and is parameterized by the *asymmetry parameter* g (or $\langle\cos\Theta\rangle$). It varies from –1 (mirror particles) to 1 (all radiation is forward scattered). Note that mirror particles (with $g < 0$) are very atypical because no grain material (with the exception of very small iron particles) gives at visual wavelengths a negative asymmetry parameter. Even pure conductors ($m = \infty$) have $g \approx -0.2$ [52].

In optically thin case the amount of observed scattered radiation is proportional to

$$I(\lambda) \propto I_\star(\lambda)[1 - g(\lambda)]\Lambda(\lambda)\tau_{\rm ext}(\lambda), \tag{16}$$

where $I_\star(\lambda)$ describes the power of the source, the quantities Λ and $1-g(\lambda)$ characterize the ability of a particle to scatter the radiation and the scattering geometry, respectively, and the last multiplier $\tau_{\rm ext}(\lambda)$ is proportional to the extinction cross-section and the number of scatterers. It is evident that the incoming scattered radiation vanishes if $I_\star(\lambda), \Lambda(\lambda)$ or $\tau_{\rm ext}(\lambda)$ approaches zero or $g(\lambda)$ tends to unity.

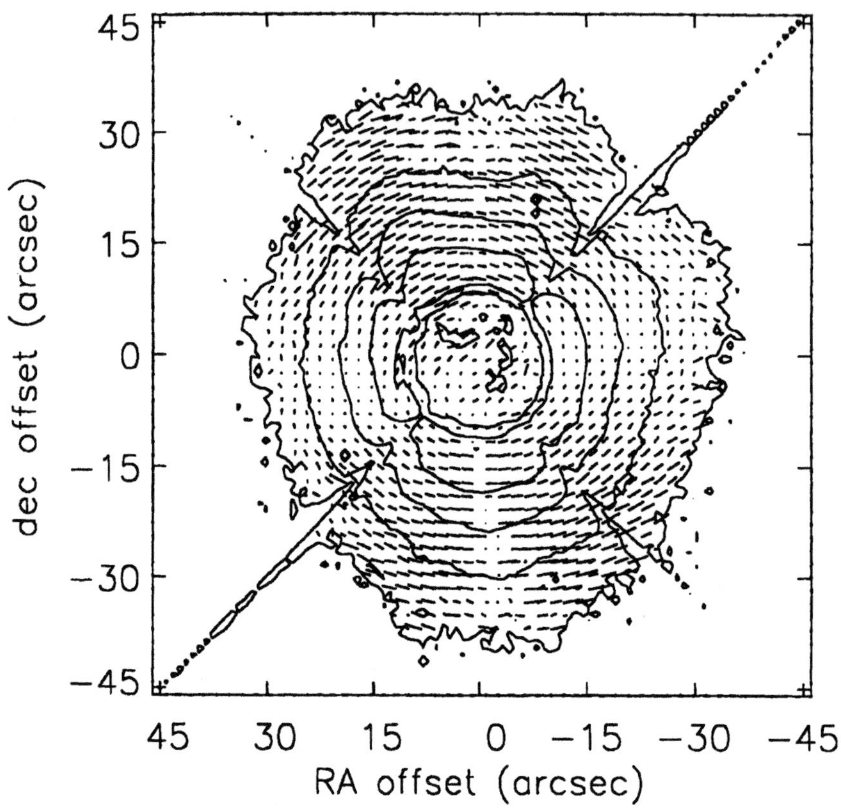

Figure 8. 2.2 μm polarimetric map constructed from a set of occulted, polarimetric images of IRC +10 216. The map is superimposed on a contour plot of an unpolarized image constructed from the same set of images (after [28]).

The standard behavior of particle albedo and asymmetry parameter is the following: $\Lambda \ll 1$ and $g \simeq 0$ for small size parameters (small sizes or large wavelengths), both characteristics grow with increasing particle size or decreasing the radiation wavelength and reach the asymptotic values for very large size parameters. The wavelength dependencies of Λ and g are shown in Fig. 9 (upper panels) for particles of astrosil. It is seen that the albedo of particles with $r_s \geq 0.05\,\mu$m is rather high (up to ~0.9) in a wide wavelength range and reduces in the far-UV. At the same time, the asymmetry parameter shows a tendency toward growth. So, we expect

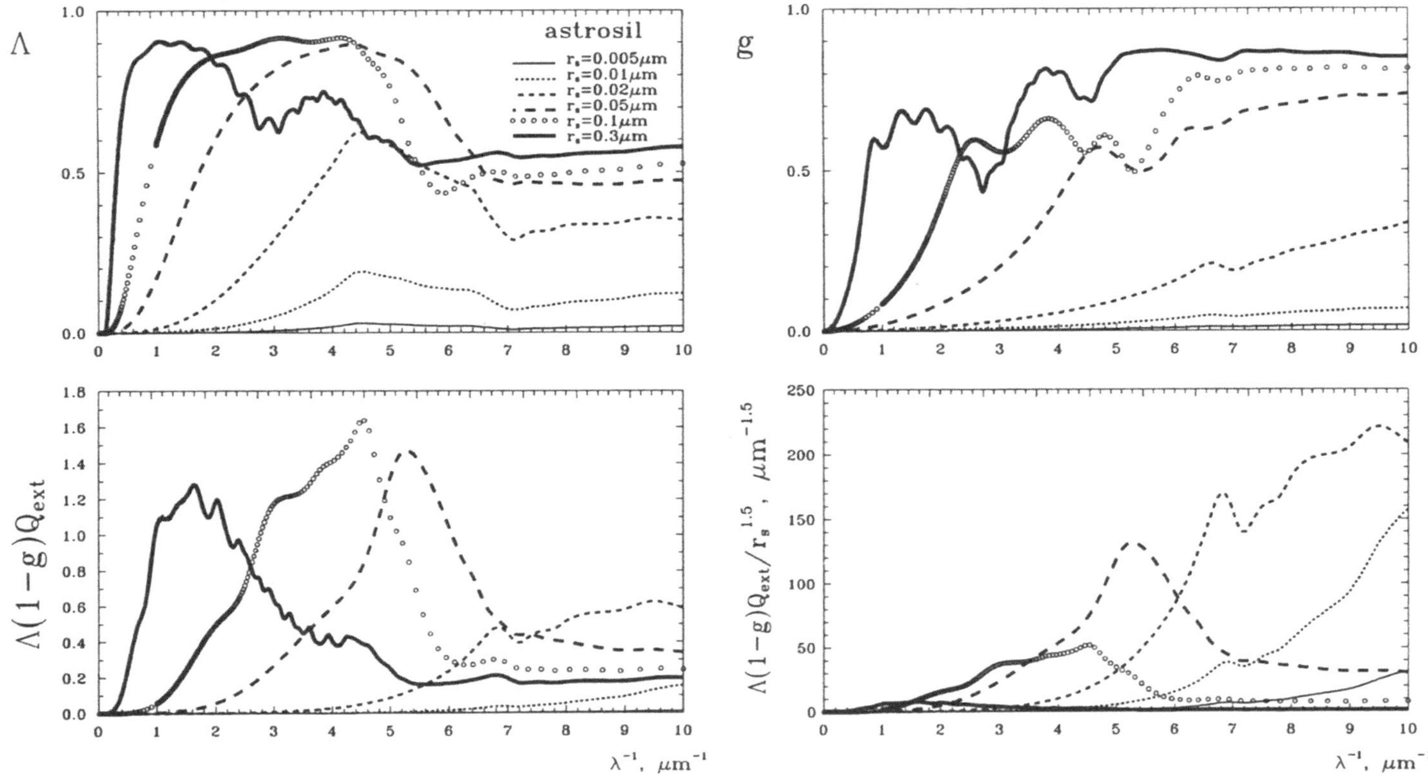

Figure 9. Wavelength dependence of the albedo, asymmetry parameter and the product of grain scattering characteristics given by Eq. (16) for homogeneous spherical particles of different sizes consisting of astronomical silicate. The lower right panel allows one to estimate the contribution of particles into scattered radiation if the size distribution is like MRN.

reduction of the role of scattered light in the UV in comparison with the visual part of spectrum. This is clearly seen at the lower left panel of Fig. 9 where the product $[1 - g(\lambda)]\Lambda(\lambda)Q_{\rm ext}(\lambda)$ as given by Eq. (16) is plotted. The contribution of a particle of given size to the scattering occurs at some wavelength which correlates with the particle size (i.e. larger particles scatter radiation at longer wavelengths). But grains with radii $r_{\rm s} \lesssim 0.01\,\mu$m hardly affect the scattered radiation. This conclusion remains valid if one considers size distributions like MRN (see the lower right panel of Fig. 9).

It is important to keep in mind that the *albedo and asymmetry parameter cannot be determined from observations separately, but only in the form of a combination.* Therefore, models with one fixed parameter and the other varying make little physical sense. The dependence of Λ on g is plotted in Fig. 10 for particles from astrosil and AC1. It is seen that some pairs of parameters correspond to *no particles.* The theoretical constraints on the albedo and asymmetry parameter were discussed by Chlewicki and Greenberg [7] who showed that some modeling results could not be represented in the optics of small particles.

In general, the radiation scattered by aligned non-spherical particles has an azimuthal asymmetry that provokes a non-coincidence of the directions of the radiation pressure force and of the wave-vector of incident radiation [59]. Another consequence of the azimuthal asymmetry is the anisotropy of the phase function in the left/right direction. The geometry of the phase function in forward/backward and left/right directions may be characterized by two asymmetry parameters $g_{||}$ and $g_{\perp}$, respectively (see [57] for details). The values of radial asymmetry factor $g_{||}$ decrease with a growth of α when the path of radiation reduces from $2a$ ($\alpha = 0°$) to $2b$ ($\alpha = 90°$). The transversal asymmetry factor $g_{\perp}$ can be rather large and even exceeds the radial one. Because the geometry of light scattering by very elongated spheroids approaches that of infinite cylinders,[2] *such particles scatter more radiation "to the side" than in the forward direction.*

At the same time, the albedo for large non-spherical particles becomes close to that of spheres. Our calculations made for particles with different absorption [62] demonstrate that the distinction of the albedo for spheres and spheroidal particles remains rather small (within $\sim 20\,\%$) if the ratio of the imaginary part of the refractive index to its real part $k/n \gtrsim 0.2 - 0.3$.

The difference in the single light scattering by a spherical particle and an aligned non-spherical particle are readily apparent in the behavior of the elements of the first column of the scattering matrix. These elements determine the scattered radiation if the incident radiation is non-polarized. In contrast to spheres, the scattering by non-spherical particles causes the

[2]In this case the scattered radiation forms the conical surface with the opening angle 2α.

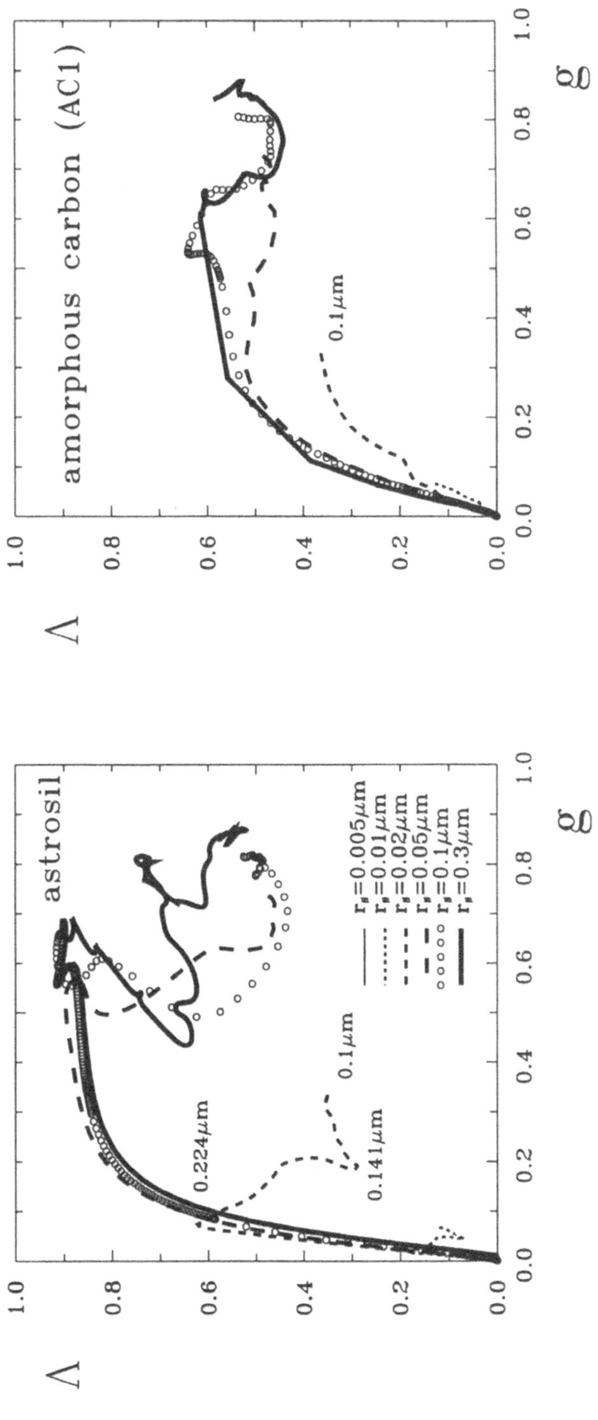

Figure 10. Albedo dependence on asymmetry parameter for homogeneous spherical particles of different sizes consisting of astronomical silicate and amorphous carbon. The values of wavelength are given in μm for particle with $r_s = 0.02\,\mu$m.

rotation of the positional angle of linear polarization and produces circular polarization after the first scattering event. This is the result of non-zero elements F_{31} and F_{41} of the scattering matrix for aligned non-spherical particles. Thus, some observational features mentioned in Sect. 6.1 may be attributed to light scattering by non-spherical grains.

The elements of the scattering matrix for spherical and spheroidal particles of the same size and refractive index are compared in Fig. 11. It shows that the major differences between spheres and spheroids as well as between prolate and oblate particles appear at large scattering angles. Such behavior is rather common (see [35] for discussion). Therefore, in order to investigate the shape effects, we need to observe an object located behind the illuminating source that occurs on occasion. However, a thin layer of the foreground dust can produce much more scattered radiation than background dust (see upper left panel in Fig. 11). So, it is rather difficult to diagnose the particle shape from scattered radiation in complex celestial objects.

7. Infrared radiation

7.1. OBSERVATIONS

As noted by Glass [17]: "The central astronomical role of dust is at its most evident in the infrared." The observed IR and submillimeter emission from interstellar clouds, circumstellar envelopes, and galaxies is generally thermal emission of dust heated by stellar radiation or shock waves. The spectrum of dust emission is blackbody. Its shape is mainly determined by the properties of heating sources and the dust distribution around them. The individual characteristics of particles manifest themselves in different temperatures of metallic and dielectric grains and in a full manner as spectral features superimposed on continuum emission. The dust features are observed in absorption and emission and are the vibrational transitions in solid materials — constituents of grain cores and mantles (usually bending or stretching modes). The IR spectral features present the most reliable method of diagnostics of dust chemical composition. As a rule, the dust bands are rather weak and in order to find them the process of subtracting the continuum is required (see Fig. 12).

The observed IR features attributed to the interstellar and circumstellar dust are collected in Table 1. This is the updated version of Table 7 published in 1986 (see [56]) which included 24 features and was significantly enlarged after the discoveries made from the Infrared Space Observatory (ISO). First of all ISO contributed into the mid-IR where numerous emission features of crystalline silicates were found. They were identified with magnesium–iron silicates: pyroxenes ($Mg_xFe_{1-x}SiO_3$) and olivines

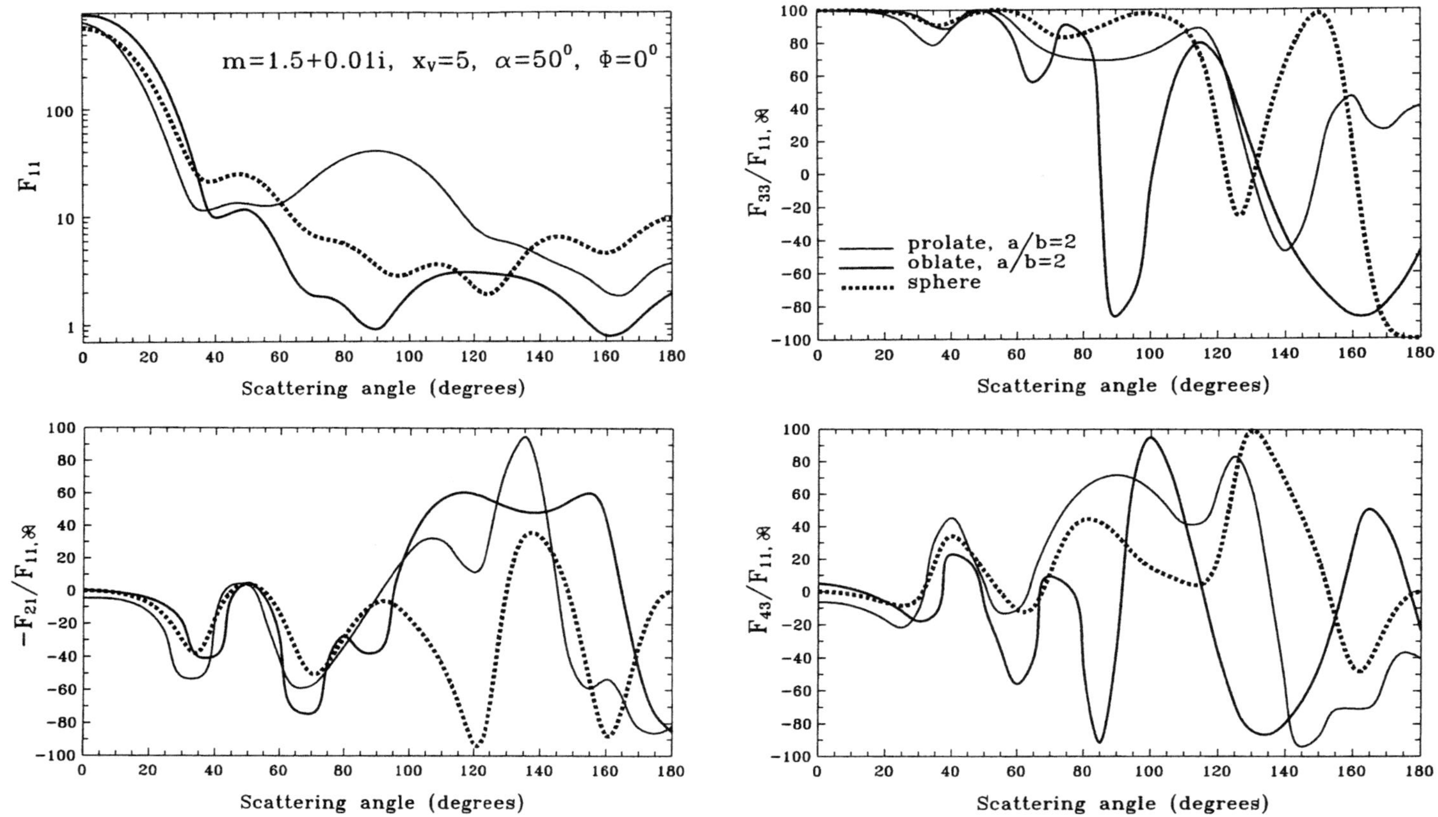

Figure 11. The element F_{11} and the ratios of elements of the scattering matrix $-F_{21}/F_{11}$, F_{33}/F_{11} and F_{43}/F_{11} for sphere, prolate and oblate spheroids, $m = 1.5 + 0.01i$, $x_V = 5$, $\alpha = 50°$, $\Phi = 0°$. The influence of variations of particle shape is illustrated. Adapted from [24].

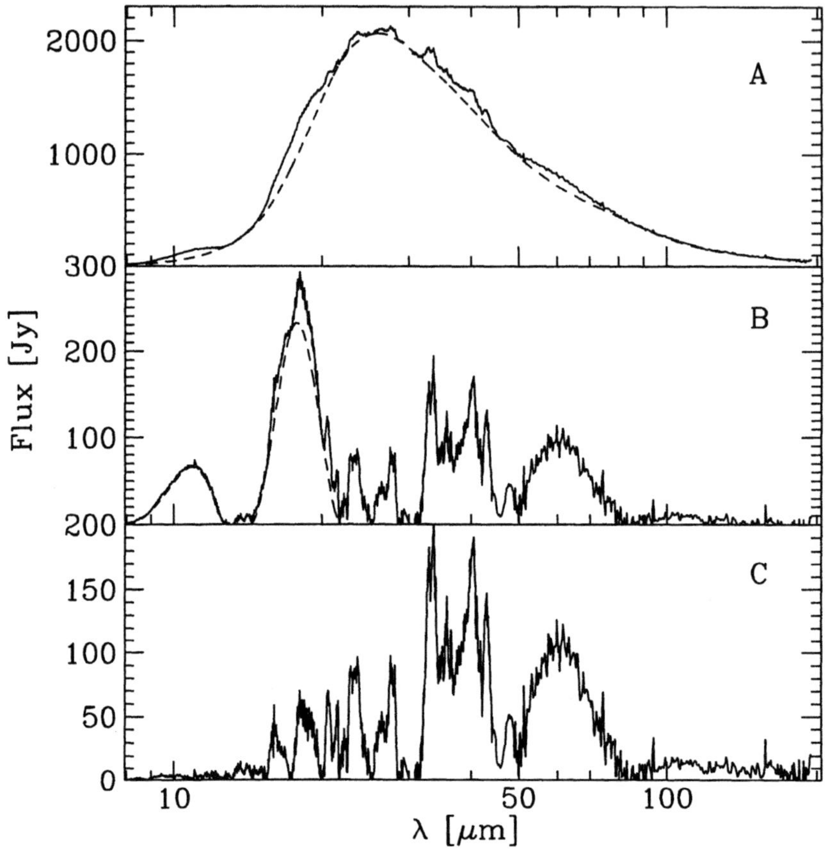

Figure 12. The ISO spectrum of the post-red supergiant star AFGL 4106 with the spline fit of continuum. The result of the subtraction of the spline fit continuum (dashed line in A) is shown in graph B. The result of the removal of the broad silicate features (dashed line in B) is given in graph C (after [39]).

($Mg_{2x}Fe_{2-2x}SiO_4$), where $0 \leq x \leq 1$. In particular, enstatite ($MgSiO_3$) and forsterite (Mg_2SiO_4) are the extreme cases of pyroxenes and olivines with $x = 1$, respectively. Note that the positions and widths of features in Table 1 vary from object to object and sometimes a feature was found in one object only. Several features like the 3.4 μm band reveal sub-features. Additional information about observational data can be found in recent publications [4], [5], [6], [9], [11], [12], [14], [15], [16], [38], [43], [46], [50], [51].

TABLE 1. : Observed IR features attributed to dust

λ, μm	$\Delta\lambda^*$, μm	A/E	Identification**	Objects
1.15	0.2	E	silicon nanoparticles?	7
1.50	0.2	E	β-$FeSi_2$?	7
2.70	0.05	A	CO_2	2
2.75		A	phyllosilicates? (O–H)	1
2.78	0.05	A	CO_2	2
2.97		A	NH_3 (N–H)	2, 9
3.07	0.7	A	H_2O (O–H)	1, 2, 8, 9
3.25	0.07	A	hydrocarbons (C–H)	2
3.28	0.05	A/E	PAH (C–H)	1, 2, 5, 7, 8
3.4	0.08	A/E	hydrocarbons, HAC	1, 2, 5, 7–9
3.47	0.10–0.15	A	hydrocarbons	2
3.473	0.03	A	H_2CO?, C_2H_6?	2
3.5	0.08	E	H_2CO	6
3.53	0.08	E	carbonaceous mater. (C–C)	6
3.53	0.025–0.27	A	CH_3OH (C–H)	2
3.9	0.07	A	H_2S	2
4.1		A	HDO?, SO_2?	2
4.27	0.02–0.05	A	CO_2	1, 2, 8
4.38		A	$^{13}CO_2$	1, 2
4.5		A	H_2O	2
4.61	0.1	A	'XCN' (–C≡N)	2, 8, 9
4.67	0.03–0.22	A	CO	2, 8, 9
4.90	0.07	A	OCS	2, 9
5.25		E	PAH	2, 5, 7, 8
5.5–5.6	0.06–0.12	A/E	metal carbonils (–C=O–)	2, 5
5.7		E	PAH	2, 5, 7, 8
5.83		A	HCOOH (–C=O–), H_2CO	2, 9
5.95		A	carbonaceous mater. (–C=O–)	1
6.0	0.7	A	H_2O (O–H)	2, 8
6.14		A	NH_3 (N–H)	2
6.2	0.2	A/E	PAH (–C=C–)	1, 2, 5, 7, 8
6.8–6.9	0.15	A/E	HAC, hydrocarbons?	2, 5, 8
6.85	0.7	A	CH_3OH	2, 9
7.24	0.1	A	HCOOH?	2
7.25		A	hydrocarbons, HAC	1, 2, 8
7.41	0.08	A	CH_3HCO?, $HCOO^-$?	2
7.67	0.06	A	CH_4	2, 5, 8
7.7	0.5	A/E	CH_4, CO	2, 5, 7, 8
7.7		E	PAH (C–C)	1, 2, 5, 7, 8
8.3	0.42	E		3, 7
8.7	0.36	E	PAH (H–C–H)	1–5, 7, 8
9.0		A	NH_3	2
9.14	0.30	E	silica?	3, 7
9.45	0.19	E		3, 7

TABLE 1. : (Continued)

λ, μm	$\Delta\lambda^*$, μm	A/E	Identification**	Objects
9.7	3	A/E	amorphous silicate	1–3, 5–9
9.8	0.17	A/E	forsterite and enstatite	3, 7
9.8	0.4	A	CH_3OH	2
10.7	0.28	A/E	enstatite	3, 7
11.2	1.7	A/E	SiC	4, 5
11.2	0.27	A/E	PAH (H–C–H)	1–5, 7, 8
11.4	0.48	E	forsterite, diopside?	3, 7
12.0	0.47	A/E	H_2O	2
12.7		E	PAH (H–C–H)	2, 5, 7, 8
13.0	0.6	E	spinel	3
13.3		E	PAH?	7
13.5	0.25	E	?	3
13.8	0.20	E	enstatite?	3
13.6		E	PAH?	7, 8?
14.2	0.28	E	enstatite?	3
14.97	0.02	A	CO_2	2
15.2	0.06	A	CO_2	2, 9
15.2	0.26	A/E	enstatite	3, 4, 7
15.8		E	PAH?	7, 8
15.9	0.43	A/E	silica?	3, 4, 7
16.2	0.16	A/E	crystalline forsterite	3, 4
16.4		E	PAH?	7
16.0		E	spinel	3
16.9	0.57	A/E	?	3, 4, 7
17.5	0.18	E	enstatite	3, 7
18.0	0.48	A/E	forsterite and enstatite	3–5, 7
18.5	3	A/E	amorphous silicate	1–3, 5–9
18.9	0.62	A/E	forsterite?	3–5, 7
19.5	0.40	A/E	cryst. forsterite and enstatite	3–5, 7
20.7	0.31	A/E	silica?, diopside?	3–5, 7
21	5	E	?	4, 5
21.5	0.35	E	?	3–5, 7
22.4	0.28	E	?	3–5, 7
23.0	0.48	E	crystalline enstatite	3–5, 7
23.7	0.79	E	crystalline forsterite	3–5, 7
23.89	0.18	E	?	3, 5, 7
24.5	0.42	E	crystalline enstatite + ?	3–5, 7
25.0	0.32	A/E	diopside?	3–5, 7
26.1	0.57	A/E	forsterite + silica?	3–5, 7
26.8	0.37	A/E	?	3–5, 7
27.6	0.49	E	crystalline forsterite	3–5, 7
28.2	0.42	E	crystalline enstatite	3–5, 7
28.8	0.24	E	?	3, 5, 7
29.6	0.89	E	diopside?	3–5, 7
30	20	E	MgS	4, 5

TABLE 1. : (Continued)

λ, μm	$\Delta\lambda^*$, μm	A/E	Identification**	Objects
30.6	0.32	E	?	3–5, 7
31.2	0.24	E	forsterite?	3, 4, 7
32.0	0.5	E	spinel	3
32.2	0.46	E	diopside?	3–5, 7
32.8	0.60	E	?	3–5, 7
33.6	0.70	E	crystalline forsterite	3–5, 7
34.1	0.12	E	crystalline enstatite + diopside?	3–5, 7
34.9	1.36	E	clino-enstatite?	3–5, 7
35.9	0.53	E	orto-enstatite?	3–5, 7
36.5	0.39	E	crystalline forsterite + ?	3–5, 7
38.1	0.57	E	?	3
39.8	0.74	E	diopside?	3–5, 7
40.5	0.93	E	crystalline enstatite	3–5, 7
41.8	0.72	E	?	3
43.0		E	clino-enstatite	3–5, 7
43–45		A/E	H_2O	2, 3–5, 7
43.8	0.78	E	orto-enstatite	3–5, 7
44.7	0.58	E	clino-enstatite, diopside?	3–5, 7
47.7	0.97	E	FeSi?, a silicate	3–5, 7
48.8	0.61	E	a silicate	3–5, 7
52.9	3.11	E	crystalline H_2O	3–5, 7
62.	20	E	crystalline H_2O	3–5, 7
65.		E	enstatite?, diopside?	3–5, 7
69.0	0.63	E	crystalline forsterite	3–5, 7
91.		E	?	4, 5, 7

*$\Delta\lambda$=FWHM (Full Width Half Max); **chemical bond responsible for given feature is shown in parentheses; A/E – absorption/emission; hydrocarbons: $-CH_2-$, $-CH_3$ groups in aliphatic solids; phyllosilicates: e.g., serpentine ($Mg_3Si_2O_5[OH]_4$) or talc ($Mg_3Si_4O_{10}[OH]_2$); metal carbonils: e.g., $Fe(CO)_4$; spinel: $MgAl_2O_4$; PAH – polycyclic aromatic hydrocarbons; HAC – hydrogenated amorphous carbon; 1 – diffuse interstellar clouds; 2 – molecular clouds and/or HII compact regions ; 3 – O-rich stars O/C > 1; 4 – C-rich stars C/O > 1; 5 – planetary nebulae and HII regions; 6 – eruptive variables; 7 – other galactic sources; 8 – galactic nuclei; 9 – comets

Linear polarization of the IR radiation has a rather complicated wavelength dependence and complex pattern. In the near-IR, the polarization is due to scattering while the polarization vectors mark the position of the illuminating source(s) [64]. The scattering efficiency of dust grains sharply drops with wavelength and the polarization mechanism is switched from scattering to dichroic extinction and thermal polarized emission. The latter dominates at the far-IR and submillimeter wavelength range. Polarization due to dichroic extinction and thermal emission is attributed to the spinning non-spherical dust grains aligned by magnetic fields. The direction of the observed polarization is *parallel* (dichroic extinction) or *perpendicular* (thermal emission) to the magnetic field as projected onto the plane of the sky. One special problem is the observed polarization profiles of dust features where both absorption and emission processes can occur in a single observational beam (see Ref. [48] for discussion).

7.2. INTERPRETATION

The modeling of the IR radiation is usually based on some radiative transfer calculations. Even if an object is optically thin at a given wavelength, the determination of the dust temperature requires the consideration of the UV–visual radiation where, as a rule, absorption by dust is large and the object's optical thickness is significant.

The IR flux at wavelength λ emerging from an optically thin medium is proportional to the total number of dust grains in the medium N, the Planck function which depends on the particle temperature $T_{\rm d}$, and the emission cross-section $C_{\rm em}(\lambda)$:

$$F_{\rm IR}(\lambda) = N\,\frac{C_{\rm em}(\lambda)}{D^2}B_\lambda(T_{\rm d}) = N\,\frac{\pi r_{\rm s} Q_{\rm abs}(\lambda)}{D^2}B_\lambda(T_{\rm d}), \qquad (17)$$

where D is the distance to the object. The right-hand side of Eq. (17) is written with the assumption that the grains are spheres of the same radius $r_{\rm s}$.

The wavelength dependence of the absorption efficiency factors $Q_{\rm abs}(\lambda)$ is shown in Fig. 13 (left panels) for spheres consisting of astrosil and amorphous carbon. At the IR wavelengths, the factors $Q_{\rm abs}$ usually increase with grain size and are larger for carbonaceous and metallic particles in comparison with those of silicates and ices. However, the contribution of the particles of different sizes into thermal radiation reverses if we take into account their size distribution like MRN (Fig. 13, right panels).

It is important to note that the IR spectrum of carbonaceous and metallic particles is almost featureless (see also Fig. 1 and Table 1). At the same time, as follows from Fig. 14, the shape of dust features and even the presence of them can tell us about the size of dust grains. If we consider spherical grains of astrosil, the 10-μm and 18-μm features disappear if the grain radius exceeds $r_{\rm s} \gtrsim 2-3\,\mu$m (Fig. 14).

The quantities $T_{\rm d}$ and $C_{\rm em}(\lambda)$ in Eq. (17) depend on the particle shape. The shape dependence on temperature of interstellar/circumstellar dust grains were analyzed in [60], [61]. It was found that the temperature of non-spherical (spheroidal) grains with aspect ratios $a/b \lesssim 2$ deviates from

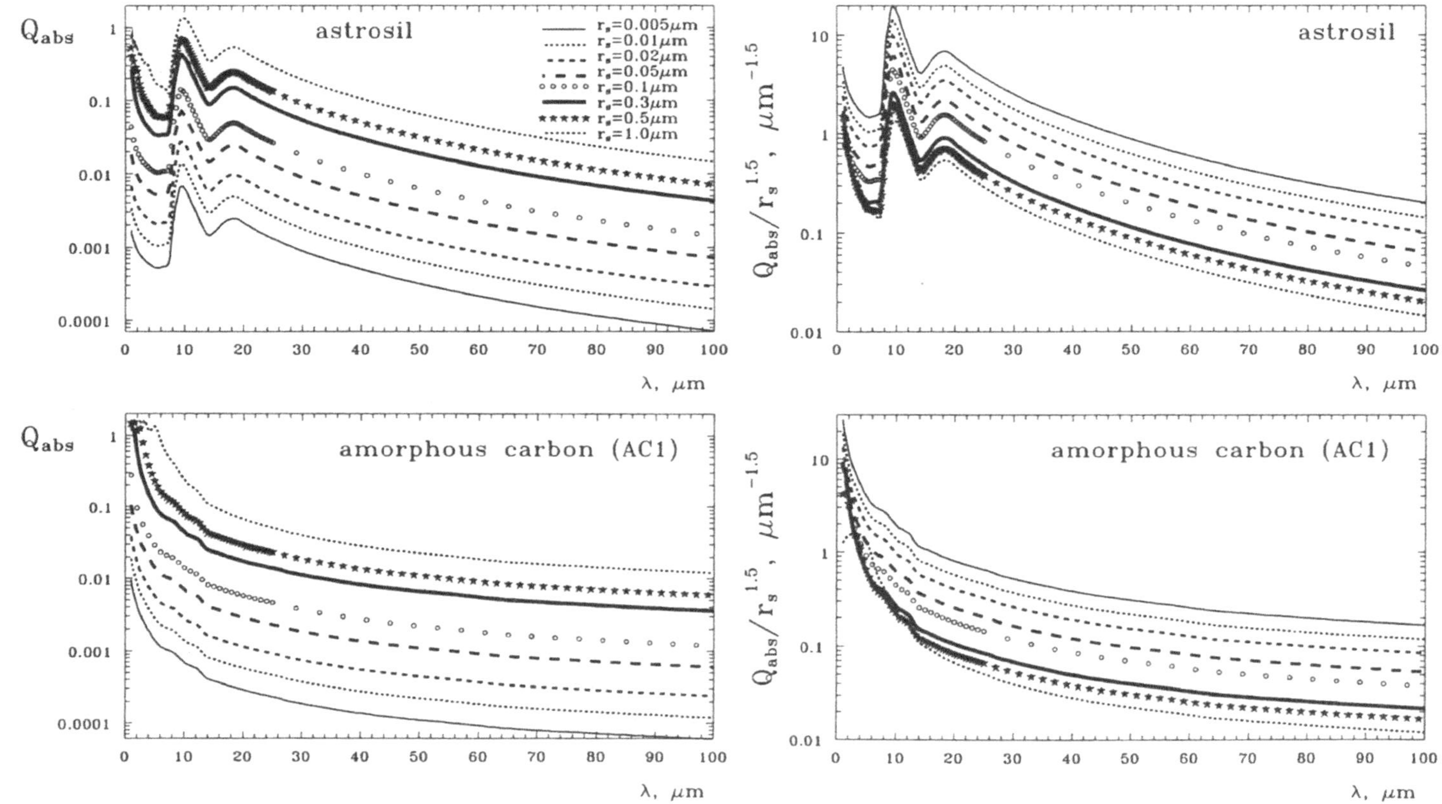

Figure 13. Wavelength dependence of the absorption efficiency factors for homogeneous spherical particles of different sizes consisting of astronomical silicate and amorphous carbon. The right panels allow one to estimate the contribution of particles into thermal radiation if the size distribution is like MRN.

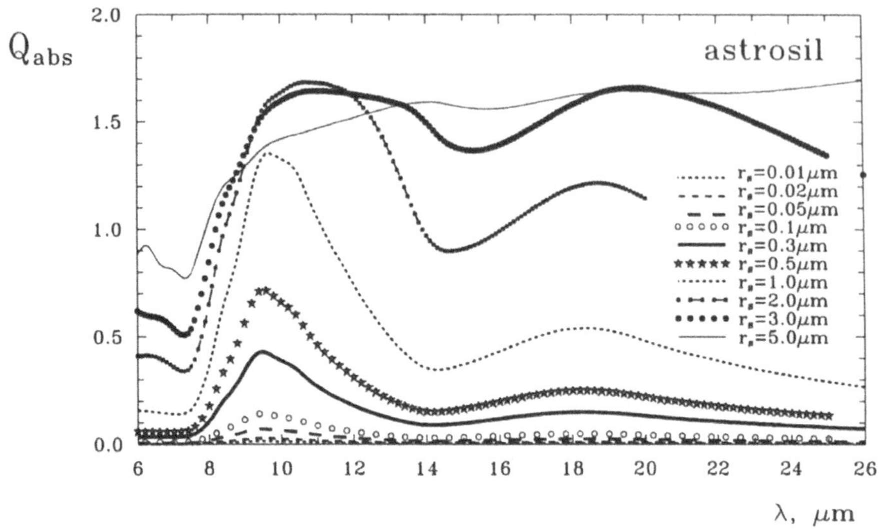

Figure 14. Wavelength dependence of the absorption efficiency factors for homogeneous spherical particles of astronomical silicate. The effect of variation of particle size is illustrated.

that of spheres by less than 10%. More elongated or flattened particles are usually cooler than spheres of the same mass and, in some cases, the temperatures may differ by even about a factor of 2.

The causes of temperature differences between dust grains with different characteristics can be clearly established with the aid the graphical method proposed by Greenberg [18]. The particles are assumed to be in an isotropic radiation field and their temperature can then be determined by solving the following equation of thermal balance (W is the dilution factor):

$$W \int_0^\infty C_{\rm abs}(\lambda)\; 4\pi B_\lambda(T_\star)\,{\rm d}\lambda = \int_0^\infty C_{\rm em}(\lambda)\; \pi B_\lambda(T_{\rm d})\,{\rm d}\lambda\,. \qquad (18)$$

If the spheroidal particles are randomly oriented in space (3D orientation), then numerical estimates show that

$$C_{\rm abs}^{\rm 3D} \approx \frac{C_{\rm em}}{4}. \qquad (19)$$

The integrals on the left- and right-hand sides of Eq.(18) then depend on the temperature T alone (for given particle chemical composition, size, and shape).

Figure 15 shows the power for the absorbed (emitted) radiation (in erg s^{-1}) for spheres and spheroids of the same volume ($r_V = 0.01\,\mu$m) composed of cellulose[3]. The method of temperature determination is indicated

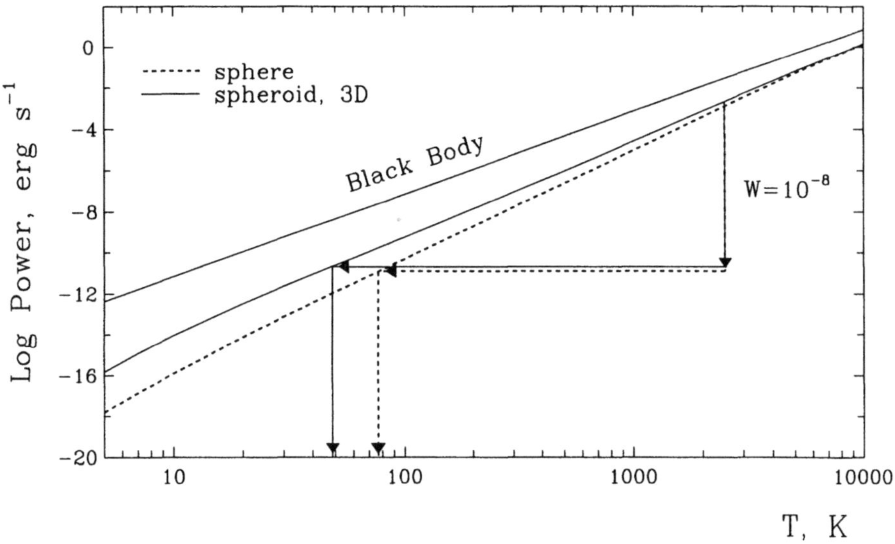

Figure 15. The power absorbed (emitted) by dust particles with $r_V = 0.01\,\mu$m in an isotropic radiation field. The straight line corresponds to a blackbody. The curves refer to spherical and prolate spheroidal ($a/b = 10$, 3D orientation) cellulose (cel1000) particles. A graphical method of determining the grain temperature is shown for $T_\star = 2500$ K and W$= 10^{-8}$. In this case, the temperatures of spherical and spheroidal particles and a blackbody are respectively, $T_{\rm d}$(sphere) $= 76.8$ K, $T_{\rm d}^{\rm 3D} = 48.7$ K, and $T_{\rm d}^{\rm BB} = 2500(10^{-8})^{1/4} = 25$ K (after [60]).

by arrows: from the point on the curve corresponding to the stellar temperature ($T_\star = 2500$ K), we drop a perpendicular whose length is determined by the radiation dilution factor and then find the point of intersection of the horizontal segment with the same curve for the power. This point gives the dust grain temperature $T_{\rm d}$ determined from the equation of thermal balance (18). It follows from Fig. 15 that different temperatures of spheres and spheroids result from different particle emissivities at low T.

If we assume the chemical composition and sizes of all particles in the medium to be the same and if we change only the grain shape, then, when spheres are replaced by non-spherical particles, the position of the maximum in the spectrum of thermal radiation is shifted longward, because the temperature of the latter is higher (Fig. 16).

[3]This is the carbon material pyrolized at 1000° C (cel1000, [26]).

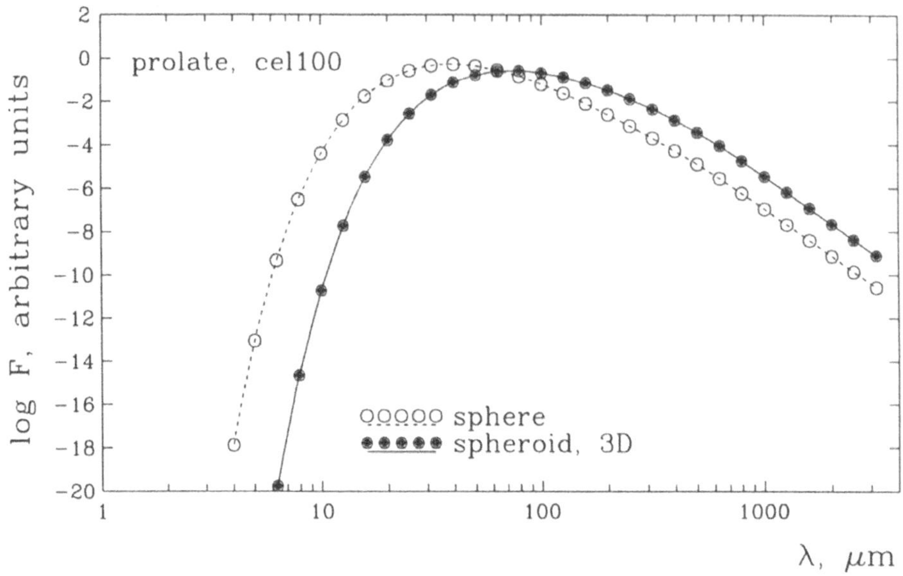

Figure 16. The normalized fluxes emerging from a medium containing the same mass of spherical and prolate spheroidal ($a/b = 10$, 3D orientation) cellulose (cel1000) particles; $r_V = 0.01\,\mu$m, $T_\star = 2500$ K, and $W = 10^{-8}$. The particle temperatures are $T_{\rm d}$(sphere) = 59.2 K and $T_{\rm d}^{\rm 3D} = 36.0$ K (after [60]).

The dust mass in an object $M_{\rm d}$ is determined from the observed flux $F_{\rm IR}(\lambda)$ and the mass absorption coefficient of a grain material $\kappa(\lambda)$

$$M_{\rm d} = \frac{F_{\rm IR}(\lambda)D^2}{\kappa(\lambda)B_\lambda(T_{\rm d})} = \frac{\rho_{\rm d} V F_{\rm IR}(\lambda)D^2}{C_{\rm abs}(\lambda)B_\lambda(T_{\rm d})}, \qquad (20)$$

where V is the particle volume and $\rho_{\rm d}$ the material density. Extensive study of the dependencies of the mass absorption coefficients on material properties and grain shape is given in [31], [41] and summarized in [19] where, in particular, it is shown that the opacities at 1 mm are considerably larger for non-spherical particles in comparison with spheres.

If $M_{\rm d}$ is determined from the observed millimeter flux, then the Rayleigh–Jeans approximation can be used for the Planck function. For the same flux $F_{\rm IR}(\lambda)$, the dust mass depends on the particle emission cross sections and temperatures, while the ratio

$$\frac{M_{\rm d}^{\rm sphere}}{M_{\rm d}^{\rm spheroid}} = \frac{C_{\rm em}^{\rm spheroid}(\lambda)}{C_{\rm em}^{\rm sphere}(\lambda)} \frac{T_{\rm d}({\rm spheroid})}{T_{\rm d}({\rm sphere})} \qquad (21)$$

shows the extent to which the values of $M_{\rm d}$ differ when changing the grain shape. The dust mass in galaxies and molecular clouds is commonly estimated from 1.3-mm observations [45]. At this wavelength, the ratio of the cross-sections for cellulose particles with $a/b = 10$ is ~ 50, and the temperature ratio is $36.0/59.2 = 0.61$, which, according to Eq. (21), gives approximately a factor of 30 larger dust mass if the particles are assumed to be spheres rather than spheroids. In other, not so extreme, cases, the dust mass overestimate is much smaller. For example, for prolate amorphous carbon particles, $M_{\rm d}^{\rm sphere}/M_{\rm d}^{\rm spheroid} \approx 1.2, 2.3$, and 6.4, for $a/b = 2$, 4, and 10, respectively.

Analysis of the polarimetric observations is based on models where dust properties and temperature, as well as the object's opacity, are varied. The main purpose of such modeling is to identify the magnetic field configurations (see [1], [22] for discussion). Unfortunately, the models usually contain many parameters and are not unique.

8. What do we really know about the cosmic dust?

Our current knowledge of cosmic dust and the possibility of extracting information about it from existing and future observations can be summarized as follows.

The study of the interstellar extinction and polarization, together with the constraints from the cosmic abundances, allows one to estimate the chemical composition, size and shape of dust grains. These three characteristics cannot be found separately but only as a combination. At the same time, it may be argued that the orientation angle of the interstellar linear polarization decisively provides the projected direction of alignment, i.e. the projected direction of the interstellar magnetic field $\mathrm{B}_{\perp}$.

Chemical composition, size and shape of dust grains can be estimated approximately from the observed intensity, colors and polarization of the scattered light. In this case, the polarization vectors may mark the position of illuminating source(s).

By investigating the thermal emission of dust, one can evaluate with confidence grain temperature, from observations in continuum, and composition,[4] from observations of dust features. Submillimeter polarization again shows the projected direction of the interstellar magnetic field $\mathrm{B}_{\perp}$. The particles' size and shape also can be estimated approximately.

The current state of different components in the modeling of dusty objects can be summed up as a rather well developed set of optical constants, light scattering theories and radiative transfer methods, which allow one to produce models of very complicated objects. However, the last stage of any modeling effort is a comparison with observations where we are restricted by the Stokes vector of incoming radiation (I, Q, U, V). *But*: the principle of

[4] or more exactly chemical bonds in solids

optical equivalency does work! This means that *all* models are ambiguous to some degree.

Acknowledgements

The author is thankful to V.B. Il'in and D.A. Semenov for valuable comments and to T.V. Zinov'eva for the assistance with compilation of Table 1. This research was partly supported by the INTAS grant (Open Call 99/652) and by grant 00-15-96607 of the President of the Russian Federation for leading scientific schools.

References

1. Aitken, D.K., Efstathiou, A., McCall, A. and Hough, J.H., Monthly Not. RAS, **329**, 647 (2002)
2. Arce, H.G., Goodman, A.A., Bastien, P., Manset, N. and Sumner, M., Astrophys. J., **499**, L93 (1998)
3. Bohren, C.F. and Huffman, D.R., Absorption and Scattering of Light by Small Particles, John Wiley and Sons, New York (1983)
4. Boogert, A.C.A., Tielens, A.G.G.M., Ceccarelli, C., Boonman, A.M.S., van Dishoeck, E.F., Keane, J.V., Whittet, D.C.B. and de Graauw, Th., Astron. Astrophys., **360**, 683 (2000)
5. Brooke, T.Y., Sellgren, K. and Geballe, T.R. Astrophys. J., **517**, 883 (1999)
6. Chiar, J.E., Tielens, A.G.G.M., Whittet, D.C.B., Schutte, W.A., Boogert, A.C.A., Lutz, D., van Dishoeck, E.F. and Bernstein, M.P., Astrophys. J., **537**, 749 (2000)
7. Chlewicki, G. and Greenberg, J.M., Monthly Not. RAS, **210**, 791 (1984)
8. Chýlek, P., Videen, G., Geldart, D.J.W., Dobbie, J.S. and Tso, H.C.W., In: Light Scattering by Nonspherical Particles, Eds., M.I. Mishchenko et al., Academic Press, San Francisco, p. 274 (2000)
9. d'Hendecourt, L., Joblin, C. and Jones A., Eds., Solid Interstellar Matter: the ISO Revolution Springer-Verlag, Berlin (1999)
10. Dolginov, A.Z., Gnedin, Yu.N. and Silant'ev, N.A., Propagation and Polarization of Radiation in Cosmic Medium, Nauka, Moscow (1979)
11. Duley, W.W., Astrophys. J., **528**, 841 (2000)
12. Fabian, D., Posch, Th., Mutschke, H., Kerschbaum, F. and Dorschner, J., Astron. Astrophys., **373**, 1125 (2001)
13. Fitzpatrick, E.L., Publ. Astron. Soc. Pacific, **111**, 163 (1999)
14. Gerakines, P.A., Whittet, D.C.B., Ehrenfreund, P., Boogert, A.C.A., Tielens, A.G.G.M., Schutte, W.A., Chiar, J.E., van Dishoeck, E.F., Prusti, T., Helmich, F.P. and de Graauw, Th., Astrophys. J., **522**, 357 (1999)
15. Gibb, E.L., Whittet, D.C.B., Schutte, W.A., Boogert, A.C.A., Chiar, J.E., Ehrenfreund, P., Gerakines, P.A., Keane, J.V., Tielens, A.G.G.M., van Dishoeck, E.F. and Kerkhof, O., Astrophys. J., **536**, 347 (2000)
16. Gordon, K.D., Witt, A.N., Rudy, R.J., Puetter, R.C., Lynch, D.K., Mazuk, S., Misselt, K.A., Clayton, G.C. and Smith, T.L., Astrophys. J., **544**, 859 (2000)
17. Glass, I.S., Handbook of Infrared Astronomy, Cambrige Univ. Press (1999)
18. Greenberg, J.M., Astron. Astrophys., **12**, 240 (1971)
19. Henning, Th., In: The Cosmic Dust Connection, Ed., J.M. Greenberg, Kluwer, p. 399 (1996)
20. Henning, Th., Il'in, V.B., Krivova, N.A., Michel, B. and Voshchinnikov, N.V., Astron. Astrophys. Suppl., **136**, 405 (1999)
21. Henyey, L.G. and Greenstein, J.K., Astrophys. J., **93**, 70 (1941)

22. Hildebrand, R.H., Davidson, J.A., Dotson, J.L., Dowell, C.D., Novak, G. and Vaillancourt, J.E., Publ. Astron. Soc. Pacific, **112**, 1215 (2000)
23. Hovenier, J.W., Ed., J. Quant. Spectrosc. Rad. Transfer, **55**, N 5 (1996)
24. Hovenier, J.W., Lumme, K., Mishchenko, M.I., Voshchinnikov, N.V., Mackowski, D.W. and Rahola, J., J. Quant. Spectrosc. Rad. Transfer, **55**, 695 (1996)
25. Il'in, V.B., Voshchinnikov, N.V., Farafonov, V.G., Henning, Th., and Perelman, A.Ya., this volume
26. Jäger, C., Mutschke, H. and Henning, Th., Astron. Astrophys., **332**, 291 (1998)
27. Jones, A.R., Progr. Energy Combust. Sci., **25**, 1 (1999)
28. Kastner, J.H. and Weintraub, D.A., In: "Polarimetry of the Interstellar Medium", Eds., W.G. Roberge and D.C.B. Whittet, ASP Conf. Ser., **97**, 212 (1996)
29. Kokhanovsky, A., Optics of Light Scattering Media: Problems and Solutions, John Wiley, Chichester (1999)
30. Kolokolova, L. and Gustafson, B.Å.S., J. Quant. Spectrosc. Rad. Transfer, **70**, 611 (2001)
31. Krügel, E. and Siebenmorgen, R., Astron. Astrophys., **288**, 929 (1994)
32. Lumme, K., Ed., J. Quant. Spectrosc. Rad. Transfer, **60**, N 3 (1998)
33. Martin, P.G., Clayton, G.C. and Wolff, M.J., Astrophys. J., **510**, 905 (1999)
34. Mathis, J.S., Rumpl, W. and Nordsieck, K.H., Astrophys. J., **217**, 425 (1977)
35. Mishchenko, M.I. and Travis, L.D., Appl. Optics, **33**, 7206 (1994)
36. Mishchenko, M.I., Hovenier, J. and Travis, L.D., Eds., Light Scattering by Nonspherical Particles, Academic Press, San Francisco (2000)
37. Mishchenko, M.I., Travis, L.D., Hovenier, J., Eds., J. Quant. Spectrosc. Rad. Transfer, **63**, N 2-6 (1999)
38. Molster, F.J. PhD Thesis, University of Amsterdam (2000)
39. Molster, F.J., Waters, L.B.F.M., Trams, N.R., van Winckel, H., Decin, L., van Loon, J.Th., Jäger, C., Henning, Th., Käufl, H.-U., de Koter, A. and Bouwman, J., Astron. Astrophys., **350**, 163 (1999)
40. Muinonen, K., In: Electromagnetic and Light Scattering by Nonspherical Particles, Eds., B. Gustafson et al., p. 219 (2002)
41. Ossenkopf, V. and Henning, Th., Astron. Astrophys. **291**, 943 (1994)
42. Sasseen, T.P., Hurwitz, M., Dixon, W.V. and Airieau, S., Astrophys. J., **566**, 267 (2002)
43. Schutte, W.A., van der Hucht, K.A., Whittet, D.C.B., Boogert, A.C.A., Tielens, A.G.G.M., Morris, P.W., Greenberg, J.M., Williams, P.M., van Dishoeck, E.F., Chiar, J.E. and de Graauw, Th., Astron. Astrophys., **337**, 261 (1998)
44. Serkowski, K., In: Interstellar Dust and Related Topics, Eds., J.M. Greenberg and D.S. Hayes, IAU Symp., **52**, p. 145 (1973)
45. Siebenmorgen, R., Krügel, E. and Chini, R., Astron. Astrophys. **351**, 495 (1999)
46. Siebenmorgen, R., Krügel, E. and Laureijs, R.J., Astron. Astrophys. **377**, 735 (2002)
47. Sihvola, A.H., Electromagnetic Mixing Formulas and Applications, Institute of Electrical Engineers, Electromagnetic Waves Series 47, London (1999)
48. Smith, C.H., Wright, C.M., Aitken, D.K., Roche, P.F. and Hough, J.H., Monthly Not. RAS, **312**, 327 (2000)
49. Snow, T.P. and Witt, A.N., Astrophys. J., **468**, L65 (1996)
50. Spoon, H.W.W., Keane, J.V., Tielens, A.G.G.M., Lutz, D., Moorwood, A.F.M. and Laurent, O., astro-ph/0202163 (2002)
51. Tielens, A.G.G.M. and Whittet, D.C.B., In: Molecules in Astrophysics: Probes and Processes, Ed., E.F. van Dishoeck, IAU Symp., **178**, p. 45 (1997)
52. van de Hulst, H.C., Light Scattering by Small Particles, John Wiley, New York (1957)
53. Videen, G. and Chýlek, P., Optics Comm., **158**, 1 (1998)
54. Videen, G., Fu, Q. and Chýlek, P., Eds., J. Quant. Spectrosc. Rad. Transfer, **70**, N 4-6 (2001)

55. Voshchinnikov, N.V., Sov. Astron., **21**, 693 (1977)
56. Voshchinnikov, N.V., In: Itogi Nauki i Tekhniki, Ser. Issledovanya Kosmicheskogo Prostranstva, **25**, 98 (1986).
57. Voshchinnikov, N.V., In: IRS 2000: Current Problems in Atmospheric Radiation, Eds., W.L. Smith and Yu.M. Timofeyev, A. Deepak Publ., p. 237 (2001)
58. Voshchinnikov, N.V., "Optics of Cosmic Dust. I", Astrophysics and Space Physics, **12**, 1 (2002)
59. Voshchinnikov, N.V. and Il'in, V.B., Sov. Astron. Lett., **9**, 101 (1983)
60. Voshchinnikov, N.V. and Semenov, D.A., Astron. Lett., **26**, 679 (2000)
61. Voshchinnikov, N.V., Semenov, D.A. and Henning, Th., Astron. Astrophys., **349**, L25 (1999)
62. Voshchinnikov, N.V., Il'in, V.B., Henning, Th., Michel, B. and Farafonov, V.G., J. Quant. Spectrosc. Rad. Transfer, **65**, 877 (2000)
63. Whittet, D.C.B., In: "Polarimetry of the Interstellar Medium", Eds., W.G. Roberge and D.C.B. Whittet, ASP Conf. Ser., **97**, 125 (1996)
64. Weintraub, D.A., Goodman, A.A. and Akeson, R.I., In: "Protostars and Planets IV", Eds., V. Mannings et al., p. 247 (2000)
65. Wilking, B.A., Lebofsky, M.J. and Rieke, G.H., Astron. J., **87**, 695 (1982)
66. Wolf, S., Voshchinnikov, N.V. and Henning, Th., Astron. Astrophys., **385**, 365 (2002)
67. Wolff, M.J., Clayton, G.C. and Gibson, S.J., Astrophys. J., **503**, 815 (1998)
68. Wriedt, T., Part. Part. Syst. Charact., **15**, 67 (1998)

INFRARED SPECTROSCOPY OF COSMIC DUST

G. J. FLYNN[1], TH. HENNING[2,3], L. P. KELLER[4] AND H. MUTSCHKE[2]
[1]*Department of Physics, SUNY, Plattsburgh, NY 12901, USA*
[2]*Astrophysikalisches Institut und Universitäts-Sternwarte Friedrich-Schiller-Universität, D-07745 Jena, Germany*
[3]*Max-Planck-Institut für Astronomie, Königstuhl 17, D-69117 Heidelberg, Germany*
[4]*NASA Johnson Space Center, Houston, TX 77051, USA*

Abstract. Cosmic dust particles are an interesting system of small particles with sizes on the nanometer- to micrometer scale. This paper reviews recent progress in infrared spectroscopy of cosmic dust. The main topics are laboratory measurements on cosmic dust analogues and the spectroscopic characterization of primitive material in meteorites and interplanetary dust grains.

1. Introduction

Cosmic dust particles actively participate in the physical processes which determine the thermal, dynamical, and chemical state of the cold and dense phases of the interstellar medium. They provide the surface for chemical reactions, dominate the opacity of the medium and contribute to heating by the photoelectric effect. The grains undergo a lifecycle from the formation in the outflows of evolved stars, their modification by energetic ions, UV radiation and attack of hydrogen atoms in the diffuse interstellar medium to the formation of ice mantles and the coagulation of particles in shielded molecular environments [19].

Information about this unique system of small particles can be obtained by astronomical spectroscopy, abundance studies, modelling of the dusty sources, and in-situ studies of primitive material in meteorites and interplanetary dust grains. These investigations have revealed a large diversity

G. Videen and M. Kocifaj (eds.), Optics of Cosmic Dust, 37–56.

of materials and grain sizes. Silicates and other oxides, carbonaceous grains, sulfides and molecular ices have been identified in interstellar space. Grain sizes range from the nanometer scale to several microns in very dense regions. In this paper, we summarize the information which has been obtained by infrared spectroscopy.

2. Astronomical Infrared Spectroscopy

The infrared wavelength range is the key wavelength region for astronomical spectroscopy of dust. Stretching, bending, rocking, wagging, and other modes of molecular bonds as well as lattice modes of solids are located in the infrared. Astronomical infrared spectroscopy started in the late-1960s when single-channel infrared detectors became available for astronomical observations. The spectral resolution of these observations was relatively low, ranging between 30 and 150 (see, e.g., [67]). In addition, infrared observations from the ground suffer always from atmospheric extinction. Therefore, spectroscopic studies concentrated mainly on the wavelength intervals 2-4 μm and 8-13 μm, with few measurements performed around 18 μm. Silicate emission and absorption bands, spectroscopic evidence for silicon carbide and various molecular ices could be detected.

The Kuiper Airborne Observatory (KAO) provided the first dust spectra at longer wavelengths [30]. The IRAS satellite delivered low-resolution data over the wavelength range 7.7-22.6 μm. The Infrared Space Observatory ISO was the first spaced-based observatory which produced spectroscopic data over the whole wavelength range of interest, from 2.5 to 200 μm. The ISO results are summarized in [17]. However, spectra could only be obtained for relatively bright objects. The spatial resolution of ISO was very limited because of the small size of the telescope. This turns out to be a disadvantage if one wants to trace the spatial distribution of circumstellar material. The situation is presently changing with more mid-infrared instruments becoming available at 10m-class telescopes.

For the investigation of cosmic dust, ISO contributed unique new information [18]. Important ISO results include the complete inventory of molecular ices (H_2O, CO, CO_2, CH_3OH, CH_4, $HCOOH$, OCN^-, NH_3) and the various emission bands associated with Polycyclic Aromatic Hydrocarbons (PAHs). The detection of crystalline silicates in various classes of objects, ranging from comet Hale-Bopp to Herbig Ae/Be stars to more evolved objects, opened the new field of astromineralogy [34]. It turned out that the crystalline silicates are predominantly the magnesium-rich silicates forsterite (Mg_2SiO_4) and enstatite ($MgSiO_3$) (see, e.g., [43], [65], [53]).

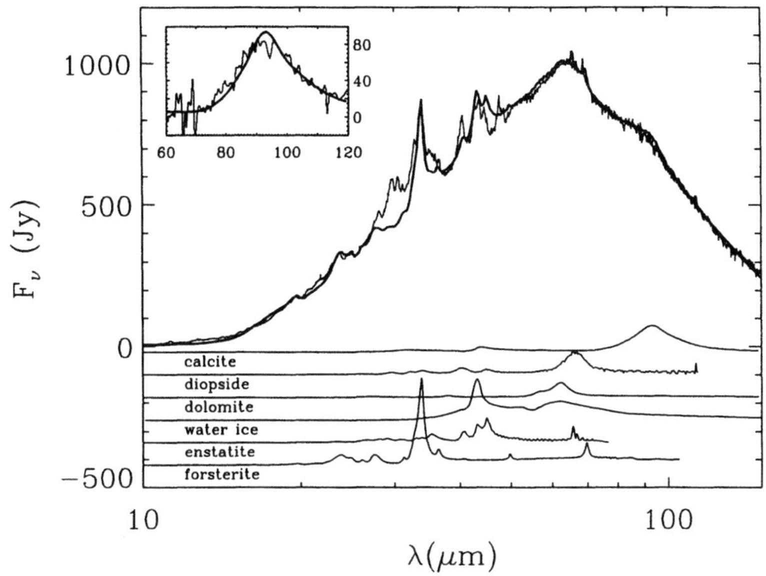

Figure 1. The rich dust spectrum of the planetary nebula NGC 6302. The observed ISO spectrum (thin line) is shown together with a model (solid line) containing different dust components. The inset shows the observed 92.6 μm feature, derived by subtracting all contributions of previously identified dust species from the observed spectrum, overplotted by the modelled contribution of calcite. After Kemper *et al.* (2002).

New dust components such as carbonates [45] (see Fig.1) and spinel were identified [22]. The identification of certain features in astronomical spectra requires that optical data of cosmic dust analogues be provided by laboratory measurements (section 4). Apart from the production and measurement of astronomically relevant materials, presolar material can be identified in meteorites and interplanetary dust grains on the basis of their isotopic anomalies (section 3). In the next step, the infrared spectra of these grains can be measured (section 4).

3. Primitive Material in IDPs and Meteorites

3.1. METEORITES

Primitive meteorites, those which experienced minimal thermal processing on their parent bodies, can preserve ancient interstellar grains that were present in the Solar Nebula 4.5 billion years ago, at the time of Solar

TABLE 1. Interstellar grains found in meteorites. Sizes after Brearley and Jones 1998

Grain Type	Size micrometer
Diamond	~0.002
SiC	<0.05 - 3
TiC	0.01 - 0.07
Graphite	0.1 - 0.3
Oxides: Mostly Al_2O_3 and rare $MgAl_2O_4$	~ 0.5 - 5
Si_3N_4	No data

System formation. Some of these ancient presolar grains can be identified because they exhibit isotopic compositions significantly different from the Solar system values. Rare, refractory interstellar grains have been isolated from primitive meteorites, usually by acid etching procedures that dissolve the silicates and other common materials in the meteorites (see review by [2]). These stardust grains include diamond, graphite, corundum, titanium carbide, and silicon carbide (see Table 1). The stardust grains occur in meteorites in trace concentrations, ranging from 30 ppb for corundum [55] up to ~ 600 ppm for the nanodiamonds. Because of the small size of the nanodiamonds, isotopic analyis of individual particles is not yet possible. The bulk carbon isotopic composition does not differ from the solar values, and the inference of an interstellar origin is based on noble gas isotopic patterns from a large number of diamonds. Therefore, we do not know if all of the nanodiamonds, only a fraction of the nanodiamonds, or even another unrecognized minor phase mixed with the nanodiamonds is the presolar material.

Since these interstellar grains have never equilibrated with Solar System material, as demonstrated by the preservation of their non-Solar isotopic signatures, each interstellar grain carries a record of the nucleosynthesis of the particular environment in which it formed. Laboratory analysis of the refractory circumstellar grains isolated from the primitive meteorites has provided significant new understanding of the chemical conditions, time scales, and stellar processes relevant to the circumstellar envelopes in which these grains formed.

However, none of the interstellar grains recovered from primitive meteorites are silicates or organic matter - the two most common materials identified in the grains in the interstellar medium. Despite extensive searches, the analyses of meteoritic silicates have not identified any grains with the non-Solar isotopic signatures that would indicate an interstellar origin. The absence of interstellar silicates in the primitive meteorites may indicate that either

1. interstellar silicates were rare or absent from the population of grains in the Solar Nebula; or,
2. interstellar silicates were too reactive to survive the nebula formation processes or the subsequent thermal and/or aqueous processing on the meteorite parent bodies.

The first alternative seems unlikely, since Si is quite abundant in the Solar System, and some of it is likely to have been carried into the Solar Nebula in the form of grains. If the second alternative is the case, then the isolation of interstellar silicate and interstellar organic grains requires the examination of even more primitive samples of interplanetary material - samples that have experienced much less thermal and aqueous processing than the most primitive meteorites.

3.2. INTERPLANETARY DUST PARTICLES

Interplanetary dust particles (IDPs), ranging from about 5 μm to over 50 μm in size, have been collected from the Earth's stratosphere by NASA aircraft since the mid-1970s [12]. Most of the IDPs are believed to be fragments from asteroids and comets, both of which are known to contribute a significant mass of dust to the inner Solar System based on the IRAS detection of dust bands associated with the Eos, Koronis and Themis families of asteroids [50] and the dust trails associated with several comets [63]. Some IDPs may be contemporary interstellar grains that are currently passing through the Solar System [26] and other IDPs may be from the moons and planets of the Solar System.

The IDPs can be divided into three groups based on the dominant mineralogy of each particle: those dominated by anhydrous silicates (olivine, pyroxene, or glass), those dominated by hydrated silicates (clay minerals), and those dominated by large Fe-sulfide grains. However, many IDPs are mixtures of two or all three of these components.

The surface texture of the IDPs ranges from highly porous, reentrant structures, usually associated with anhydrous IDPs (see Figure 2a), to flat, platy structures, usually associated with hydrated (see Figure 2b) or Fe-sulfide IDPs. The porous, reentrant structures are likely to effect the optical properties of the IDPs [46].

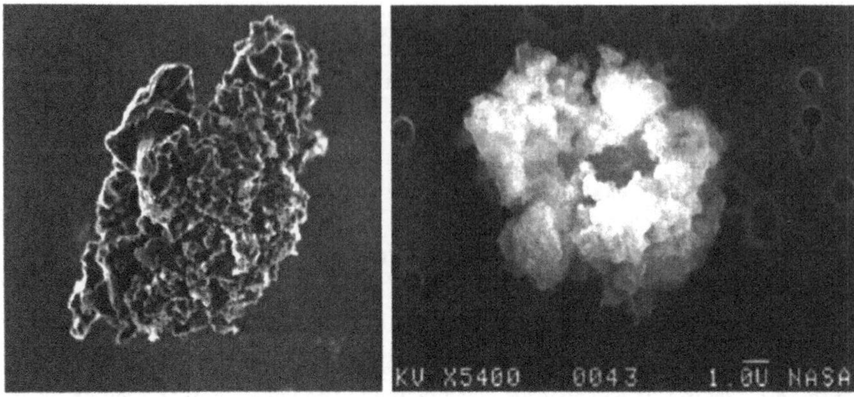

Figure 2. IDPs have surface textures varying from porous, reentrant structures (left image) to flat, platy structures (right image).

The porous, reentrant IDPs must have "bulk densities", the mass divided by the volume of the envelope that would surround the particle, that are much lower than the densities of their mineral constituents. Measurements of the bulk densities of 23 IDPs gave a mean of 1.1 gm/cc, with a bimodal distribution peaking at 0.6 gm/cc and 1.9 gm/cc [24]. Since the silicate minerals that dominate most IDPs have densities near 3 gm/cc, these bulk density measurements demonstrate that most IDPs have significant porosity. Transmission Electron Microscope (TEM) images of ultramicrotome slices of IDPs frequently show pore spaces in the interior (see Figure 3). Fraundorf *et al.* (1982) [29] speculated that these pores were once filled with ices that sublimed either before the particles reached Earth or during heating on atmospheric entry. If so, then the optical and dynamical properties of the IDPs while they are in space may differ significantly from the properties of IDPs measured in the laboratory after collection from the stratosphere.

Many of the anhydrous IDPs have characteristics that indicate they are significantly more primitive than the most primitive meteorites. The presence of moderately volatile minor elements at concentrations 2 to 4 times those in the most primitive CI meteorites [25] and the observation of unequilibrated mineral assemblages [29] indicate that many IDPs are extremely primitive. Some of these IDPs also exhibit localized (on the micrometer or sub-micrometer scale) non-Solar isotopic ratios - particularly enrichments in deuterium and ^{15}N [51]. Non-Solar oxygen isotopic ratios have also been measured in some silicate grains in primitive IDPs, indicating that these IDPs contain preserved interstellar silicates [52], which have

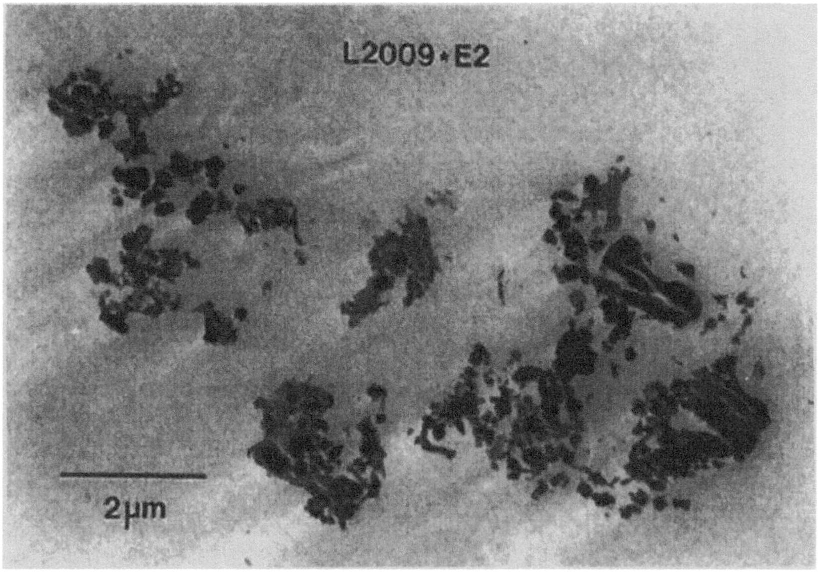

Figure 3. Transmission electron microscope image of an IDP showing the submicron mineral grains and pore spaces.

never been found in meteorites. Messenger *et al.* [52] estimated that at least 1% of the silicates in the primitive IDPs are interstellar, compared to less than 10^{-5} of the silicates in meteorites. These measurements demonstrate the presence of ~micrometer-size interstellar fragments of silicate and H- and N-bearing organic compounds in some IDPs. Thus, the primitive IDPs are the best candidates, among the extraterrestrial material available for laboratory characterization, in which to find and characterize preserved interstellar grains.

4. Laboratory Data of Cosmic Dust Analogues

4.1. PRODUCTION OF COSMIC DUST ANALOGUES

A large variety of different cosmic dust analogues have been produced and spectroscopically characterized [32]. The materials include silicon and iron particles, silicates, metal oxides, carbonates, silicon and titanium carbide, iron and magnesium sulfides, iron silicides, silicon nitrides, carbonaceous solids (carbon black, hydrogenated amorphous carbon, graphite, nanodiamonds), and a large number of molecular ices. The reader can find comprehensive reviews and extensive compilations of references for silicates in the papers by Colangeli *et al.* [16] and Henning [35], for carbides in the

Figure 4. Laser ablation system of the Jena laboratory, used for the production of small carbonaceous and oxide particles.

paper by Henning & Mutschke [37], for carbonaceous solids in the papers by Henning & Schnaiter [38] and Pendleton & Allamandola [60], and for molecular ices in the paper by Ehrenfreund & Schutte [20]. Main production methods for cosmic dust analogues are condensation of particles from the gas phase (laser-induced pyrolysis, microwave plasma processes, combustion), evaporation, sputtering or melting of material and subsequent quenching, and sol-gel reactions. Crystalline silicates can be produced by condensation from the gas phase [64] or by annealing amorphous precursors [21], [31], [13]. Figure 4 displays the laser ablation system of the Jena group which has been frequently used to produce carbonaceous particles and oxide grains.

4.2. FROM OPTICAL CONSTANTS TO FLUXES

A primary goal of laboratory measurements is the acquisition of optical constants or absorption and scattering cross sections of well-characterized materials. The necessary basic knowledge indicating how optical data of small particles can be measured has been recently summarized by Henning & Mutschke (2000) [36] and will not be repeated here. The optical constants are directly related to the phase velocity and attenuation of waves in the material. Compilations of frequency-dependent optical constants of solids

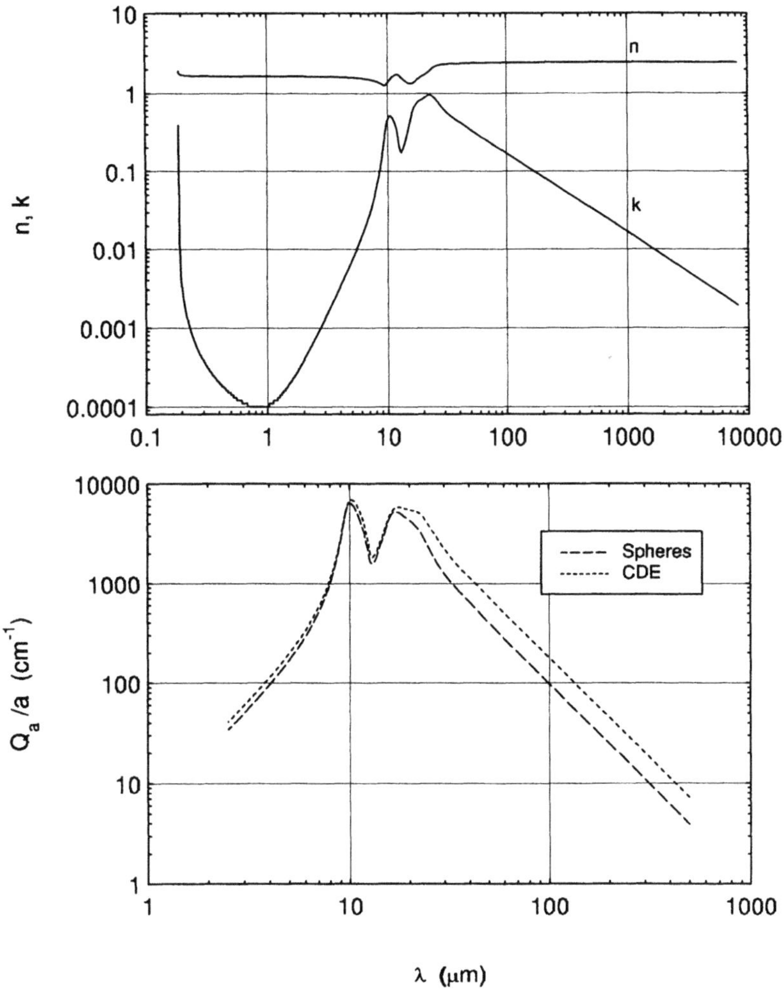

Figure 5. Optical constants n and k and absorption efficiency Q_a divided by radius for spheres and a distribution of ellipsoids for amorphous $Mg_{2.4}SiO_{4.4}$.

can be found in the various editions of the "Handbook of Optical Constants" [57], [58], [59]. An electronic database [40], especially dedicated to cosmic dust analogues, can be accessed via the web page http://www.astro.uni-jena.de. An example of a measurement of optical constants and calculated absorption efficiencies for an amorphous magnesium silicate produced by the sol-gel method is given in Figure 5.

The absorption and scattering cross sections of small particles depend not only on the optical constants (or the dielectric function), but also on the particle size and morphology and a surrounding mantle if it exists. The reason for this behaviour is the occurrence of geometrical resonances (sometimes called surface modes) which dominate the absorption if the band strength is large enough to produce negative values of the real part of the dielectric function. We should note that many crystalline solids, including the astronomically interesting crystalline silicates, show an anisotropic optical behaviour. In these cases, measurements with polarized light along the different axes of the crystal have to be carried out. In addition, the optical data may depend on the temperature, leading to a change of the width and wavelength position of sharp bands of crystalline grains or a change of the far-infrared continuum. In the case of simple particle shapes or agglomerates, the absorption and scattering cross sections can be calculated from the optical constants [5]. This fact recently has been used to constrain the particle shape of circumstellar olivine particles [23]. The comparison between the calculated mass absorption coefficients and the astronomical spectra of optically thin circumstellar envelopes indicates that both sphere-like particles and ellipsoids elongated along the z-axis are necessary to obtain a convincing band assignment.

Astronomers obviously do not measure optical constants or cross sections, but fluxes. In the case of optically thin dust configurations, absorption cross sections can be derived from astronomical spectra under certain simplifying assumptions [39], [56]. For complicated source structures with higher optical depth, radiative-transfer calculations are necessary in order to compare measured and calculated fluxes [33]. A nice example of radiative-transfer (temperature) effects can be found in the ISO spectra. The relatively low temperatures of circumstellar crystalline silicates are a simple explanation for the presence of strong emission features in the ISO spectra beyond 20 μm in comparison with the weak 11.3 μm feature.

4.3. NEW LIGHT SOURCES

The IDPs, which are the most primitive extraterrestrial materials available for laboratory analysis, range from 5 μm to about 25 μm in size. The anhydrous IDPs are aggregates of submicron and micron-size grains. Thus a bulk analysis examines many different types of grains at once. To analyze a single, isolated mineral for comparison with an interstellar absorption or emission feature, it is necessary to prepare ~100 nm thick sections by ultramicrotomy, and to analyze small, isolated, mono-mineralic areas of those thin sections. Traditional laboratory infrared spectrometers produce very noisy spectra from samples that are this small and thin.

TABLE 2. Synchrotron and Free Electron Laser Infrared Facilities

Infrared Beamlines	Status in 2002
Synchrotron Facilities NSLS (Brookhaven, USA)	6 Operational
ALS (Berkeley, USA)	3 Operational
Aladdin (Wisconsin, USA)	1 Operational
NIST (Gaithersburg, USA)	1 Operational
SRC (Daresbury, UK)	1 Operational
LURE (Paris, France)	2 Operational
MAX-Lab (Lund, Sweden)	1 Operational
UVSOR (Okasaki, Japan)	2 Operational
DAFNE (Rome, Italy)	1 Operational
ANKA (Karlsruhe, Germany)	1 Operational
BESSY-2 (Berlin, Germany)	1 Operational
SPring-8 (Nishi-Harima, Japan)	1 Operational
SRRC (Hsinchu, Taiwan)	1 Planned
CLS Saskatoon (Canada)	2 Planned
NSRL (Hefei, China)	1 Planned
CAMD (Baton Rouge, USA)	1 Planned
DELTA (Dortmund, Germany)	1 Planned
SRC (Daresbury, UK)	1 Planned
ESRF (Grenoble, France)	1 Planned
SOLEIL, (France)	1 Planned
SLS (Villigen, Switzerland)	1 Planned
JLab (Newport News, USA)	1 Planned
FEL Facilities JLab (Newport News, USA)	1 Operational
Stanford Picosecond FEL (Palo Alto, USA)	1 Operational
FELIX-FOM (Rijnhuizen, The Netherlands)	1 Operational

The high intensity of the infrared beamlines at synchrotron light sources and free electron lasers (FEL) allows the analysis of very small areas (down to 3x3 μm at 3000 cm^{-1}) or very thin samples ($\sim$80 nm thick sections). These synchrotron and FEL infrared sources have IR fluxes 1,000 or more times the flux obtained from a typical globar IR source. Using synchrotron based micro-FTIR, infrared spectra have been obtained from ultramicrotome sections, $\sim$80 nm thick, of silicates in IDPs.

4.4. SPECTROSCOPY OF IDPS AND METEORITES

The examination of IDPs has allowed comparison of the infrared absorption features of silicate, sulfide, and organic material in the IDPs with the infrared emission and absorption features in interstellar and circumstellar environments.

Silicate Features:

Interstellar grains in molecular clouds and the diffuse interstellar medium exhibit two strong absorption features assumed to result from silicates: ~9.5 μm absorption, attributed to Si-O stretching vibrations, and, ~18 μm absorption, attributed to Si-O-Si bending vibrations. Both features are quite broad, indicating an "amorphous," glassy or radiation damaged, silicate. Over the past several decades there have been numerous attempts to match the interstellar 9.5 μm and 18 μm absorption features with laboratory spectra of naturally occurring and synthetic silicates.

Bradley (1994) [10] suggested that ~0.5 μm glass subunits, containing nanometer size Fe-Ni-metal and Fe-sulfides, which he called GEMS (for Glass with Embedded Metal and Sulfides), found in interplanetary dust were good candidates for interstellar silicate. If the GEMS are not interstellar silicate, then perhaps the grain formation process in our Solar System was sufficiently similar to the grain formation processes in interstellar or circumstellar regions that glass subunits in the primitive IDPs will provide a spectral match to the IR spectral features seen in the interstellar medium, allowing identification of the minerals responsible for the features.

Bradley *et al.* (1999) [11] measured the infrared absorption of silicates in an IDP named L2011*B6, which is an 12x8 μm fragment from an interplanetary dust particle that was so weak that it broke up into tens of fragments during impact collection. L2011*B6 is dominated by Fe-sulfide, but contains two lobes of GEMS as well as some carbonaceous material associated with the GEMS (see Figure 6). Other fragments from L2011*B6 have deuterium and ^{15}N excesses, indicating the particle contains some preserved interstellar material [51]. The spectrum of GEMS in L2011*B6 provides an excellent match to the 9.5 μm feature of interstellar grains (see Figure 7).

Bradley *et al.* (1999) [11] measured the IR spectra of glass with embedded metals and sulfide (GEMS) in several interplanetary dust particles, and found that not all GEMS are the same. Many GEMS contain small olivine or pyroxene grains as well as Fe-sulfides and Fe-metal. Bradley (1994) [10] has suggested the GEMS started as crystalline silicate, which became amorphous by radiation damage, but sometimes small crystals survived in the interior of the glass. These GEMS contain the spectral signatures of the crystalline silicate or other mineral inclusion as well as the glass. The infrared spectra of some of the GEMS that contain some crystalline silicate

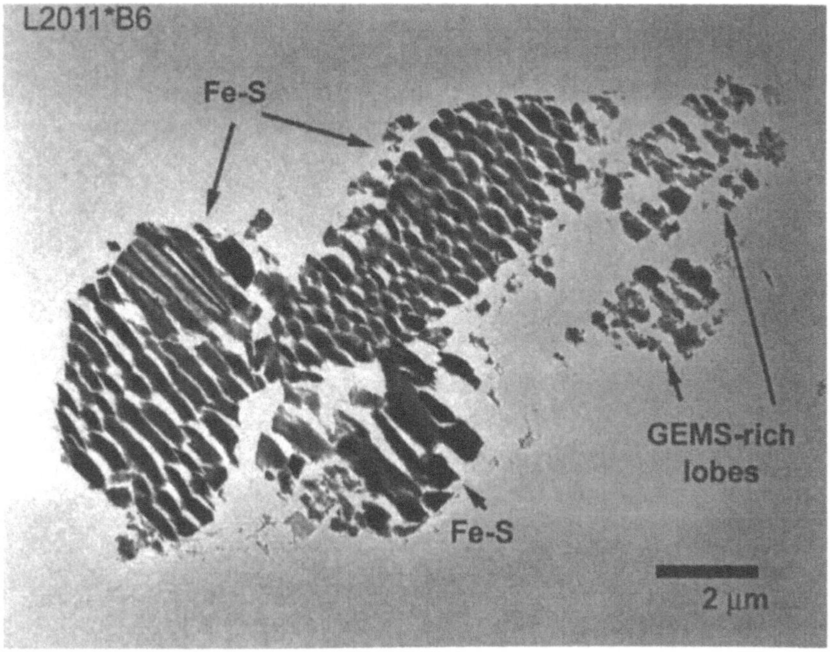

Figure 6. Transmission electron microscope image of an ultramicrotome section of L2011*B6, a 12x8 μm fragment from a larger anhydrous IDP. This particle is dominated by two Fe-sulfide lobes and to GEMS-rich lobes.

is an excellent match to the spectral features of comet Halley, comet Hale-Bopp, and the dust surrounding the young star HD163296 (see [11]).

Sulfide Features:

The ISO spectra of some cold, dense molecular clouds and in evolved stars show an excess flux between 20 μm and 26 μm. This feature cannot be from amorphous silicate, which has its peak absorption near 18 μm, or crystalline silicates that exhibit much narrower peaks in the infrared. The feature was attributed to FeO based on calculated spectra derived from experimentally determined optical constants combined with theoretically modeled grain shapes [6]. But FeO (wüstite) is thermodynamically metastable and extremely rare in meteorites.

Sulfur is highly depleted in the gas phase in cold molecular clouds. The solid S-carrier in the molecular clouds is unknown, but Fe-sulfide, particularly pyrrhotite, is the main sulfur mineral in interplanetary dust and meteorites. Keller *et al.* (2002) [44] measured the infrared spectra of a pyrrhotite standard as well as Fe-sulfides in two IDPs. The pyrrhotite standard is an

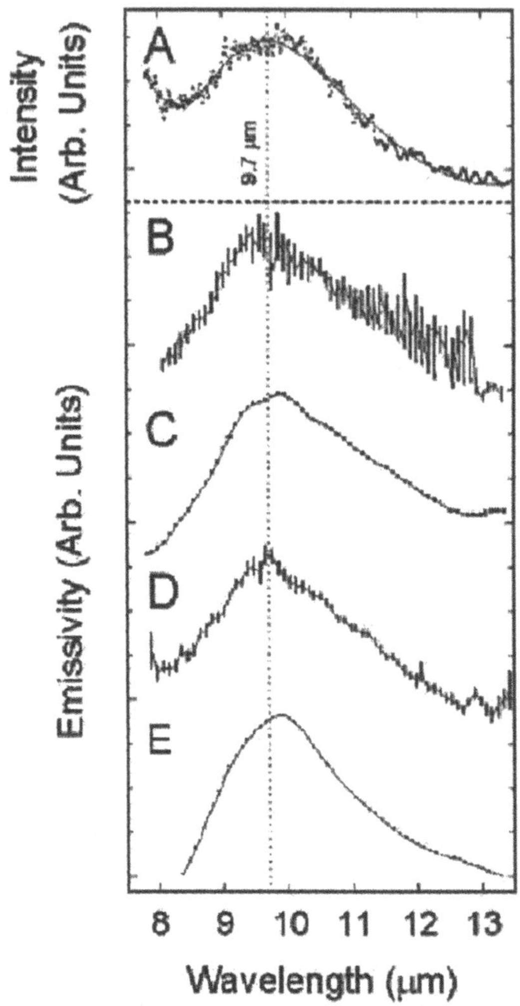

Figure 7. Comparison of the 10 μm Si–O stretch bands of GEMS with those of astronomical silicates (after Bradley *et al.* 1999): (A) GEMS in IDP L2011*B6; (B) Elias 16 molecular cloud; (C) Trapezium molecular cloud; (D) Pre main-sequence T Tauri YSO DI Cephei; (E) Post main-sequence M type supergiant μ-Cephei.

excellent match to the ISO feature (see Figure 8) while the Fe-sulfides in the two IDPs have narrower features. These differences may be due to a difference in composition - pyrrhotite has a range of $Fe_{1-x}S$, with $0<x<0.2$. Thus Fe-Ni sulfides appear to be a common, previously unrecognized, dust species in many circumstellar environments.

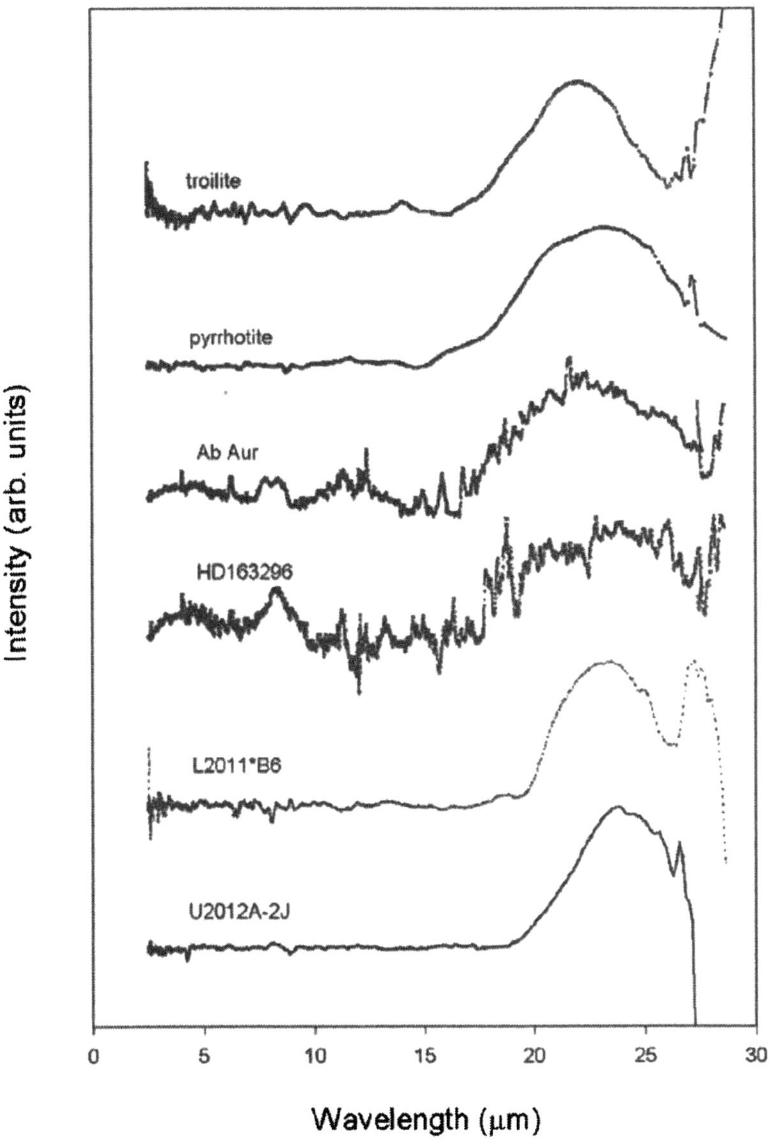

Figure 8. Comparison of the infrared absorption of Fe sulfides (pyrrhotite, troilite) and sulfide-rich IDPs (L2011*B6, U2012A-2J) with the absorption detected by ISO in two Herbig stars.

The identification of abundant Fe-sulfides in circumstellar environments may be particularly significant for astrobiology because Llorca and Casanova (2000) [49] and Cody et al. (2000) [15] have recently suggested sulfide grains may have played a catalytic role in the synthesis of organic molecules in the solar nebula and on the early Earth.

Organic Feature:

C-H stretching vibrations, which occur in the 2800 to 3000 cm^{-1} range, have been detected in interstellar grains, demonstrating the presence of organic matter. It has been suggested that many interstellar grains are amorphous silicates covered by a mantle of organic matter [48].

Spectropolarimetric data, probing the line of sight through the interstellar medium to the Galactic center source Sagittarius A IRS 7, show that this feature is largely unpolarized [1]. This speaks against a model where the carrier material is associated with the aligned silicate component of interstellar dust, as it is the case in silicate core-organic mantle models. The simplest explanation of this observation is that the carrier of the feature resides in a population of small, nonpolarizing grains, physically separate from the silicates. Another important result of the ISO observations of the *diffuse* interstellar medium is that the carbonaceous material is poor in oxygen and that no strong evidence for the presence of water ice exists [66], [62], [14].

Flynn *et al.* [27] [28] measured the C-H stretching absorptions in 9 anhydrous interplanetary dust particles, and detected C-H stretching absorptions at 2960 and 2870 cm^{-1} from $C\text{-}H_3$ and at 2925 and 2854 cm^{-1} from $C\text{-}H_2$ in an aliphatic hydrocarbon. Features at the same energies have been detected in the interstellar grains (see, e.g., Sandford *et al.* (1991) [61] and the recent detailed review in [60]), but the relative strengths of the $C\text{-}H_3$ and $C\text{-}H_2$ features are different from those in the IDPs. The simplest aliphatic molecules are chains of $C\text{-}H_2$ groups terminated on both ends by $C\text{-}H_3$. Thus, in this simple case, the ratio of $C\text{-}H_2$ to $C\text{-}H_3$ absorption features is a measure of the length of the aliphatic molecule. The more intense $C\text{-}H_3$ absorption, relative to $C\text{-}H_2$, in the interstellar feature shown in [60] compared to the IDP spectra [27] suggests that the aliphatic chain is shorter in the interstellar material than in the IDPs.

Interstellar Grains from Meteorites:

Although interstellar and circumstellar grains are present in primitive meteorites in only trace amounts, they can be concentrated using acid extraction techniques that dissolve the major minerals including the silicates, oxides, carbonates and sulfides. Thus, enough material can be extracted to allow infrared analysis using conventional infrared spectrometers. Infrared spectra of SiC grains extracted from the Murchison meteorite depend on the grain size, with small SiC having an absorption feature near 11.3 μm, consistent with a feature seen in carbon stars, while SiC grains greater than 5 μm in size have an absorption centered around 10 μm [4].

The nanodiamonds extracted from meteorites are only a few nanometers in size, so infrared spectra are aquired from groups of these diamonds. The first infrared spectrum of meteoritic nanodiamonds was published by [47]. While diamonds generally show a broad absorption near 21 μm [42], the meteoritic nanodiamonds also show features near 3200 cm^{-1}, attributed to O-H, in the 2800 to 3000 cm^{-1}, attributed to C-H stretching modes, and near 1100 cm^{-1} [54] [3]. However [54] and [8] suggest these latter features are either partially or completly artifacts from the chemical extraction process. Searches for isolated diamond in the meteorites have found only individual nanodiamonds [7] that are too small for the measurement of infrared spectra.

5. Conclusions

Astronomical infrared spectroscopy performed with the Infrared Space Observatory over a large wavelength region opened a new chapter in the investigation of the nature of cosmic dust. With the SIRTF mission, an infrared space telescope scheduled for launch in 2003, we will be able to obtain data from fainter objects. The SOFIA airborne observatory and the ground-based Herschel telescope will provide data with high spectral and spatial resolution.

Enormous progress has been made in the production, analysis, and spectroscopic characterization of cosmic dust analogues. Future work will concentrate on the formation of particles from the gas phase, the modification of grains by high-energy ions, UV radiation, and reactions with hydrogen and other atoms, and the growth process of the particles by grain-grain collisions. The spectroscopic characterization of very small grains and large molecules is another area of strong interest for laboratory astrophysics.

The combination of intense synchrotron and FEL based infrared sources with modern infrared spectroscopic analysis instruments has made it possible to analyze, frequently on a grain-by-grain scale, the mineralogical constituents of IDPs and meteorites. This provides laboratory spectra for comparison with astronomical observations of the emission and absorption features of the interstellar grains. Finally, the Stardust mission will return primitive grains to Earth for direct analysis.

6. Acknowledgments

We want to thank Johann Dorschner, Cornelia Jäger and Gwyn Williams for useful discussions. This work was supported by the DFG grant "Research Group Laboratory Astrophysics Chemnitz/Jena" and a NASA Cosmochemistry Grant NAG 5-4843 (GJF).

References

1. A.J. Adamson, D.C.B. Whittet, A. Chrysostomou, J.H. Hough, D.K. Aitken, G.S. Wright, and P.F. Roche, Astrophys. J. **512**, 224 (1999).
2. E. Anders and E. Zinner, Meteoritics **28**, 490 (1993).
3. A.C. Andersen, U.G.Jorgensen, F.M. Nicolaisen, P.G. Sorensen, and K. Glejbjol, Astron. Astrophys. **330**, 1080 (1998).
4. A.C. Andersen, C. Jäger , H. Mutschke, A. Braatz, D. Clement, Th. Henning, U.G. Jorgensen, and U. Ott, Astron. Astrophys. **343**, 933 (1999).
5. C.F. Bohren and D.R. Huffman: "Absorption and Scattering of Light by Small Particles" (Wiley, New York, 1983).
6. J. Bouwman, A. de Koter, M.E. van den Ancker, L.B.F.M. Waters, Astron. Astrophys. **360**, 213 (2000).
7. F. Banhart, Y. Lyutovich, A. Baatz, C. Jäger, T. Henning, J. Dorschner, and U. Ott, Meteoritics Planet. Sci. **33**, S4, A12
8. (1998). A. Braatz, J. Dorschner, Th. Henning, C. Jäger, and U. Ott, Meteoritics Planet. Sci. **33**, S4, 21 (1998).
9. A. J. Brearley and R. H. Jones: "Chondritic Meteorites", in "Planetary Materials", Ed. by J. J. Papike (Min. Soc. of America, Washington, 1998), p. 3-01.
10. J. P. Bradley, Science **265**, 925 (1994).
11. J. P. Bradley, L. P. Keller, T. P. Snow, M. S. Hanner, G. J. Flynn, J. C. Gezo, S. J. Clemett, D. E. Brownlee, and J. E. Bowey, Science **285**, 1716 (1999).
12. D. E. Brownlee: "Microparticle Studies By Sampling Techniques", in "Cosmic Dust", Ed. by J. A. M. McDonnell (John Wiley & Sons, New York, 1978), p. 295.
13. J.R. Brucato, L. Colangeli, V. Mennella et al., Astron. Astrophys. **348**, 1012.
14. J.E. Chiar, A.G.G.M. Tielens, D.C.B. Whittet *et al.* Astrophys. J. **537**, 749 (2000)
15. G. Cody et al., Science **289**, 1337 (2000).
16. L. Colangeli, Th. Henning, J.R. Brucato et al., Astron. Astrophys. Rev. (2002), in press.
17. P. Cox and M.F. Kessler (eds.): "The Universe as seen by ISO". ESA SP-427, Vol. 1 and 2 (1999).
18. L. dHendecourt, C. Joblin, and A. Jones (eds.): "Solid Interstellar Matter: The ISO Revolution" (EDP Sciences, Les Ullis, 1999)
19. J. Dorschner and Th. Henning, Astron. Astrophys. Rev. **6**, 271 (1995).
20. P. Ehrenfreund and W.A. Schutte: "Infrared Observations of Interstellar Ices", in "Astrochemistry: From Molecular Clouds to Planetary Systems", Ed. by Y.C. Minh and E.F. van Dishoeck (IAU Symp. 197, 2000), p.135.
21. D. Fabian, C. Jäger, Th. Henning, J. Dorschner, H. Mutschke, Astron. Astrophys. **364**, 282 (2000)
22. D. Fabian, Th. Posch, H. Mutschke, F. Kerschbaum, and J. Dorschner, Astron. Astrophys. **373**, 1125 (2001).

23. D. Fabian, Th. Henning, C. Jäger, H. Mutschke, J. Dorschner, and O. Wehrhan, Astron. Astrophys., **378**, 228 (2001).
24. G. J. Flynn and S. R. Sutton, Proc. of Lunar and Planet. Sci. **21**, 541 (1991).
25. G. J. Flynn, S. Bajt, S. R. Sutton, M. Zolensky, K. L. Thomas, and L. P. Keller: "The Abundance Pattern of Elements Having Low Nebular Condensation Temperatures in Interplanetary Dust Collected from the Stratosphere", in "Analysis of Interplanetary Dust", Ed. by B. A. S. Gustafson and M. S. Hanner (AIP Press, New York, 1996), p. 291.
26. G. J. Flynn, Nature **387**, 248 (1997).
27. G.J. Flynn, L. P. Keller, and M. A. Miller, Lunar Planet. Sci. XXIX, Abs. No. 1157 (1998).
28. G.J. Flynn, L.P. Keller, C. Jacobsen, S. Wirick, and M. A. Miller,"Organic Carbon in Interplanetary Dust", in "A New Era in Bioastroomy", ASP Conf. Series **213**, 191 (2000). P. Fraundorf, Geochim. Cosmochim. Acta **45**, 915 (1981).
29. P. Fraundorf, D. E. Brownlee, and R. M. Walker: "Laboratory Studies of Interplanetary Dust", in "Comets", Ed. by L. L. Wilkening (U. of Arizona Press, Tucson, 1982), p. 383.
30. M.R. Haas, J.A. Davidson, and E.F. Erickson (eds.): "Airborne Astronomy Symposium on the Galactic Ecosystems: From Gas to Stars to Dust" (ASP, San Francisco, 1994)
31. S.L. Hallenbeck, J.A. Nuth, and P.L. Daukantas, Icarus **131**, 198 (1998).
32. Th. Henning: "Infrared Spectroscopy of Cosmic Grains - Contributions of Laboratory Astrophysics", in "ISO beyond the Peaks" , Ed. by A. Salama, M.F. Kessler, K. Leech, B. Schulz (ESA SP-456, 2000), p. 329.
33. Th. Henning: "Frontiers of Radiative Transfer", in "The Formation of Binary Stars", Ed. by H. Zinnecker and R. D. Mathieu (ASP, San Francisco, 2001), p.567.
34. Th. Henning (ed.): "Astromineralogy" (Springer, Berlin, 2002), in press.
35. Th. Henning: "Cosmic Silicates - A Review", in "Solid State Astrochemistry", Ed. by V. Pirronello, J. Krewlowski (Kluwer, Dordrecht, 2002), in press.
36. Th. Henning and H. Mutschke: "Optical Properties of Cosmic Dust Analogs", in "Thermal Emission Spectroscopy and Analysis of Dust, Disks, and Regoliths" (ASP Conf. Ser. 196, 2000), p. 253
37. Th. Henning and H. Mutschke, Spectrochimica Acta A **57**, 815 (2001).
38. Th. Henning and M. Schnaiter: "Carbon - From Space to the Laboratory", in "Laboratory Astrophysics and Space Research", Ed. by P. Ehrenfreund, H. Kochan, C. Krafft, V. Pirronello (Kluwer, Dordrecht, 1998), p.249.
39. Th. Henning, J. Gürtler, J. Dorschner, Astrophys. Space Sci., **94**, 333 (1983).
40. Th. Henning, V.B. Ilin, N.A. Krivova, B. Michel, N. Voshchinnikov, Astron. Astrophys. Suppl. **136**, 405 (1999).
41. H. Hill, L. D'Hendecourt, C. Perron, and A. Jones, Meteoritics Planet. Sci. **32**, 713 (1997).
42. H.G.M. Hill, A.P. Jones, L.B. D'Hendecourt, Meteoritics Planet. Sci. **32**, A67 (1998)
43. C. Jäger, F.J. Molster, J. Dorschner, Th. Henning, H. Mutschke, and L.B.F.M. Waters, Astron. Astrophys. **339**, 904 (1998).
44. L. P. Keller, S. Hony, J.P. Bradley, J. Molster, L.B.F.M. Waters, J. Bouwman, A. de Koter, D.E. Brownlee, G.J. Flynn, T. Henning, and H. Mutschke, Nature (2002), in press.
45. C. Kemper, C. Jäger, L.B.F.M. Waters, Th. Henning, F.J. Molster, M.J. Barlow, T. Lim, and A. de Koter, Nature **415**, 295 (2002).
46. M. Kocifaj, I. Kapinisjy, and F. Kundracik, J. Quantitative Spectroscopy & Radiative Transfer **63**, 1 (1999).
47. R.S. Lewis, E. Anders, and B.T. Draine, Nature **339**, 117 (1989).
48. A. Li and J.M. Greenberg, Astron. Astrophys. **323**, 566 (1997).
49. J. Llorca and I. Casanova: Meteoritics & Planet. Sci. **35**, 841 (2000).
50. F. J. Low, D. A. Beintema, T. N. Gautier, F. C. Gillett, C. A. Beichman, G. Neu-

genbauer, E. Young, H. H. Aumann, N. Boggess, R. G. Walker, and P. R. Wesselius, Astrophys. J. **278**, L19 (1984).
51. S. Messenger, Nature **404**, 968 (2000)
52. S. Messenger, L. P. Keller, and R. M. Walker: "Discovery of Abundant Interstellar Silicates in Cluster IDPs" Lunar & Planetary Science XXXIII, Abs. 1887 (2002)
53. F.J. Molster, L.B.F.M. Waters, A.G.G.M. Tielens, Astron. Astrophys. **382**, 222 (2002).
54. H. Mutschke, J. Dorschner, Th. Henning, C. Jäger, U. Ott, Astrophys. J. Lett. **454**, L157 (1995)
55. L. Nittler, C.M. Alexander, X. Gao, R.M. Walker, E. Zinner, Astrophys. J. **483**, 475 (1997).
56. V. Ossenkopf, Th. Henning, and J.S. Mathis, Astron. Astrophys. **261**, 567 (1992).
57. E.D. Palik (ed.): "Handbook of Optical Constants" (Academic Press, Orlando, 1985).
58. E.D. Palik (ed.): "Handbook of Optical Constants II" (Academic Press, San Diego, 1991).
59. E.D. Palik (ed.): "Handbook of Optical Constants III" (Academic Press, San Diego, 1998).
60. Y.J. Pendleton and L.J. Allamandola, Astrophys. J. Suppl. **138**, 75 (2002).
61. S.A. Sandford, L.J. Allamandola, A.G.G.M. Tielens, K. Sellgren, M. Tapia, and Y. Pendleton, Astrophys. J. **371**, 607 (1991).
62. W.A. Schutte, K.A. van der Hucht, D.C.B. Whittet et al., Astron. Astrophys. **337**, 261 (1998).
63. M. V. Sykes, L. A. Lobofsky, D. M. Hunten, and F. Low, Science **232**, 1115 (1986).
64. A. Tsuchiyama, Mineral. J. **20**, 59 (1998).
65. L.B.F.M. Waters et al.: "ISO Spectroscopy of Circumstellar Dust", in: "Thermal Emission Spectroscopy and Analysis of Dust, Disks, and Regoliths", Ed. M.L. Sitko, A.L. Sprague, D.V. Lynch (ASP Conf. Ser. 196, 2000), p.3.
66. D.C.B. Whittet, A.C.A. Boogert, P.A. Gerakines et al., Astrophys. J. **490**, 729 (1997).
67. S.P. Willner et al., Astrophys. J. **155**, L81 (1969).

EXPERIMENTAL LIGHT SCATTERING MATRICES RELEVANT TO COSMIC DUST

O. MUÑOZ [1], H. VOLTEN[2,3], AND J.W. HOVENIER[3].
[1]*Instituto de Astrofisica de Andalucia, Granada, Spain*
[2]*FOM-institute AMOLF, Amsterdam, The Netherlands*
[3]*Astronomical institute "Anton Pannekoek", Amsterdam, The Netherlands*

Abstract. A brief overview is given of experimental scattering matrices of natural particles that seem to be most relevant to ensembles of cosmic dust particles. The matrices have been obtained by letting laser radiation at 441.6 nm and 632.8 nm be scattered by collections of small, randomly oriented particles over angles from about 5° – 173°. The samples considered include feldspar, red clay, quartz, loess, Pinatubo and Lokon volcanic ashes, Sahara sand, olivine, Allende meteorite and fly ash. The elements of the scattering matrices as functions of the scattering angle show a number of interesting similarities and differences. These are briefly discussed, especially in relation to particle shape.

1. Introduction

An important optical characteristic of an ensemble of cosmic dust particles is its scattering matrix. This is a real 4×4 matrix, which may be implicitly defined as follows [1]. The Stokes vector of the light scattered in a particular direction by the ensemble is proportional to the product of the scattering matrix and the Stokes vector of the incident light. Here the scattering plane acts as a reference plane for the Stokes vectors. The particles are often assumed to be in random orientation, which implies that the scattering angle suffices to describe the angular dependence of the scattering matrix. Furthermore, cosmic dust particles differ in size and shape, so that size/shape ensembles should be considered.

To gain insight into the scattering matrices of cosmic dust particles a variety of computations have been performed, while early experiments were

G. Videen and M. Kocifaj (eds.), Optics of Cosmic Dust, 57–70.

mostly conducted in the microwave region to simulate scattering at shorter wavelengths [2, 3, 4]. In both approaches, however, model dust particles are used that must be fabricated, either mentally or literally. This means that certain assumptions must be made with regard to microphysical properties of the particles under study like their shapes and refractive indices at various wavelengths.

An alternative starting point for studies of light scattering by cosmic dust particles is provided by 4×4 experimental scattering matrices that are determined in the laboratory for ensembles of natural dust particles and for the visible part of the spectrum, by using lasers. A number of such scattering matrices have been reported in various papers [5, 6, 7, 8, 9]. In this chapter we bring together those experimental scattering matrices of ensembles of randomly oriented natural particles (as functions of the scattering angle) that seem to be most relevant to investigations of cosmic dust. The measurements were conducted in the past few years with the light scattering apparatus in Amsterdam [10]. Here we focus on the experimental results and refer the reader to the original papers for details and more extensive comparisons and discussions.

2. Experimental method

A detailed description of the experimental apparatus that we used to measure the scattering matrices is given by Hovenier [10]. A brief summary is given here. We use either a HeNe laser (632.8 nm, 5 mW) or a HeCd laser (441.6 nm, 40 mW) as a light source. The laser light passes through a polarizer and an electro-optic modulator. The modulated light is subsequently scattered by an ensemble of randomly oriented particles located in a jet stream produced by an aerosol generator. The scattered light may pass through a quarter-wave plate and an analyzer (both optional) and is then detected by a photomultiplier tube which moves in steps along a ring around the ensemble of particles. In this way a range of scattering angles is covered from approximately 5° (nearly forward scattering) to about 173° (nearly backward scattering). Another photomultiplier is placed at a fixed position and used for normalization purposes. We employ the polarization modulation in combination with lock-in detection to determine all elements F_{ij} of the 4×4 scattering matrix of the ensemble as a function of the scattering angle, θ. All matrix elements (except F_{11} itself) are normalized to F_{11}, that is, we consider F_{ij}/F_{11}, with $i, j = 1$ to 4 with the exception of $i = j = 1$. All scattering functions or phase functions, $F_{11}(\theta)$, considered in this chapter are plotted on a logarithmic scale and are normalized to 1 at 30°. Due to reciprocity $F_{21}(\theta)/F_{11}(\theta) = F_{12}(\theta)/F_{11}(\theta)$. Further, we verified that $F_{43}(\theta)/F_{11}(\theta) = -F_{34}(\theta)/F_{11}(\theta)$ and all other off diagonal

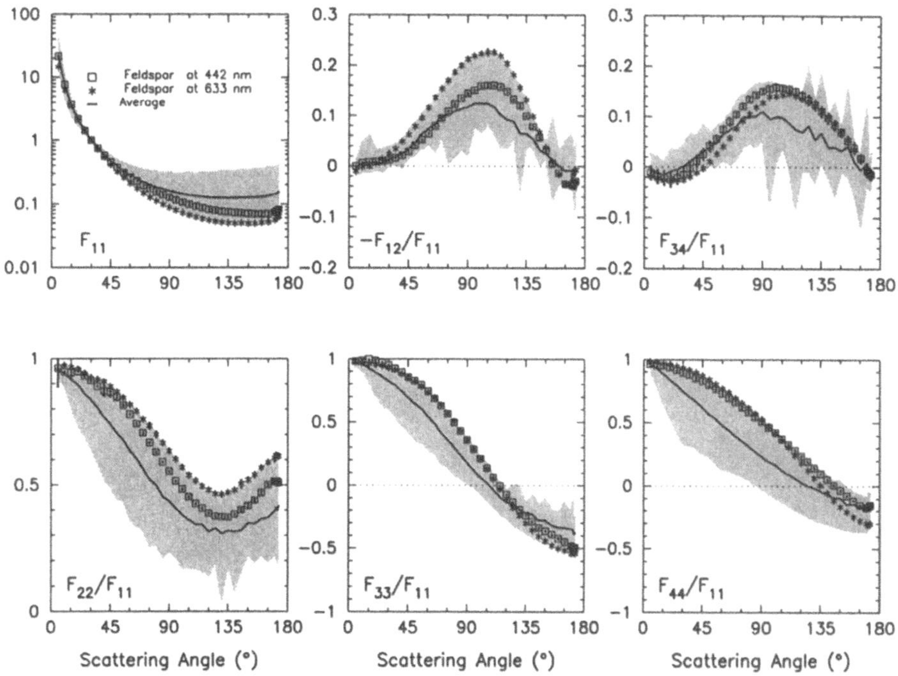

Figure 1. Measured scattering matrix elements $F_{11}(\theta)$, and element ratios $-F_{12}(\theta)/F_{11}(\theta)$, $F_{34}(\theta)/F_{11}(\theta)$, $F_{22}(\theta)/F_{11}(\theta)$, $F_{33}(\theta)/F_{11}(\theta)$, and $F_{44}(\theta)/F_{11}(\theta)$ for feldspar at 441.6 nm (squares) and 632.8 nm (stars). The data are shown together with their error bars. In case no errors are shown they are smaller than the symbols. The domains occupied by the best values of the measurements for seven samples of mineral aerosols at both wavelengths are indicated in grey. The corresponding average aerosol scattering matrix is also shown as a function of the scattering angle (solid lines).

elements divided by $F_{11}(\theta)$ vanish identically within experimental errors. Thus $-F_{12}(\theta)/F_{11}(\theta)$ is the degree of linear polarization for incident unpolarized light. All matrices presented in this chapter have been verified to obey the Cloude test [11].

3. Mineral aerosol particles

Volten *et al.* [8] presented measured scattering matrices as functions of the scattering angle at wavelengths of 441.6 nm and 632.8 nm for seven distinct samples of mineral aerosols in random orientation. These samples,

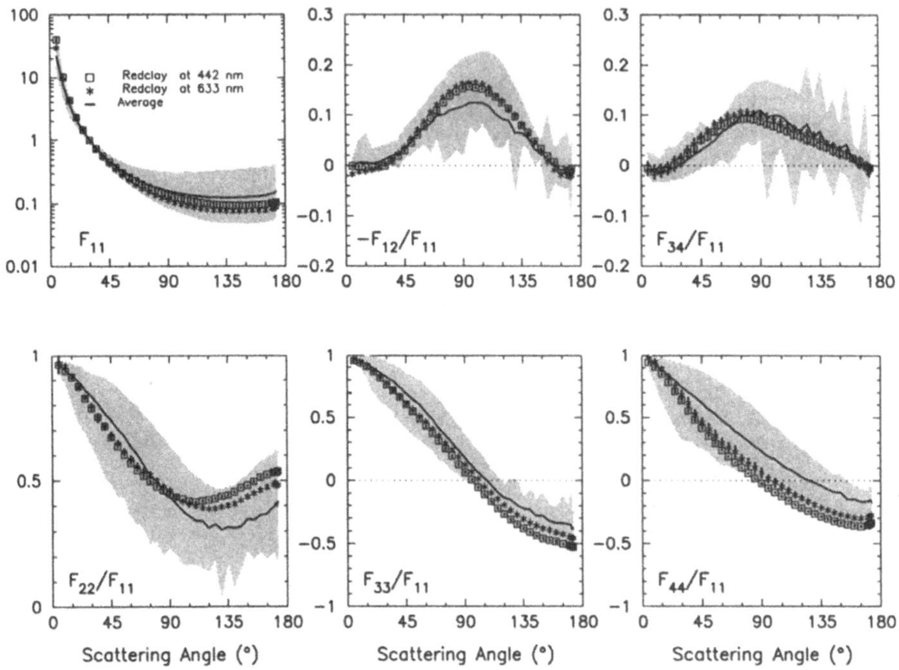

Figure 2. Same as Figure 1, but for red clay.

i.e., feldspar, red clay, quartz, Pinatubo volcanic ash, loess, Lokon volcanic ash, and Sahara sand represent a wide variety of particle sizes and compositions (mainly silicates). Figures 1–7 show the measured scattering matrix elements $F_{11}(\theta)$, as well as element ratios $-F_{12}(\theta)/F_{11}(\theta)$, $F_{22}(\theta)/F_{11}(\theta)$, $F_{33}(\theta)/F_{11}(\theta)$, $F_{34}(\theta)/F_{11}(\theta)$, and $F_{44}(\theta)/F_{11}(\theta)$ for the seven samples. The shapes of the particles can be described as compact and irregular. Most of them, except for the quartz particles, are inhomogeneous in composition. Evidently, it would be impossible or very difficult to model such natural particles in a satisfactory way. The size of each particle can be described in terms of the radius of a sphere having the same projected surface area. The size distribution of each sample can then be characterized by an effective radius, r_{eff}, and effective standard deviation, σ_{eff} [12]. The following values for r_{eff} and σ_{eff}, respectively, were determined: 1.0 μm and 1.0 for feldspar; 1.5 μm and 1.3 for red clay; 2.3 μm and 1.5 for quartz; 3.0 μm and 3.5 for Pinatubo volcanic ash; 3.9 μm and 1.6 for loess; 7.1 μm and 1.6 for Lokon

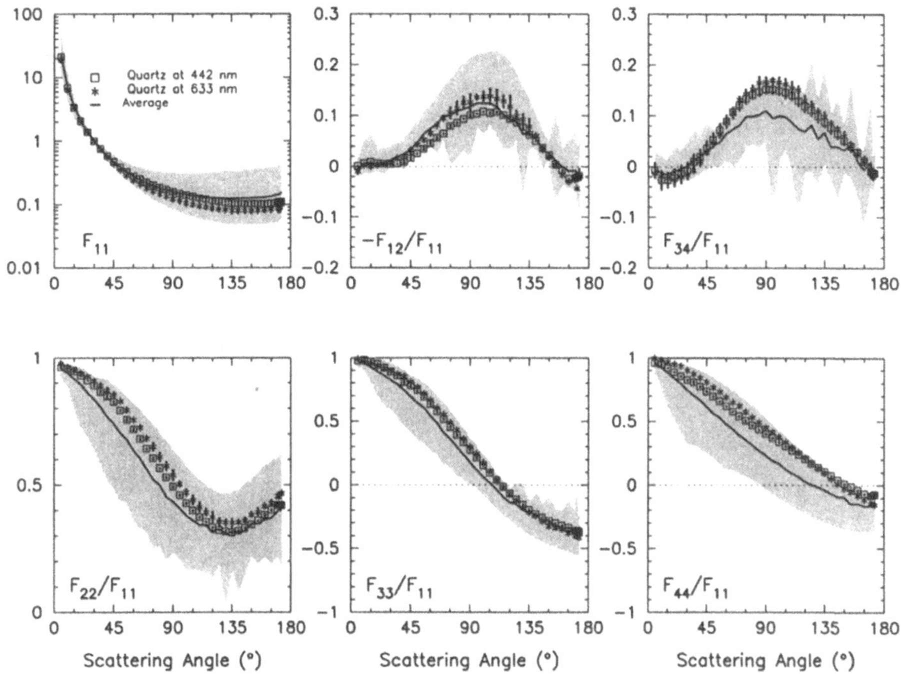

Figure 3. Same as Figure 1, but for quartz.

volcanic ash; and 8.2 μm and 2.0 for Sahara sand.

Figs. 1–7 show that all phase functions, $F_{11}(\theta)$, are rather smooth with a steep forward peak and virtually no structure for side scattering and backscattering. The curves for $-F_{12}(\theta)/F_{11}(\theta)$ in Figs. 1–7 are all more or less bell-shaped and show that some negative polarization occurs at very large scattering angles. This is very interesting in view of the fact that several types of cosmic dust particles exhibit negative polarization at small phase angles, i.e., at large single scattering angles (see e.g. [13, 14]). The curves for $F_{22}(\theta)/F_{11}(\theta)$ in Figs. 1–7 prove directly that the scattering behavior of the pertinent particles is very different from that of homogeneous optically inactive spheres, since in the latter case $F_{22}(\theta) \equiv F_{11}(\theta)$. Theoretically [15] our measured scattering matrices must obey the inequality

$$\left| \frac{F_{33}(\theta)}{F_{11}(\theta)} - \frac{F_{44}(\theta)}{F_{11}(\theta)} \right| \leq 1 - \frac{F_{22}(\theta)}{F_{11}(\theta)}, \tag{1}$$

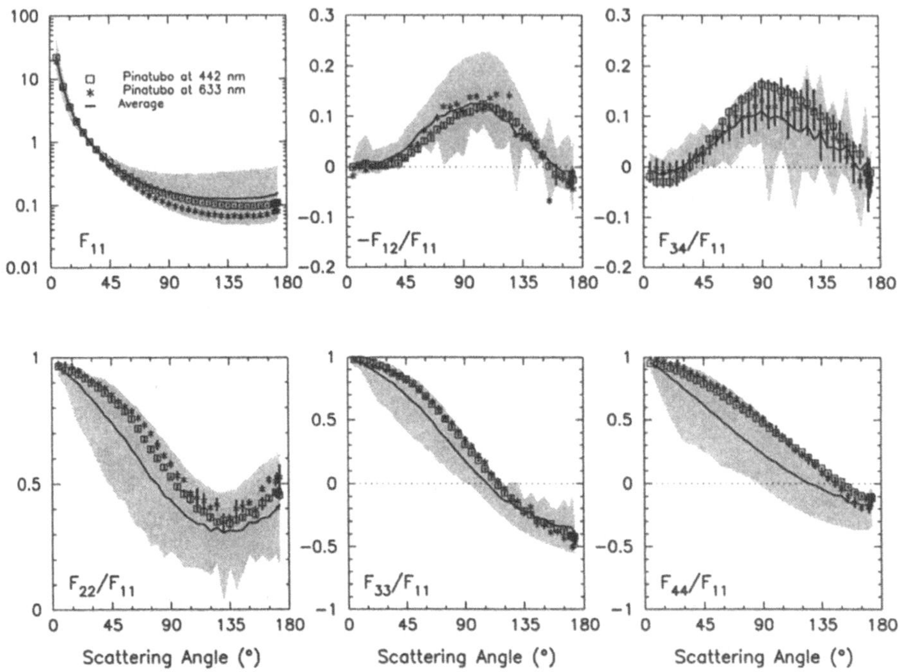

Figure 4. Same as Figure 1, but for Pinatubo volcanic ash.

but except at small angles the experimental upper bound for the left-hand side of Eq. (1) is always much lower for the seven samples at both wavelengths. It may be noted that the overall shapes of the curves for $F_{34}(\theta)/F_{11}(\theta)$ in Figs. 1–7 are similar to the shapes for $-F_{12}(\theta)/F_{11}(\theta)$ of the same samples. In fact, the differences may be called small i.e. generally smaller than 0.1.

The experimentally determined scattering functions and element ratios as functions of the scattering angle for the distinct samples of mineral aerosols are generally found to agree well in their overall trends and behavior, independent of the wavelength considered. Therefore, averages were taken over all 7 samples and both wavelengths to construct (unweighted) average mineral aerosol scattering matrix elements $\langle F_{11}(\theta)\rangle$ and element ratios $\langle F_{ij}(\theta)\rangle/\langle F_{11}(\theta)\rangle$. Since no scattering cross sections were available, all measured matrices were averaged giving them equal weights. The resulting matrix can be used for several purposes, e.g., (i) when a scattering matrix

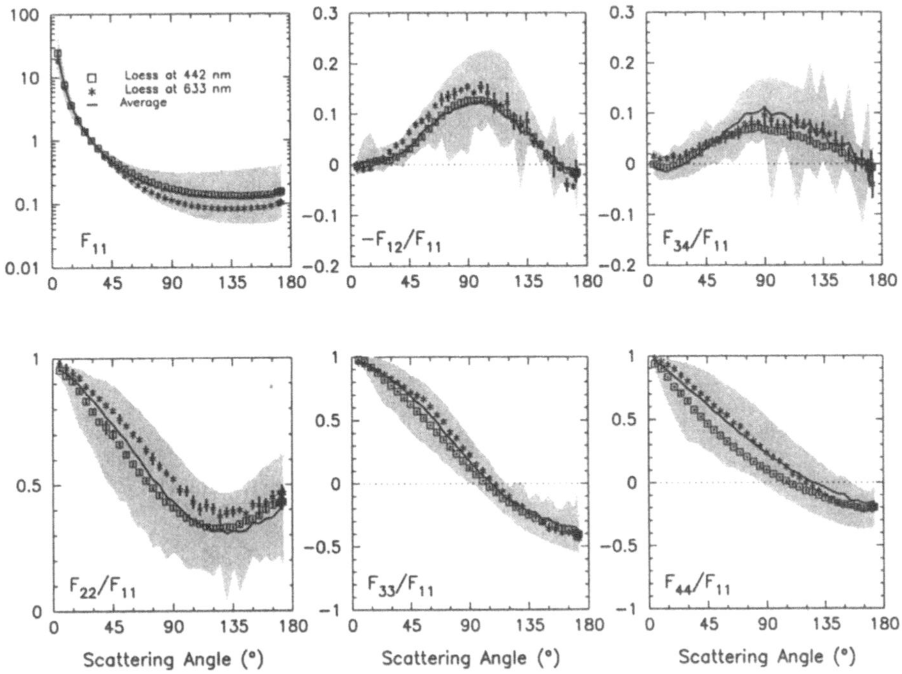

Figure 5. Same as Figure 1, but for loess.

for mineral particles is needed (e.g. for multiple scattering computations) and not much information on the particles exist and (ii) as a reference matrix for comparison with other scattering matrices [cf. Section 4]. The latter is the reason why we also included the average scattering curves in Figs. 1–7 together with the domains occupied by the best values of the measurements for all seven samples at both wavelengths. This may provide more insight in the scattering behavior of one sample with respect to the rest. For example, it is striking that for the sample of feldspar, which has the smallest effective radius, when illuminated at 632.8 nm $F_{11}(\theta)$ shows the lowest values and $F_{22}(\theta)/F_{11}(\theta)$ shows the highest values within the domain at virtually all scattering angles. In contrast, the Sahara sand has the highest effective radius, and when illuminated at 441.6 nm the situation for $F_{11}(\theta)$ and $F_{22}(\theta)/F_{11}(\theta)$ is reversed. It is premature to conclude that the size of the particles (apart from their shape) determines the height of these values, but it is clear that size must play an important role. For the

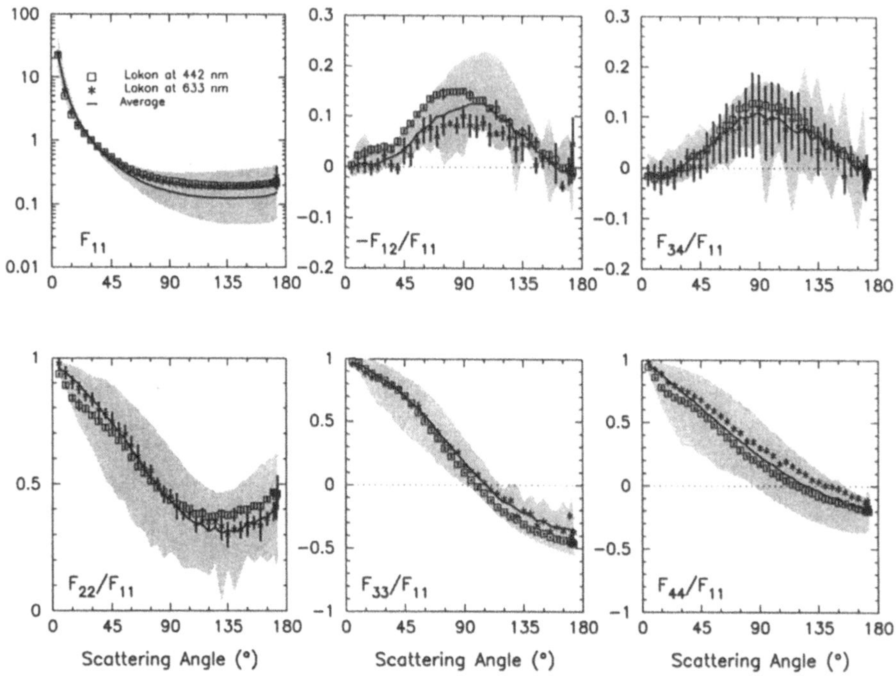

Figure 6. Same as Figure 1, but for Lokon volcanic ash.

other elements the situation is more complex.

4. Olivine and Allende Meteorite Particles

Olivine is of particular interest for cosmic dust particles, since infrared spectra have provided strong evidence for the presence of this mineral in, e.g., comets [16, 17, 18, 19, 20] and circumstellar matter [21, 22]. Therefore, Muñoz *et al.* [7] measured the entire scattering matrix as a function of the scattering angle of randomly oriented compact particles of natural Mg-rich olivine samples and a sample consisting of a ground piece of Allende meteorite, which also contains olivine, but has a higher amount of iron than the olivine samples. The olivine sample we will discuss here is the same as sample M presented by Muñoz *et al.* [7] and has $r_{\text{eff}} = 2.6\mu\text{m}$ and $\sigma_{\text{eff}} = 2.2$. For the Allende meteorite sample we have $r_{\text{eff}} = 0.8\mu\text{m}$ and $\sigma_{\text{eff}} = 1.8$. The measured scattering matrix elements are shown in Figs. 8 and 9. For

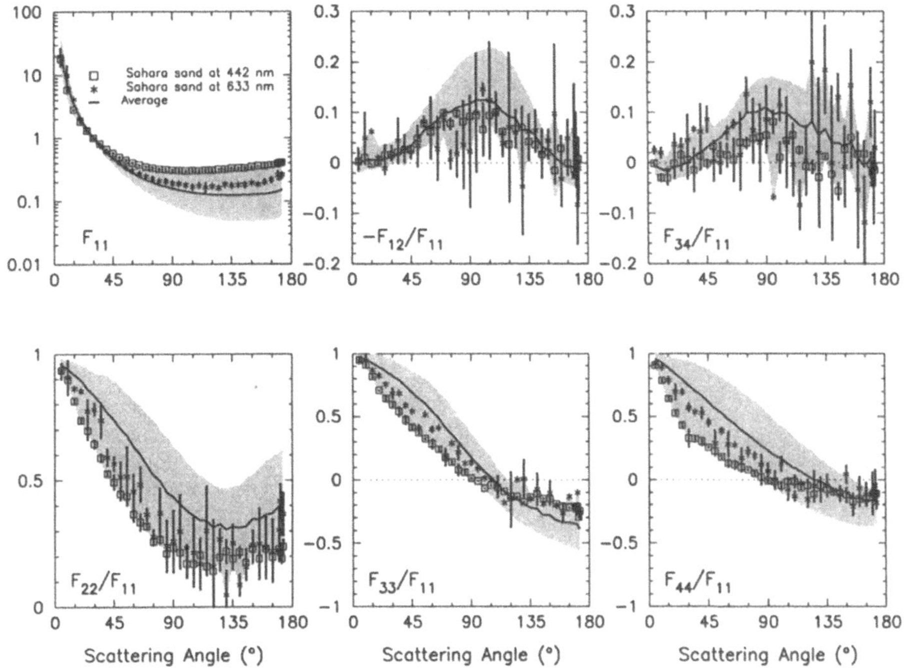

Figure 7. Same as Figure 1, but for Sahara sand.

comparison, the elements of the average mineral scattering matrix discussed in Sect. 3 are also shown in Figs. 8 and 9 together with the domains covered by the best values of all measurements for mineral aerosols mentioned in Sect. 3. Figs. 8 and 9 show that the measured angular distributions of the scattering matrix elements of the samples of olivine and Allende meteorite are not very different from those of the mineral aerosol samples reported by Volten *et al.* [8]. In fact, the deviations from the average curves are small.

5. Fly Ash Particles

In the preceding sections we have considered particles with a more or less compact structure. For a typical example of such particles see the upper panel of Fig. 10. But in the literature on cosmic dust particles, aggregated or fluffy particles also play an important role. As an example of this type of particle Muñoz *et al.* [9] used a sample of polydispersed randomly

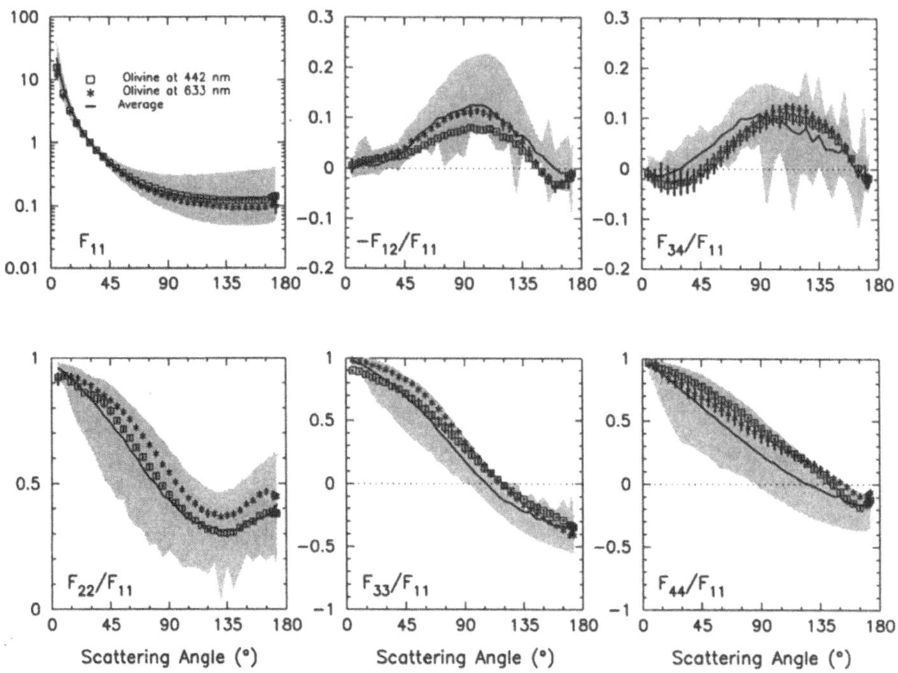

Figure 8. Same as Figure 1, but for olivine.

oriented particles of fly ash. This fly ash consists of aggregates of nearly spherical particles (see Fig. 10, lower panel), and originates from the inorganic fraction, mainly clays, of the combustion of powdered coal in an electric power plant. The main components are SiO_2 and Al_2O_3. The sizes of the aggregates are estimated to be characterized by $r_{\text{eff}} = 3.7\mu$m and $\sigma_{\text{eff}} = 3.3$. The measured scattering matrix elements of the fly ash sample are shown in Fig. 11 together with the domains covered by the best values of all measurements for the seven mineral aerosol samples considered in Sect. 3. Apparently, considerable differences occur between the results for fly ash on the one hand and the results for the mineral aerosols on the other hand. For example, for the fly ash, $F_{11}(\theta)$ shows a strong increase at backscattering angles. Furthermore, the curves for $-F_{12}(\theta)/F_{11}(\theta)$ do not show the bell-shape presented by the compact mineral particles. The degree of linear polarization for incident unpolarized light at 441.6 nm for fly ash particles is positive at almost all scattering angles with a strong in-

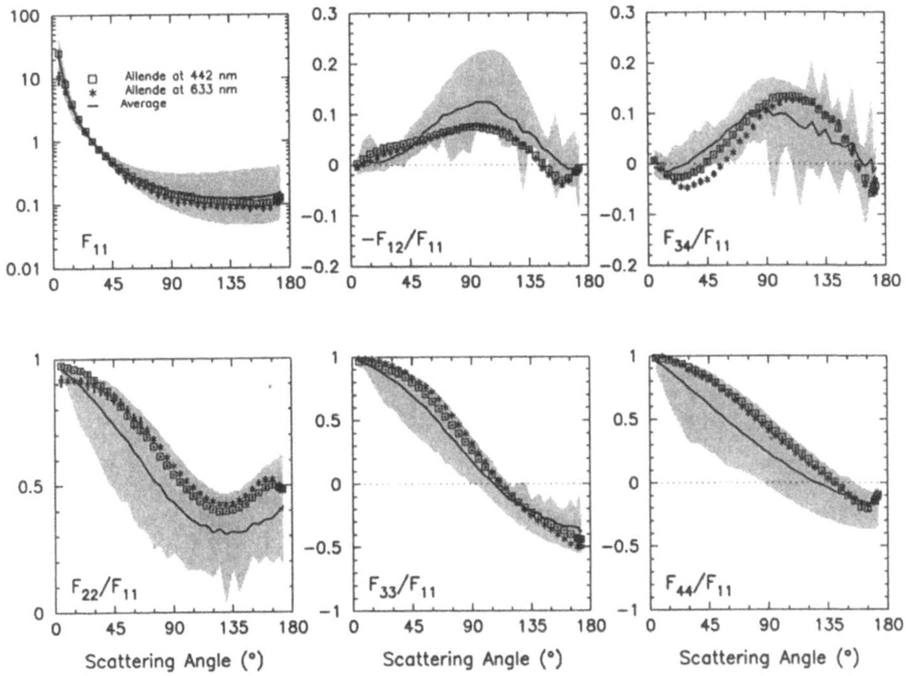

Figure 9. Same as Figure 1, but for Allende meteorite.

crease at backscattering positions. Such an increase is not observed for the compact mineral particles. At 632.8 nm the behavior of $-F_{12}(\theta)/F_{11}(\theta)$ is even more dissimilar to the general behavior for compact mineral particles, showing negative values at most scattering angles. Moreover, the shapes of $-F_{12}(\theta)/F_{11}(\theta)$ and $F_{34}(\theta)/F_{11}(\theta)$ for the fly ash are not at all similar. The increase of $F_{22}(\theta)/F_{11}(\theta)$ near backscattering angles for fly ash is spectacular. A sudden decrease of $F_{33}(\theta)/F_{11}(\theta)$ and $F_{44}(\theta)/F_{11}(\theta)$ occurs for the fly ash near backscattering at 632.8 nm. These differences in scattering behavior in comparison with the mineral aerosol particles must be mainly due to shape and structure effects, since the refractive indices and the sizes of the fly ash particles lie within the range of values of the samples studied by Volten *et al.* [8]. Indeed, some of the main differences are consistent with results of computations for bispheres [23] and aggregates of spheres [24].

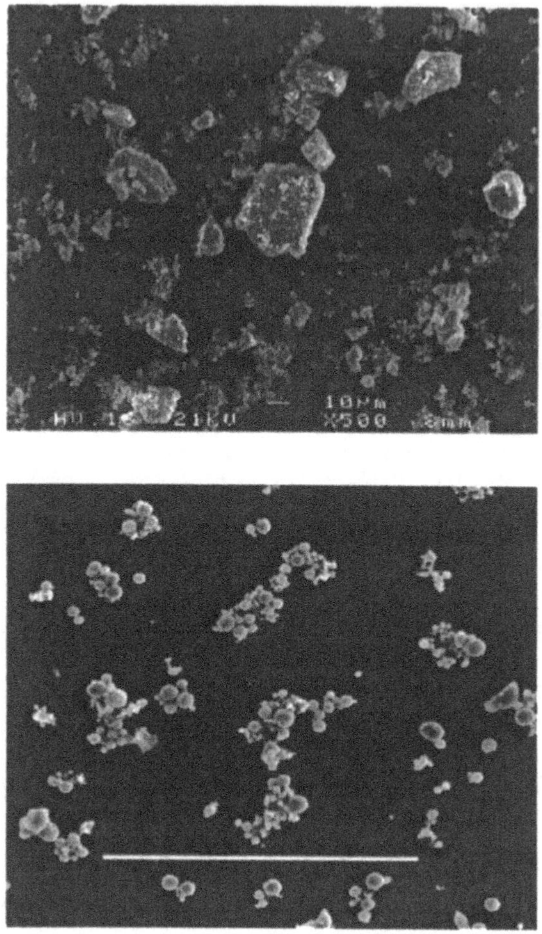

Figure 10. SEM photograph of olivine (upper panel) and fly ash particles (lower panel). The white bar denotes 10 μm in the upper panel and 100 μm in the lower panel. The pictures are not representative for the size distributions.

6. Conclusions

A number of experimental scattering matrices at optical wavelengths for size/shape ensembles of randomly oriented natural particles are now available for studies of cosmic dust particles. An example is provided by Moreno *et al.* [25]. Compact mineral particles with sizes in the micrometer range exhibit a fairly similar scattering behavior, although interesting small differences occur. The scattering patterns of aggregated mineral particles, as presented by fly ash, appear to be markedly different, but more experiments are needed for general statements on this issue.

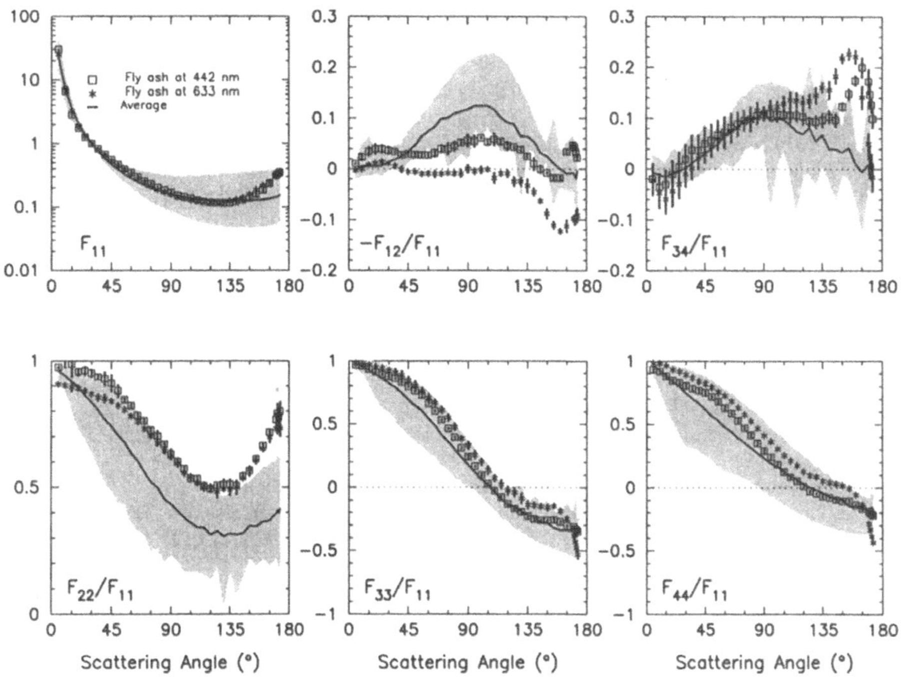

Figure 11. Same as Figure 1, but for fly ash.

7. Acknowledgments

We would like to thank R. D. Schuiling from the University of Utrecht for providing the olivine and fly ash samples and D. Heymann from the Rice University for the Allende meteorite sample. We are grateful to several people of the Free University, particularly to J. Bouma for technical support, M. Konert for measuring the size distributions of the samples, and S. Kars for providing SEM photographs of our samples. We enjoyed many fruitful discussions with J. F. de Haan and W. Vassen also from the Free University in Amsterdam. Comments by W. J. van der Zande and M. I. Mishchenko on an earlier version of this paper are gratefully acknowledged.

References

1. H. C. van de Hulst, *Light scattering by small particles*, Wiley, New York (1957) (also Dover New York, 1981) .
2. M. I. Mishchenko, J. W. Hovenier, and L. D. Travis, (Eds.), *Light Scattering by Nonspherical Particles,* Academic Press, San Diego, Calif., (2000).
3. G. Videen, Q. Fu, P. Chylek (Eds.), *Light scattering by Nonspherical Particles, Halifax contributions*, Adelphi (2000).
4. G. Videen, Q. Fu, P. Chylek (Eds.) *Light scattering by Nonspherical Particles*, Special issue *J. Quant. Spectrosc. Radiat. Transfer,* **70,** 373, (2001).
5. A. C. Holland, and G. Gagne, *Appl. Opt.*, **9**, 1113 (1970).
6. F. Kuik, P. Stammes, and J.W. Hovenier, *Appl. Opt.*, **30**, 4872 (1991).
7. O. Muñoz, H. Volten, J. de Haan, W. Vassen, and J. W. Hovenier, *Astron. Astrophys.*, **360,** 777, (2000).
8. H. Volten, O. Muñoz, E. Rol, J. F. de Haan, W. Vassen, J. W. Hovenier, K. Muinonen, and T. Nousianen, J. Geophys. Res., **106** 17,375 (2001).
9. O. Muñoz, H. Volten, J. de Haan, W. Vassen, and J. W. Hovenier, *J. Geophys. Res.*, **106,** 22833, (2001).
10. J.W. Hovenier, Measuring scattering matrices of small particles at optical wavelengths, In: *Light Scattering by Non-Spherical Particles*, Mishchenko M.I., Hovenier J.W., Travis L.D. (Eds.), Academic Press, San Diego, Calif., (2000).
11. J. W. Hovenier, and C. V. M. van der Mee, *J. Quant. Spectrosc. Radiat. Transfer,* **55,** 6,49 (1996).
12. J. E. Hansen, and L. D. Travis, Space Sci. Rev., **16,** 527 (1974).
13. K. Lumme, Scattering properties of interplanetary dust particles, in *Light Scattering by Nonspherical Particles,* M. I. Mishchenko, J. W. Hovenier, and L. D. Travis, (Eds.) pp. 555-582, Academic Press, San Diego, Calif., (2000).
14. See the chapter on "Photometric and Polarimetric Opposition Effects in Solar System Bodies,"
15. J. W. Hovenier, H.C. van de Hulst, C. V. M. van der Mee, *J. Quant. Spectrosc. Radiat. Transfer,* **55,** 649, (1986).
16. H. Campins, E. V. Ryan, ApJ **341**, 1059 (1989)
17. M. S. Hanner, D. K. Lynch, R. W. Russel, ApJ **425**, 274 (1994)
18. L. Colangeli, V. Mennella, C. Di Marino, A. Rotundi, and E. Bussoletti, *Astron. Astrophys.,* **293,** 927, (1995).
19. L. Kolokolova, K. Jockers, *PSS* **45**, No 12, 1543 (1997)
20. J. R. Brucato, L. Colangeli, V. Mennella, P. Palumbo, E. Bussoletti, *PSS* **47**, 773 (1999)
21. A.G.G.M. Tielens, L.B.F.M. Waters, F.J. Molster and K. Justtanont, *Astrophys. and Space Sci.*, **255**, 415 (1998).
22. C. Jager, F.J. Molster, J. Dorschner, T. Henning, H. Mutschke, and L.B.F.M. Waters *Astron. Astrophys.*, **339,** 904, (1998)
23. M. I. Mishchenko, D. W. Mackowski, and L. D. Travis, Appl. Opt., **34**, 4589 (1995).
24. D. W. Mackowski, and M. I. Mishchenko, J. Opt. Soc. Am. A Opt. Image Sci. **11**, 2266 (1996).
25. See the chapter on "Monte Carlo Modelling of Dusty Cometary Atmospheres Including Polarization," F. Moreno, O. Muñoz, and A. Molina, this book.

LIGHT SCATTERING TOOLS FOR COSMIC DUST MODELING

V.B. IL'IN AND N.V. VOSHCHINNIKOV
St. Petersburg State University, Russia
V.G. FARAFONOV
St. Petersburg University of Aerocosmic Instrumentations, Russia
TH. HENNING
Max Planck Institute for Astronomy, Heidelberg, and Friedrich Schiller University Jena, Germany
AND
A.YA. PERELMAN
St. Petersburg Technical Forest Academy, Russia

Abstract.
Because cosmic dust grains vary significantly in both morphology and chemical composition, it is necessary to develop different light scattering tools to analyze their scattering properties and to reconcile these properties with observations. We present a set of recently developed tools which includes a database of optical constants of materials of astronomical interest, exact and approximate methods and numerical codes using various models of a non-spherical inhomogeneous scatterer, a database of optical properties of non-spherical particles, a new approach to find a solution of ill-posed inverse problems in optics, and an original polarized radiation-transfer code applicable to 3D media populated by aligned non-spherical scatterers.

1. Introduction

Dust has been detected in practically all astronomical objects, and its role in the physics of objects and formation of their observed spectra has been found to be important. Evolutionary modeling of objects should not only explain observational manifestations of dust (infrared radiation, bands, polarization, etc.), but also include clear ideas concerning possible realizations

G. Videen and M. Kocifaj (eds.), Optics of Cosmic Dust, 71–88.

of the dust model used for the given physical conditions and its consistency with the general scenario of evolution of dust in galaxies.

To simulate the observed effects of dust, one usually needs the following tools (see, e.g., [1, 2] for more details):

i) a database of the optical constants of materials which can be considered as cosmic dust analogues;

ii) light-scattering codes to transform these data into the optical properties of particles; and

iii) radiative-transfer codes to convert these properties into the observational quantities for a selected model of the object.

In this paper we briefly consider the state of the art in the field and present some tools recently developed by us. Sect. 2 describes our database of optical constants and gives examples of possible use of the data it contains. Several methods and codes to solve the light-scattering problem for inhomogeneous (layered) non-spherical particles are introduced in Sect. 3, where we also discuss a new approach to solve the inverse problem of particle sizing and creating database of optical properties of non-spherical scatterers. Finally, Sect. 4 illustrates possible applications of the tools by making use of an original polarized radiation transfer code.

2. Optical constants

The optical constants – refractive index, dielectric function, etc. are macroscopic quantities that characterize the interaction of radiation with solids. Materials generally can be divided into two groups: those whose optical properties are typical of dielectrics (ices, silicates, etc.) and those typical of metals (metals, carbon, etc.). The optical constants can depend on temperature, and for metals, also on the size of granular structure. Note that when modeling cosmic dust, one may need the data both in narrow spectral features and in continuum for a wide wavelength range.

Many terrestrial analogues of cosmic solids have been studied extensively in laboratories (in particular during the last 10 years in Jena and Leiden), and for different extraterrestrial and artificial materials the optical constants have been determined as well. These data are generally in the form of tables and graphics in papers, and free World Wide Web (WWW) resources of optical constants are generally limited by several collections of refractive indices for a few materials. We attempt to develop a universal site that could help visitors find required data in the literature and WWW and would be as useful as possible in other aspects: the Jena-Petersburg database of optical constants (JPDOC) has free access via the Internet (*http://www.astro.spbu.ru/JPDOC/entry.html*) and contains *references* to the papers, *data* files, and *links* to the Internet resources related to measurements and calculations of the optical constants in the wavelength interval from X-rays to radio domain. The materials considered are amorphous and crystalline silicates of different kinds, various ices, oxides, sulfides, carbides, carbonaceous species from amorphous carbon to graphite and diamonds and some other materials of astrophysical interest. We have comprised over 1000 references to the papers, books, and dissertations where the refractive index, reflectivity, transmittance, etc. were derived. Data accessible via JPDOC are those obtained in the laboratory in Jena and supplemented with some other data available to us (see [3] for more details).

Most of the materials studied in Jena are synthetic compounds prepared especially for the purpose of spectroscopic investigation. They include silicates in both amorphous and crystalline state, oxides of magnesium, iron, and aluminum, sulfides, and carbon in different forms. Chemical and physical analytical methods were generally applied to confirm the homogeneity, composition, and crystal structure of the products prior to the spectroscopic measurements. Further, some natural crystals (oxides and silicates) have been included in the studies. If necessary, data have been determined for the different crystallographic axes. For some of the compounds, data are available at cryogenic temperatures. In the following we provide two examples of the data and their possible applications.

Amorphous carbonaceous materials are known to show a great diversity of optical properties due to the variability in their microstructure. Especially in the infrared range, the optical constants can differ by orders of magnitude according to the conducting or insulating electrical behaviour of the material. The amorphous carbon data contained in the database cover a wide range of these properties as is illustrated in Fig. 1. Calculations of the absorption and scattering cross-sections of small particles using these data should take into account that for the more conducting materials the results depend on the particle geometry (and size) to a large extent [4].

Silicate minerals of the olivine and pyroxene classes have been shown to be present in outflows of evolved stars as well as in comets and protoplanetary disks. The positions of the infrared emission bands produced by these materials are diagnostics for the chemical composition, especially the iron content which can constrain the condensation processes in the environments. On the other hand, they are influenced by the particle shape due to the occurrence of surface modes [5]. We have used the infrared optical constants of forsterite contained in the database for calculating the absorption cross-section of spherical and non-spherical particles in the Rayleigh approximation. The spectra contain resonances due to surface modes which shift very strongly in dependence on the semiaxis ratio of the particles (see Fig. 2–3), which may be important for the identification of emission features in astronomical spectra [6].

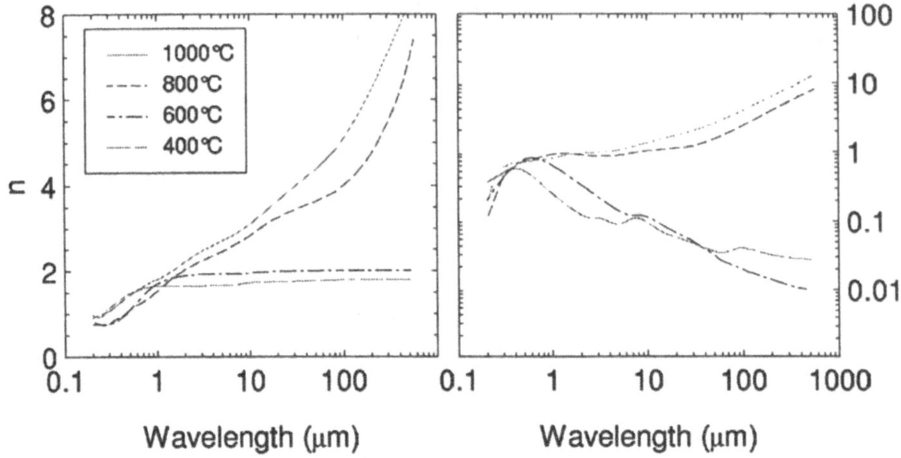

Figure 1. Complex refractive index ($m = n + k\,i$) of hydrogenated amorphous carbon prepared by pyrolysis (annealing) of cellulose at different temperatures.

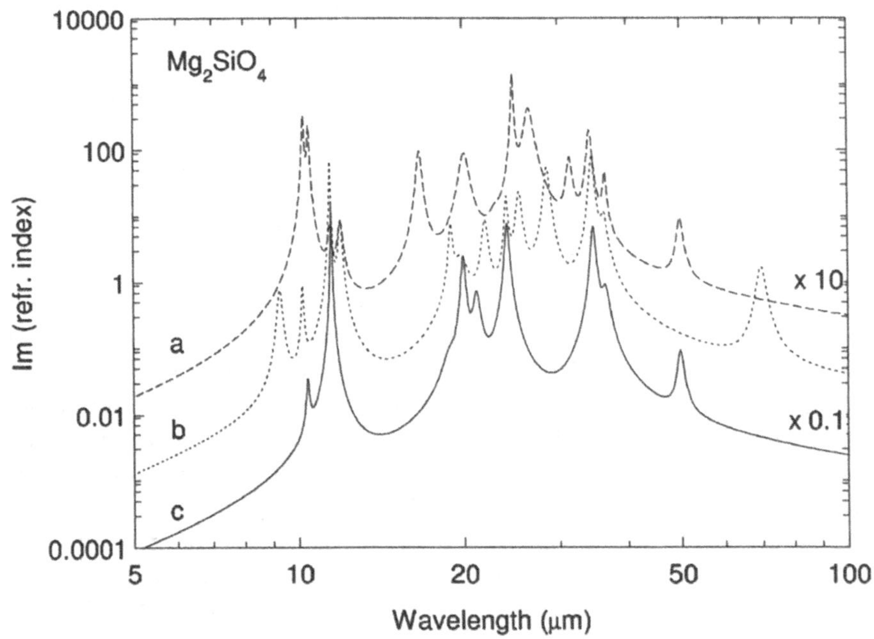

Figure 2. Imaginary part of the refractive index for crystalline forsterite (Mg_2SiO_4) in the three different crystallographic directions.

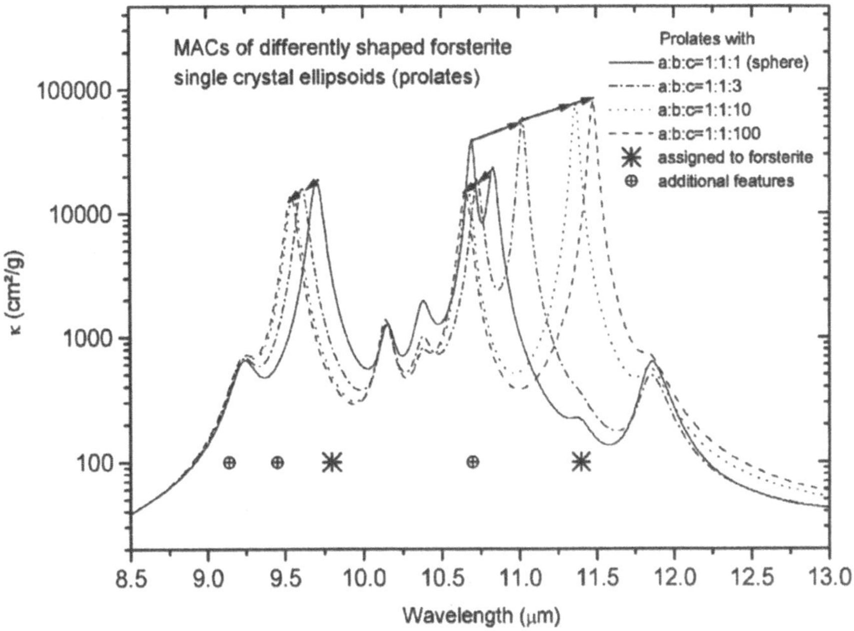

Figure 3. Mass-normalized absorption cross-section (MAC) of prolate spheroidal forsterite particles (rotationally averaged) with different semiaxis ratios. The dots and asterisks below the spectra indicate positions of observed emission bands (after [7]).

3. Light scattering

3.1. INTRODUCTORY REMARKS

The efficiency of interaction of radiation with a particle depends on its chemical composition, size, shape, structure, etc. [8]. We have discussed the composition and its relation to optical constants in Sect. 2; we address other primary characteristics of scatterers in this section.

3.1.1. *Size*

Particle size expressed in units of the wavelength of incident radiation, together with chemical composition, most strongly affects the optical properties of a scatterer. The simplest model that allows one to consider the influence of size and composition variations is that of a homogeneous sphere. The Mie theory is the solution to the light scattering from a sphere and is applicable in the entire range of parameter values. Note that the interaction of large molecules/very small grains (e.g., polycyclic aromatic hydrocarbons, PAH) with radiation is not described by the macroscopic quantities and requires quantum mechanical consideration.

3.1.2. *Shape*

Cosmic dust grains certainly differ from homogeneous spheres. Deviations of the real shapes from the spherical one can be of different kinds – from small-scale surface fluctuations to large-scale shape defects. The asphericity characterized by the ratio of the maximum to minimum dimension of a scatterer looks to be most essential in astronomy and can be well modelled using spheroidal particles as their shapes can vary from needles or disks to spheres. Thus, the spheroidal model allows one to analyze the effects of composition, size, and, to some extent, shape.

There are many methods to solve the light-scattering problem for a homogeneous spheroid. First among them should be mentioned the separation of variables method (SVM) where the wave equation is formulated and the spheroidal wave functions are used in the solution to expand the incident, internal and scattered fields. Another popular approach – the extended boundary condition method (EBCM) coupled with the T-matrix method (TMM) consists of forming surface integral equation, but utilizes the simpler spherical wave functions in the expansion. For spheroids, the SVM and to a lesser degree the TMM conform more readily to the system geometries, and as a result, the methods are much more fast and exact than other (more universal) approaches like the discrete dipole approximation (DDA) and finite difference time domain (FDTD) methods [8, 9]. Note that medium scale defects of the scatterer surface (like cavities) can be considered using the model of the Chebyshev particles and the TMM (see, e.g., [10] and references therein). Scatterers with random surface fluctuations were studied, for instance, in [11].

3.1.3. *Structure*

Solid particles in cosmic conditions hardly are homogeneous. Their internal structures can include layers, large and small (in comparison with the particle size) inclusions, and so on. In principle, scatterers of any structure can be treated by the methods using the representation of the scattering problem in the form of volume integral equation (e.g., the widely used discrete dipole approximation, DDA). The back sides of this universality are strong demands for computer memory and speed which often make computations impossible. Layered scatterers could be treated by the SVM and TMM, which would allow really extensive calculations, but the theoretical methods are not yet well developed. If the dimensions of inclusions are smaller than the wavelength of incident radiation and their volume fraction is not large, one can calculate the effective refractive index of a composite particle using a rule of the effective medium theory (EMT; see, e.g., [12] for a review) and then apply any light scattering method for homogeneous particles.

3.1.4. *Applicability of scatterer models*

For astronomical applications the models of homogeneous and core-mantle spheres (and infinite cylinders) are primarily used. Models for homogeneous spheroids (and ellipsoids) have been applied only seldomly. Thus, the combined effects of shape and structure of cosmic dust grains have barely been studied. It should be added that although inhomogeneous non-spherical aggregates were utilized in a number of papers on cosmic dust modeling, very little attention has been paid on these effects.

Below we present several methods and codes developed to solve the light scattering problem for *inhomogeneous* (multilayered) *non-spherical* particles. Both exact and approximate approaches are considered as they are useful in astronomy (in particular the Rayleigh approximation in the infrared). The ranges of applicability of the methods that we extend to treat layered particles are shown in Fig. 4 for the case of homogeneous dielectric spheroids. For layered particles, the ranges are generally similar.

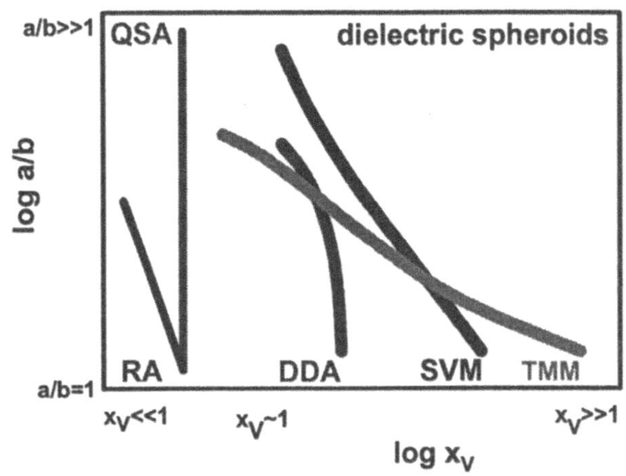

Figure 4. Applicability ranges (domains left to the curves) of different methods in the case of dielectric prolate spheroids. Calculations were made with codes based on the Rayleigh (RA) and quasistatic (QSA) approximations, the discrete dipole (DDA), separation of variables (SVM) and T-matrix (TMM) methods.

3.2. TOOLS FOR LAYERED PARTICLES

The model of multilayered particles deserves special attention in astronomy. To begin with, layers of different composition can form during dust grain evolution in the interstellar medium. Particles with many relatively thin layers can be used to model grains with a radially changing refractive index, appearing, for instance, in the case of fractal-like, fluffy aggregates typical of interplanetary media. The model with a large number of very thin layers of several cyclically changing materials opens an efficient way to study effects of composition inhomogeneity expected in cosmic dust grains. Note that the layered model also makes it possible to obtain the corresponding light-scattering solution using fast and exact methods like the SVM and TMM and useful approximations like the RA and can therefore have wide applicability.

3.2.1. *Layered spheres (nMie, PCHA)*
Let us start by considering light scattering tools available for multilayered particles, specifically layered spheres. Aside from the standard code for such spheres (n-layered Mie theory, nMie) based on the well known algorithm [13], we have developed a program to calculate scatter in more general case of a sphere with a homogeneous core and a layered mantle of relative thickness Δ that can be changed from nearly 0 up to 1. The latter code realizes the piecewise continuous hyperbolic approximation (PCHA) described in detail in [14]. Here we should note that in this approach the layers must have the radial dependence of the refractive index $m(r)$ in the form of a hyperbolic function (which is not very restictive as the number of the layers is not limited) and the solution to the light scattering problem is obtained analytically [15]. The specific characteristics of the approach (speed, accuracy, wide range of applicability, etc. – see discussion in [14]) make it very useful in finding solutions of inverse problems (see Sect. 3.3.1).

3.2.2. *Layered spheroids (nSVM)*
Our new approach for layered spheroids extends the SVM solution earlier developed for homogeneous and core-mantle spheroids [16]. This solution differs from the classical one of Asano & Yamamoto [17] by a proper choice of the scalar potentials of the electromagnetic fields that makes computations more efficient. In our approach the fields are divided in two parts

$$\vec{E} = \vec{E}_{\rm A} + \vec{E}_{\rm N}, \quad \vec{H} = \vec{H}_{\rm A} + \vec{H}_{\rm N}, \tag{1}$$

where the axisymmetric parts $(\vec{E}_{\rm A}, \vec{H}_{\rm A})$ do not depend on the azimuthal angle φ and averaging of the non-axisymmetric parts $(\vec{E}_{\rm N}, \vec{H}_{\rm N})$ gives zero. Special scalar potentials are selected for each of the parts – for the axisymmetric one, p and q being proportional to the azimuthal components of the fields

$$p = \vec{E}_{{\rm A},\varphi} \cos\varphi, \quad q = \vec{H}_{{\rm A},\varphi} \cos\varphi, \tag{2}$$

and for the non-axisymmetric one, a superposition of the Debye potential V and the z-component of the Hertz vectors U, e.g. in the case of TE mode

$$\vec{E}_{\rm N} \; = \; \nabla \times \left(U\vec{i}_{\rm z} + V\vec{r}\right), \tag{3}$$

$$\vec{H}_{\rm N} = \frac{1}{i\mu k}\nabla\times\nabla\times\left(U\vec{i}_z + V\vec{r}\right). \tag{4}$$

Both p, q and U, V potentials satisfy the scalar Helmholtz equation.

The scalar potentials are expanded in terms of the spheroidal wave functions

$$p(\vec{r}) = \sum_{l=1}^{\infty} a_l\, R_{1l}(c,\xi)\, S_{1l}(c,\eta)\cos\varphi, \tag{5}$$

$$U(\vec{r}) = \sum_{m=1}^{\infty}\sum_{l=1}^{\infty} a_{ml}\, R_{ml}(c,\xi)\, S_{ml}(c,\eta)\cos m\varphi, \tag{6}$$

where R_{nl} and S_{nl} are the radial and angular spheroidal functions, respectively, $c = kd/2$ and d is the focal distance of the spheroid. Substitution of the expansions into the corresponding equations for the potentials and orthogonality of the wave functions lead to a system of linear algebraic equations for the expansion coefficients of the scattered field potential $\vec{a}_{\rm sca}$

$$B_1 B_2^{-1}\vec{a}_{\rm sca} = \vec{a}_{\rm in}, \tag{7}$$

where $\vec{a}_{\rm in}$ are known coefficients for the incident radiation and the matrices B_1, B_2 can be found from the matrix equation

$$\begin{pmatrix} B_1 \\ B_2 \end{pmatrix} = \begin{pmatrix} -B_{31}^1 & -B_{33}^1 \\ B_{11}^1 & B_{13}^1 \end{pmatrix}\cdot\ldots\cdot\begin{pmatrix} -B_{31}^{n-1} & -B_{33}^{n-1} \\ B_{11}^{n-1} & B_{13}^{n-1} \end{pmatrix}\cdot\begin{pmatrix} B_{31}^n \\ B_{11}^n \end{pmatrix}, \tag{8}$$

where B_{ij}^n are matrices whose elements are sums of products of the radial spheroidal functions and their derivatives and integrals of the angular spheroidal functions, n indicates dependence only on the parameters of n-th layer and i, j show the kinds of the radial functions used. It is essential that the integrals can be calculated analytically [16]. In principle Eq. (8) can be converted into the recursive form, but we find it less efficient. Note also that the only known SVM scheme for layered spheroids [18] includes non-linear equations for the scattered field expansion coefficients $\vec{a}_{\rm sca}$, while in our scheme these equations are linear.

The formalism we suggest is described in detail in [19]; development of the numerical code is in progress. It should be mentioned that the code will include a new method of calculation of the spheroidal wave functions [20]. This method essentially expands the applicability of the SVM and has for the first time allowed a real comparison of the optical properties of very elongated spheroids and infinitely long cylinders (see [21] for more details). Based on the results obtained for homogeneous and core-mantle spheroids, we expect that the new code will work for spheroids with layer

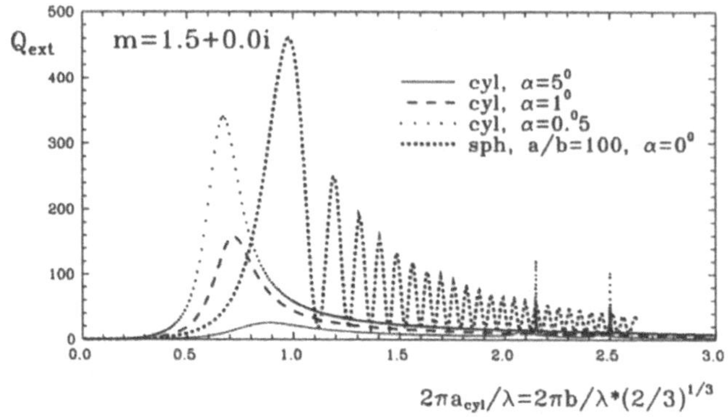

Figure 5. Extinction efficiency factors (TM mode) for prolate spheroids ($Q_{\rm ext} = C_{\rm ext}/G(\alpha)$) and infinitely long cylinders ($Q_{\rm ext} = C_{\rm ext}/(2a_{\rm cyl}\sin\alpha)$) as a function of the size parameter $2\pi a_{\rm cyl}/\lambda = 2\pi b/\lambda(2/3)^{1/3}$, where α is the angle between the (main) particle axis and the wave vector of incident radiation, $G(\alpha)$ the geometrical cross-section of a spheroid, b its minor semiaxis, $a_{\rm cyl}$ the radius of a cylinder.

boundaries of high eccentricity. Note that all SVM schemes including ours are applicable only to the case of confocal boundaries, otherwise one must apply another approach, for instance, the TMM.

3.2.3. *Layered axisymmetric particles (nTMM)*

We extend the TMM for homogeneous particles [22] to handle layered particles. Both features of the approach – the deviation of the fields into two parts and the special choice of the scalar potentials for the fields are analogous to those of the nSVM (see Eqs. (1)–(4)). However, the potentials are expanded in terms of simpler spherical wave functions

$$p(\vec{r}) = \sum_{l=1}^{\infty} a_l \, j_l(kr) \, P_l^1(\cos\theta) \cos\varphi, \tag{9}$$

$$U(\vec{r}) = \sum_{m=1}^{\infty}\sum_{l=1}^{\infty} a_{ml} \, j_l(kr) \, P_l^m(\cos\theta) \cos m\varphi, \tag{10}$$

where j_l denotes the Bessel or Hankel functions and P_l^m the associate Legendre functions, $r = |\vec{r}|$. Substitution of the expansions into the equations written in the form of surface integrals leads to the system of linear algebraic equations for the expansion coefficients of the scattered field potential equivalent to Eq. (7). The matrices B_1 and B_2 can be found from an expression similar to Eq. (8), where B_{ij}^n are the matrices whose elements are sums of integrals of products of the spherical functions and their derivatives (the integrals can be calculated only numerically). Note that in contrast to the nSVM, the nTMM is applicable to particles with any axisymmetric coaxial boundary layers. Figure 6 illustrates how large the difference may be in the internal structure and the optical properties in the case of spheroidal boundaries.

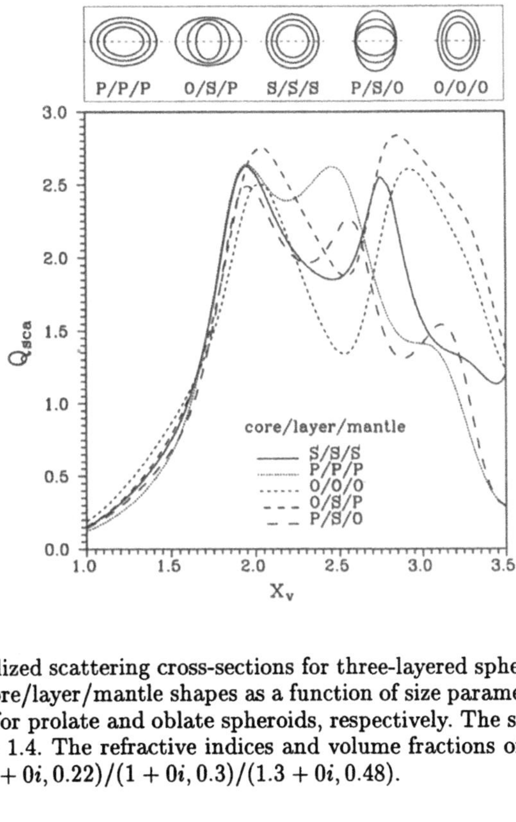

Figure 6. Normalized scattering cross-sections for three-layered spheroids with different combinations of core/layer/mantle shapes as a function of size parameter $x_{\rm v}$. S stands for spheres, P and O for prolate and oblate spheroids, respectively. The semiaxis ratios of all spheroids is $a/b = 1.4$. The refractive indices and volume fractions of core/layer/mantle are $(m, \delta_{\rm v}) = (2.5 + 0i, 0.22)/(1 + 0i, 0.3)/(1.3 + 0i, 0.48)$.

Comparison with the DDA code of Draine & Flatau shows that accuracy of the nTMM is higher and, for these particle sizes, requires an order of magnitude less memory and computational time [27]. The computational time for the nTMM, t, grows more slowly that of DDA with an increase in size. Although $t \propto n$, where n is the number of layers, but for large n, the number of dipoles in the DDA should be increased to reproduce correctly the scatterer structure and this enlarges computational time as well.

An analytical study of the applicability of TMM modifications was done in [20]. For multilayered particles, the conditions formulated for homogeneous particles should be applied to each boundary layer (there are also some additional but less important conditions – see [27] for more details). Our comparison of the nTMM and DDA shows that for spheroids, the ranges of applicability of these methods do not differ significantly with the size parameter (and refractive index), unless the shape of the boundary layers strongly deviate from spherical.

3.2.4. *Layered ellipsoids (nDDA, nRA, nQSA, nEMT)*
DDA codes [24] easily can be modified to treat multilayered ellipsoidal particles. Such a routine is useful to verify developed approaches and to compare them with a good representative of the wide class the volume integral methods. Faster SVM and TMM approaches have not yet been developed for such particles and we will suggest several useful approximations applicable for layered ellipsoids. Primary attention is paid to the quasistatic approximation (QSA) which is a generalization of the Rayleigh (RA) and Rayleigh-Gans approximations (RGA) [25].

The classical RA is known for ellipsoids with confocal coaxial boundary layers. We formulate it as well as the QSA in the general case of *nonconfocal* boundaries [26]. In the QSA the field inside a particle is represented by the incident one (like in the RGA), taking into account the particle polarizability (like in the RA). The problem is to solve the Laplace equation with the corresponding boundary conditions. Because of the non-confocality of boundaries, each layer of the particle is to be divided into a number of sublayers with its own ellipsoidal coordinate system defined inside. For any sublayer, approximate boundary conditions can be formulated (see [26] for more details). As a result one obtains the following expression for the polarizability along an axis of the coordinate system connected with the particle

$$\alpha = V \frac{\bar{\varepsilon} - 1}{(\bar{\varepsilon} - 1)L + 1}, \quad \bar{\varepsilon} = A_2/A_1, \tag{11}$$

where V is the volume of the particle, L the corresponding geometrical factor of its surface (changing from 1/3 for a sphere to about 1 for strongly elongated particles), and

$$\begin{pmatrix} A_1 \\ A_2 \end{pmatrix} = \begin{pmatrix} 1 & L_n \\ \epsilon_n & \epsilon_n(L_n - 1) \end{pmatrix} \cdot \Psi \cdot \begin{pmatrix} (\epsilon_1 - 1)L_1 + 1 \\ -(\epsilon_1 - 1)\delta_1 \end{pmatrix}. \tag{12}$$

Here $\epsilon_j = \varepsilon_j/\varepsilon_{j+1}$, and $\varepsilon_j, L_j, \delta_j$ are the permittivity, the geometrical factor and the fraction of volume inside the external boundary of j-th layer, respectively, and

$$\Psi = \prod_{j=2}^{n-1} (\Omega_j \Gamma_j) \Omega_1, \quad \text{with} \quad \Omega_j = c_j \prod_{l=1}^{n_{\rm sub}} \Lambda_{jl}, \tag{13}$$

where c_j are some constants, the 2×2 matrices Ω_j and Γ_j depend only on the parameters of the j-th layer $(\varepsilon_j, L_j, \delta_j)$, and the 2×2 matrices Λ_{jl} depend on these parameters for the l-th sublayer [26].

Our analysis of the QSA in the case of homogeneous spheroids demonstrates that the condition of its applicability is $|m-1|x_\text{v} \ll 1$, whereas that of the RA is known to be $|m-1|x_\text{d} \ll 1$, where x_d and x_v are the size parameters related to the maximum dimension and the radius of the equivolume sphere, respectively. The difference of these ranges becomes essential for scatterers of large eccentricity (see Fig. 4) and numerical calculations confirm this conclusion [23]. For layered particles, the convergence is similar as can be seen from Fig. 7. Comparison of the nQSA with other methods available for layered particles (SVM, TMM, DDA, etc.) shows that the nQSA is many orders of magnitude faster than the nDDA and much faster and more universal than the nSVM and nTMM.

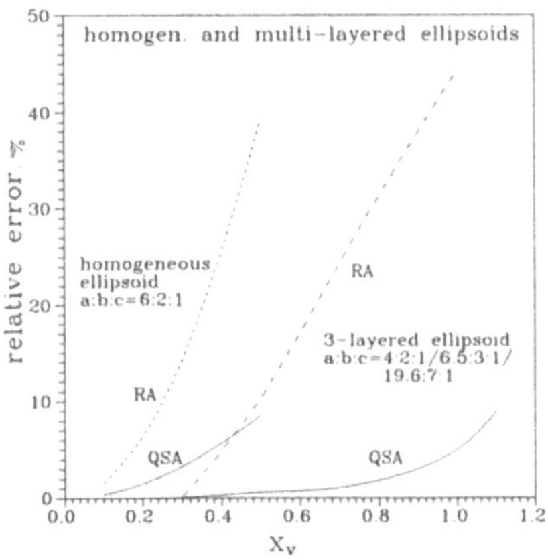

Figure 7. Relative errors of the scattering cross-sections derived with the nRA and nQSA for homogeneous ellipsoids with semiaxis ratio $a : b : c$ = 6:2:1 and refractive index m =1.5 and for three-layered ellipsoids with semiaxis ratios, refractive indices and volume fractions of core/layer/mantle $(a : b : c,\ m,\ \delta_\text{v})$ = (4:2:1, 1.3+0i, 0.22) / (6.5:3:1, 1.7+0.03i, 0.34) / (20:7:1, 1+0i, 0.44).

An effective dielectric function ($\bar{\varepsilon}$) appears in the formulation of the RA and QSA for layered scatterers (see Eq. (11) and [25] for more details). For small particles ($|m-1|x_\text{v} < 1$), it was used to produce a new effective medium theory (EMT) mixing rule. This rule takes into account the internal structure of particles and, as our calculations demonstrate, can give significantly more accurate results than other mixing rules (see Fig. 8 and [25] for discussion).

3.3. SOME OTHER TOOLS

Two other tools need to be developed to apply the scattering theories to observations: a database of optical properties of non-spherical scatterers

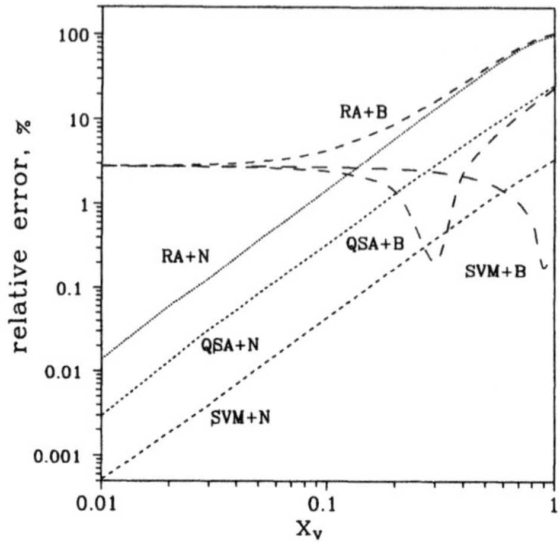

Figure 8. Relative errors of the scattering cross-sections calculated with a SVM code for homogeneous particles with the permittivity according to the new EMT rule (SVM+N) in the case of two-layered confocal spheroids (mantle: a/b =4:1, m =1.33+0.01i; core: a/b =6.7:1, m =1.7+0.03i; the volume fraction of the core is 0.33). Curves marked by RA+N and QSA+N give the errors of the nQSA and nRA. Results obtained with the standard Bruggeman rule of the EMT are denoted by SVM+B, QSA+B, and RA+B, respectively.

and a new method of solution of inverse problems. For additional tools created and results obtained within the project, we refer the reader to other chapters of this book (for instance, studies of internal fields and heating of dust grains are described in [28, 29], coherent backscattering is considered in [30] and so on) and to the annual reports available at the project site in the internet [31].

3.3.1. *Database of optical properties (DOP)*

Light-scattering codes are used to calculate optical properties of particles (cross-sections, scattering matrix, etc.). For many applications it is necessary to understand general trends in the data. This understanding often can be gained from consideration of results obtained by simplified scattering models. Optical properties derived from such models have been discussed in some books (e.g., [2, 32]) and in principle can be calculated, since various light-scattering codes are freely available via the internet. Nevertheless, there is a definite necessity of a WWW database devoted to systematic consideration of the optical properties of various model scatterers and related topics. As of this writing, our database includes:

i) original light scattering codes for non-spherical and/or inhomogeneous particles;

ii) reviews of exact and approximate light scattering methods, including discussion of their ranges of applicability;

iii) a database containing about 7000 references to papers on various aspects of light-scattering theory and applications;

iv) a library of graphics illustrating light scattering by particles of different sizes/shapes/structures (a part of the data will be in tabular form to serve as benchmarks);

v) a special tool for on-line calculations of selected optical characteristics of different scatterers (homogeneous and core-mantle spheres, infinitely long cylinders, spheroids, etc.);

vi) a self-training algorithm of determination of the optical properties of fractal-like clusters;

vii) a collection of links to related WWW resources.

The database will be available via the Internet at *http://www.astro.spbu.ru/DOP* upon its completion scheduled at the end of 2002.

3.3.2. *Inverse-problem solution*

Determination of characteristics of cosmic dust grain ensembles from their observational manifestations is often an ill-posed inverse problem. Because of existing uncertainties and the difficulty, special tools seldomly have been applied and the problem is often solved through direct fitting. However, there are good examples of proper techniques in [33, 34], where the size distribution of dust grains in the diffuse interstellar medium is determined through least squares and maximum entropy methods, respectively (using the assumption that the particles were homogeneous spheres). Problems of this sort can be formulated as the Fredholm integral equation of the first kind

$$F(\lambda) = \int_0^\infty K(a,\lambda) f(a) \mathrm{d}a, \tag{14}$$

where F is an observed quantity, K the kernel, and f an unknown function. For example, for the interstellar extinction, we have $F = A_\lambda \propto \tau_\lambda \propto \langle C_{\rm ext}\rangle$, and in the case of spheres $K \propto C_{\rm ext}(2\pi a/\lambda, m_\lambda)$, where a is the radius and m_λ the refractive index. We suggest a more physically consistent way of obtaining a solution of the inverse problem than that applied in [33, 34]. Our approach is based on a combination of the PCHA (see Sect. 3.2.1) and the method of optical equivalence (MOE).

In the MOE, a mixture of particles of complex shapes, structures, etc. is replaced by a disperse ensemble of simplified model particles whose optical properties are *approximately equivalent* to those of the mixture at the wavelengths considered. The PCHA allows one to treat inhomogeneous spheres with a great variety of refractive index profiles and hence provides enough freedom (i.e. includes a sufficient number of free parameters) to realize the MOE replacement, in contrast with the Mie theory for homogeneous spheres where no free parameters are available. Our numerical experiments [36] show the following: *i)* in all considered cases of inhomogeneous spheres only two parameters are principal – the relative thickness of the inhomogeneous mantle Δ, and the position where the refractive index reaches its maximum value (in the case of nonmonotonous profiles), and *ii)* variations of these parameters provide a good possibility of establishing the optical equivalence. When the selected free parameters are approximately settled, one can derive the size (or mass) distribution function. Note that although the structure and shape of the model of inhomogeneous particles may be far from those typical of real scatterers, the size distribution of these scatterers can be restored. The advantages of using the PCHA arises from the fact that the approximation has a simple realization and results can be calculated as quickly as those using the Mie theory. The described approach has been applied to "antireflection" particles (particles with peculiarly low backscattering efficiency) [35] and fluffy aggregates of cosmic dust grains [37].

4. Radiative transfer

In almost all cases, one needs to consider the transfer of (polarized) radiation to simulate observations of dusty objects. Various radiative transfer programs used in astronomy are listed in Table 7 of [38]. The most popular approach to solve the polarized radiative-transfer equation is the Monte-Carlo method. This is usually applicable to objects of any geometry, but not very large optical thickness.

An original Monte-Carlo code was developed in Jena (see [39] and references therein). It can treat polarized radiative transfer in a three-dimensional anisotropic media (e.g., populated by aligned non-spherical particles) in a self-consistent way (cf., e.g., [40]). Figure 9 illustrates possible applications of tools developed in our project. A light scattering code (SVM) was applied to calculate the scattering matrix and other required optical properties of aligned spheroidal particles whose refractive indices are typical of astronomical silicate in the visual. Then the polarized radiative-transfer code was used to produce the intensity and polarization maps (see [39] for more details) which resemble observed ones (e.g., maps for the objects IRAS 19500-1709 or 17106-3046 published in [41]).

5. Acknowledgments

The authors are thankful to G.Videen for many suggestions that greatly improved the manuscript and to B.T.Draine for the DDA code (ver.5a9). This work was supported by the INTAS grant 99/652. V.I. and N.V. ac-

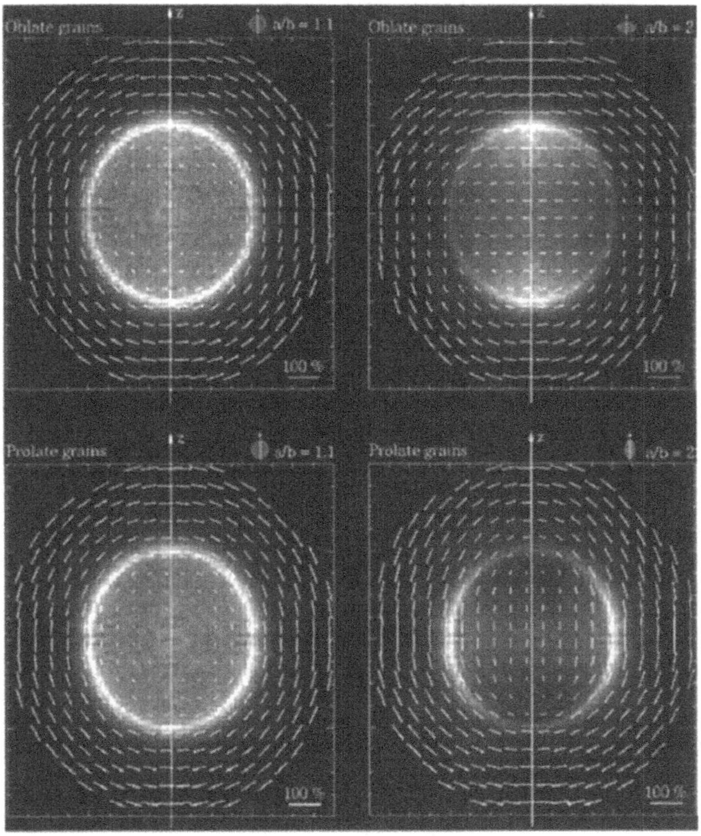

Figure 9. Intensity maps with overlaid polarization pattern of a spherical shell containing perfectly aligned dust grains (see [41] for description of the model used).

knowledge the support within the Russian federal program "Astronomy" and by the RFBR grant for the scientific school of V.V.Sobolev.

References

1. Martin, P.G. (1978) *Cosmic Dust*, Oxford Univ. Press, Oxford.
2. Bohren, C.F., and Huffman, D.R. (1983) *Absorption and Scattering of Light by Small Particles*, J.Willey & Sons, NY.
3. Henning, Th., Il'in, V.B., Krivova, N.A. et al. (1999) *Astron. Astrophys. Suppl.*, **136**, 405.
4. Jäger, C., Mutschke, H., Dorschner, J., and Henning, Th. (1998) *Astron. Astrophys.*, **332**, 291.
5. Henning, Th., and Mutschke, H. (2000) in *Thermal Emission Spectroscopy and Analysis of Dust, Disks, and Regoliths*, eds. M.L. Sitko et al., ASP Conf. Ser. **196**, 253.
6. Fabian, D., Henning, Th., Jäger, C. et al. (2001) *Astron. Astrophys.*, **378**, 228.

7. Molster, F.J. (2000) Ph.D. Thesis, Univ. Amsterdam.
8. Mishchenko, M.I., Hovenier, J.W., and Travis, L.D. (2000) *Light Scattering by Nonspherical Particles*, Academic Press, San Diego.
9. Hovenier, J.W., Lumme, K., Mishchenko, M.I. et al. (1996) *J. Quant. Spectr. Rad. Transf.*, **55**, 695.
10. Wiscombe, W.J., and Mugnai, A. (1986) *NASA Reference Publ.*, **1157**, 1.
11. Lumme, K., and Rahola, J. (1998) *J. Quant. Spectr. Rad. Transf.*, **60**, 439.
12. Kolokolova, L., and Gustafson, B.A.S. (2001) *J. Quant. Spectr. Rad. Transf.*, **70**, 611.
13. Wu, Z.P., and Wang, Y.P. (1991) *Radio Sci.*, **26**, 1393.
14. Perelamn, A.Ya., Zinov'eva, T.V., and Moseev, I.G. (2002) this volume.
15. Perelman, A.Ya. (1996) *Appl. Opt.*, **35**, 5452.
16. Farafonov, V.G., Voshchinnikov, N.V., and Somsikov, V.V. (1996) *Appl. Opt.*, **35**, 5412.
17. Asano, S., and Yamamoto, G. (1975) *Appl. Opt.*, **14**, 29.
18. Gurwich, I., Kleiman, M., Shiloah, N., and Cohen, A. (2000) *Appl. Opt.*, **39**, 470.
19. Farafonov, V.G. (2001) *Opt. Spectrosk.*, **90**, 743.
20. Farafonov, V.G., and Voshchinnikov, N.V. (2002) *J. Comp. Math. & Math. Phys.*, submitted.
21. Voshchinnikov, N.V., and Farafonov, V.G. (2002) *Meas. Sci. & Techn.*, **13**, 246.
22. Farafonov, V.G., Il'in, V.B., and Henning, Th. (1999) *J. Quant. Spectr. Rad. Transf.*, **63**, 205.
23. Farafonov, V.G., Il'in, V.B., and Prokopjeva, M.S. (2002) *Opt. Spectrosk.*, in press.
24. Draine, B.T., and Flatau, P.J. (1999) *Preprint Princeton Obs.*, POPe-785.
25. Posselt, B., Farafonov, V.G., Il'in, V.B., and Prokopjeva, M.S. (2002) *Meas. Sci. & Techn.*, **13**, 256.
26. Farafonov, V.G. (2002) *Opt. Spectrosk.*, **92**, 829.
27. Farafonov, V.G., Il'in, V.B., and Prokopjeva, M.S. (2002) *Opt. Spectrosk.*, **92**, 608.
28. Babenko, V.A., and Petrov, P.K. (2002) this volume.
29. Astafyeva, L.G. (2002) this volume.
30. Mishchenko, M.I., Tishkovets, V.P., and Litvinov, P.V. (2002) this volume.
31. INTAS project 99/652 website, *http://www.astro.spbu.ru/staff/ilin2/INTAS*.
32. Wickramasinghe, N.C. (1973) *Light Scattering Functions for Small Particles with Applications in Astronomy*, Chapman & Hall, London.
33. Mathis, J.S., Rumpl, W., and Nordsieck, K.H. (1977) *Astrophys. J.*, **217**, 425.
34. Kim, S.H., Martin, P.G., and Hendry, P.D. (1994) *Astrophys. J.*, **422**, 164.
35. Shifrin, K.S., Perelman, A.Ya., and Kokorin, A.M. (1986) *Opt. Spectrosk.*, **59**, 361.
36. Perelman, A.Ya., and Zinov'eva, T.V. (2002) *Opt. Spectrosk.*, **92**, 481.
37. Perelman, A.Ya., and Zinov'eva, T.V. (2001) *Opt. Atmosph. & Ocean*, **14**, 669.
38. Voshchinnikov, N.V. (2002) *Astrophys. & Space Phys. Rev.*, **12**, 1.
39. Wolf, S., Voshchinnikov, N.V., and Henning, Th. (2002) *Astron. Astrophys.*, **385**, 365.
40. Whitney, B.A., and Wolff, M. (2002) *Astrophys. J.*, in press.
41. Gledhill, T.M., Chrysostomou,A., Hough, J.H., and Yates, J.A. (2001) *Mon. Not. Roy. Astr. Soc.*, **322**, 321.

FRACTAL AGGREGATES IN SPACE

[1]GERHARD WURM and [2]MARTIN SCHNAITER
[1]*Westfälische Wilhelms-Universität, Institut für Planetologie, Wilhelm-Klemm-Straße 10, 48149 Münster, Germany*
[2]*Forschungszentrum Karlsruhe, Institut für Meteorologie und Klimaforschung, Postfach 3640, 76021 Karlsruhe, Germany*

Abstract. Individual cosmic dust particles come in very complex shapes but there are a variety of environments where dust particles are aggregates of smaller grains. These aggregates often can be characterized by a fractal particle structure. Within this fractal framework, morphologically averaged particle properties depend only on a few parameters. Despite the complexity of the individual particles, fractal dimension, aggregate size, and building-block size are sufficient parameters for many applications. This paper reviews the process of aggregation to generate fractal aggregates with focus on some general features of those aggregates (mass to surface ratios, aspect ratios). Some implications for aerodynamical and optical properties (gas- grain coupling times, polarization, extinction) are discussed.

1. Introduction

Most of the information on cosmic dust particles is gathered by astronomical observations. However, the process of deducing the information from the data is ill-defined, with a rather large number of parameters to fit a particular dust population. If restrictions on the parameters and the knowledge of the influence of others are known *a priori* for a given environment, remote-sensing analysis can benefit strongly. This paper focuses on the fractal morphology of aggregated cosmic dust particles and using these fractal properties as a parameter for particle characterization.

G. Videen and M. Kocifaj (eds.), Optics of Cosmic Dust, 89–102.

Ever since interplanetary dust particles have been captured in the stratosphere and later analyzed [1], aggregate particles in space have become visually accessible (see Fig.1). Though compact and not as open-structured as the aggregates we will refer to as fractals, a great body of work has been invested to characterize the optical behavior of such aggregates to interpret astrophysical observations [2,3]. With building blocks on the order of 100nm, we see similarity to relevant particle sizes in the interstellar medium that might indicate some common history, at least for certain types of aggregates. As observational wavelengths in the visual are comparable to this size there is little doubt that the aggregate morphology is one of the major parameters in determining the scattering properties.

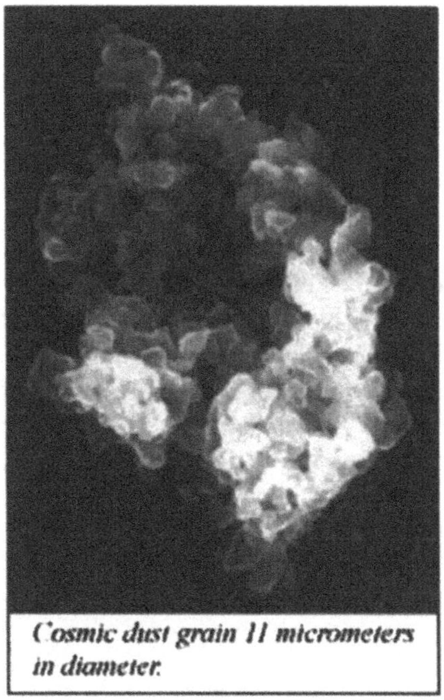

Figure 1: Interplanetary Dust Particle (IDP) captured in the stratosphere (taken from the NASA web site). Clearly visible is the aggregated structure with sub-micrometer-size building blocks.

From observations of the extinction and polarization by the interstellar medium, it is clear that interstellar particles themselves are not spherical but must have an elongated structure [4,5]. It is still unclear where this asymmetry originates. Crystalline particles e.g. naturally condense into elongated shapes and are certainly present in cosmic environments [5]. These asymmetries might be modeled in terms of spheroids or cylinders [6,7]. However, the asymmetry might naturally result from a fractal-like structure in a self-consistent and therefore determined manner. If so, the latter would be a rather elegant step of unification of cosmic dust models.

Observations of the edge of protoplanetary disks (so far one) might indicate the existence of particles of many micrometers is size [8]. The origin of these large particles has yet to be determined. In later stages of planet formation, impacts between larger bodies can create compact aggregates [9,10]. However this happens predominantly in the inner parts of protoplanetary disks. If these fragments are the characteristic particles seen at the edges of the disk, there must be an efficient transport mechanism [11]. At early stages of the protoplanetary disks and even earlier stages within dense molecular cloud cores, particles evolve by collisional growth, and this growth results in less compact, but fractal-like structures [12,13]. These aggregates also evolve naturally at the edges of the disk but should hardly be visible in extinction as larger particles [14].

Because of their population densities, fractal aggregates must play a significant role in many astrophysical environments and in the following paragraphs we outline simulated production mechanisms.

2. Fractal Grain Growth

A current paradigm in models of planet formation begins with the growth of larger bodies by sticking collisions. The basic idea is that two (dust) particles collide and are able to stick to each other creating a larger grain. If such a process happens over and over again, km-size bodies would eventually form from initially micron-size dust grains. The details of this process determine the growth timescales and particle properties. We note especially that the making of km-sized planetesimals from micron-size dust has to span 9 orders of magnitude in size. This is a task not easily accomplished.

A number of different growth processes for dust particles is discussed in the literature [15,16]. Here we concentrate on the two classical cases, namely ballistic cluster-cluster aggregation (BCCA) and ballistic particle-cluster aggregation (BPCA), which are probably the most relevant for astrophysical applications. The basic mechanism in BCCA is that, on average, two particle clusters of the same size collide on ballistic trajectories with each other and form a rigid connection on their first contact point. Since aggregates are unable to penetrate each other significantly, this process leads to very fluffy structures as seen in Fig. 2.

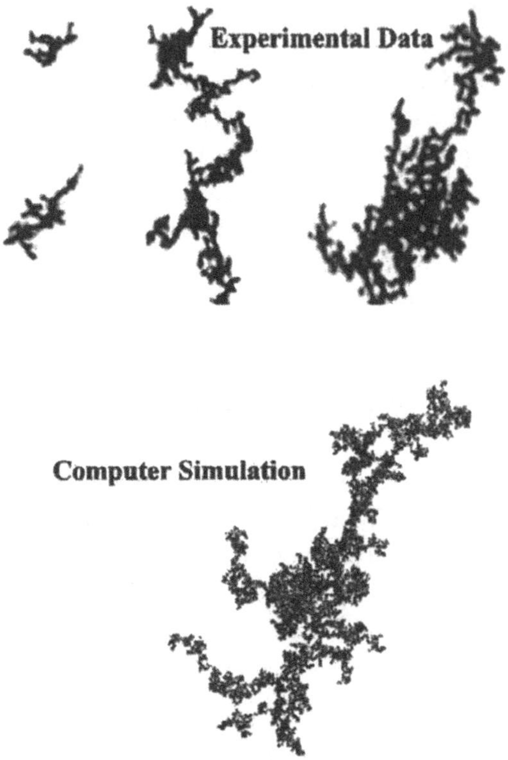

Figure 2: Examples of aggregates formed in a BCCA process. The upper part represents aggregates observed by long distance microscopy in experiments. Individual monomers in these aggregates are approximately 2 micrometer in diameter, which corresponds to the thickness of the chains in the aggregate [17]. The lower part of the image is a computer simulation created by R. Sablotny [18]. The fluffy structure, as well as the similarity between theory and experiment, is clearly visible.

Before continuing further we must make the standard disclaimer that the particles we are studying are not actually fractals, since they are not infinite in extent. They do, however, have many of the same characteristics as fractals, including their production algorithms, in which we simply truncate the procedure. It is, perhaps, more correct, although much more awkward to refer to them as fractal-like. The particle qualities we discuss are very much the same as those for fractals, and we will not discuss the distinctions here.

Introducing the fractal dimension d_f, BCCA results in $d_f \approx 2$. The fractal dimension here simply refers to the power in the mass-size relation of an aggregate with mass M and size R

$$M \propto R^{d_f}. \quad (1)$$

With $d_f \approx 2$ the mass of such a particle thus depends on its overall size as the surface does.

BPCA is characterized by a collision of a growing cluster (aggregate) of particles (monomers) with individual monomers on ballistic trajectories. The individual monomers are better suited to penetrate deeper into the aggregate and close large gaps. Therefore the aggregates might still be characterized as porous but are much more compact as seen in Fig. 3 [19]. In terms of fractal dimension the process leads to $d_f = 3$: the value for a compact solid body. In this chapter, we would like to restrict the name fractal to aggregates with a fractal dimension much less then 3, especially BCCAs.

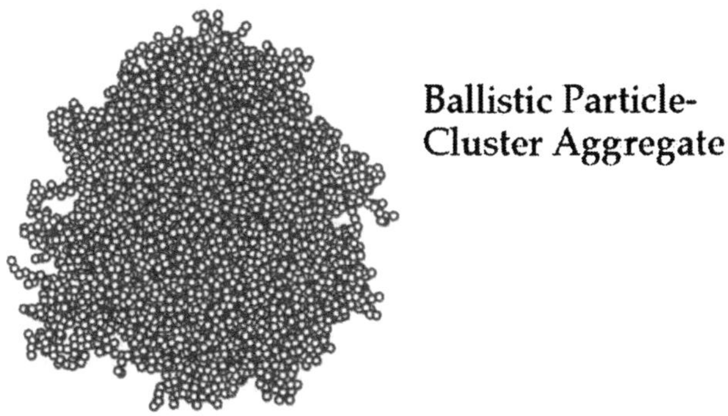

Figure 3: Computer generated BPCA by H. Kimura [19]. Though porous, the particle shows a rather compact morphology with a fractal dimension of $d_f = 3$.

In principle a relation similar to Eq. 1 (with another fractal dimension) can provide the cross section of the aggregate [20]. However, if d_f is close to or larger than 2, the cross section A will approximately be proportional to R^2 as already indicated above. The difference between BPCA and BCCA therefore shows an immediate consequence, namely the difference in the mass to surface ratio depends on the size:

$$\frac{M}{A} = \text{const} \qquad \text{(BCCA)} \quad (2)$$

$$\frac{M}{A} \propto R \qquad \text{(BPCA).}$$

In gaseous environments, the consequence is that the gas-grain coupling time τ does not depend on size for aggregates generated in a BCCA process. This can be seen in equation (20) in Blum *et al.* 1996 [21]:

$$\tau = \varepsilon \frac{M}{A} \frac{1}{\rho_g v_g}. \qquad (3)$$

Here, ε=0.58 is a constant, ρ_g is the gas density, and v_g is the mean thermal velocity of the gas molecules.

Since it is the difference in τ that, in general, generates collision velocities between small particles embedded in gas, collision velocities remain constant throughout BCCA growth. It is always only the small statistical deviation from the average value of τ which determines the collision velocity, but it is not the size difference. On the other hand by the same reasoning we must consider that collision velocities increase with size or size difference in the case of BPCA or compact aggregates.

An example where this is of importance is the growth of planetesimals in protoplanetary disks by compact particles. It is often assumed that particles "rain out" to the midplane of the disk very early, since the larger particles are faster, sweep up the smaller particles, and attain even higher velocities. This cannot happen during the first stages where BCCA dominates the growth and therefore determines aggregate morphology. Grains have to grow to several cm in size before they get compact and runaway growth of the larger bodies occurs [12].

2.1 Non-Sphericity, Asymmetry, and Aspect Ratios

There are several reasons, why particles can have spherical or non-spherical shapes. Starting at a rather fundamental point of amorphous condensation of solid matter, dust grains form as spheres in many cases (see Fig. 4). There might be small deviations or irregularities but these are minor compared to the elongations needed to describe e.g. interstellar dust particles.

There are other modes though; e.g., dust particles crystallize and already commence as non-spherical, very elongated particles [5]. And even if starting from a grain condensed as a sphere, destructive cosmic events, like high-energy collisions between two dust grains, might alter the shape of a single grain from spherical to an arbitrary, non-spherical structure. It is not easy to predict the importance of these effects in a quantitative way. That offers a wide field for speculation.

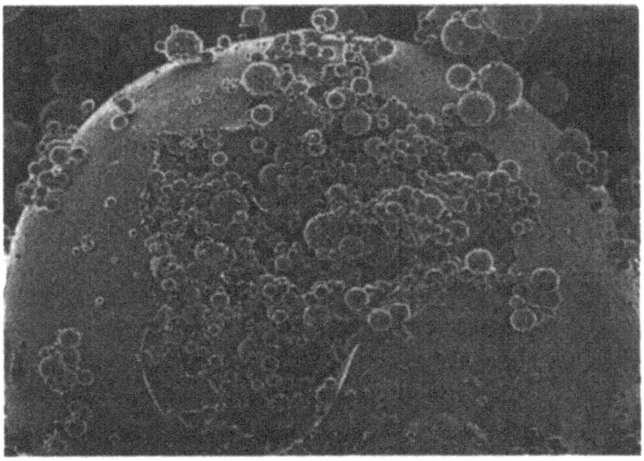

Figure 4: Fly ash as found in exhausts of power plants providing an illustration of the sphericity of condensed particles. The original picture is taken from the cover of a Post Card of the Bauhaus University in Weimar, Germany.

We outline yet another origin of asphericity. Fractal aggregates themselves might diminish the asphericity for the individual solid-grains building blocks, but might offer a determined aspect ratio within the fractal framework at the same time. It has been shown in calculations, as well as in experiments that particles grown in a CCA process show a characteristic average aspect ratio close to 2 [17, 22]. This appears to be determined by the fractal dimension of the process. The asymmetry v_R, or aspect ratio, can be measured by the ratio of the two radii of gyration of an aggregate

$$v_R = \frac{R_l}{R_s}, \tag{4}$$

where R_l and R_s, are the radii of gyration R defined with respect to the large and small principal axis:

$$R = \frac{\sqrt{\sum_i r_i^2}}{N} \tag{5}$$

Here, the distance of the ith of N spherical-building-block monomers to the axis of the aggregate is given by r_i. With respect to BCCA the aspect ratio is $v_R \approx 2$ [17]. For Brownian coagulation with a fractal dimension as low as $d_f = 1.3$ the aspect ratio can be as large as $v_R \approx 4$ [23].

3. Fractal Aggregates in Astrophysical Environments

One environment where fractal dust aggregates are undoubtedly generated and play a significant role is during the first phases of planet formation as already indicated above. The standard model of planet formation assumes sub-micron-size dust particles embedded in a gaseous disk around a young stellar object to collide with each other, stick together, and grow [23-25]. Since collision velocities are of the order of mm/s at the start, the growth commences as a BCCA process. Only when the aggregates have grown to approximately cm-size is the impact energy large enough to restructure the aggregates into compact aggregates [12]. This raises the question of how dust grains show up in observations of a protoplanetary disk in Orion [8], which, when interpreting extinction measurements, are several micrometers in size. Even if they grow to these size ranges, the building blocks of the fractal aggregates should dominate the extinction measurements in the visible [14,26]. Since those building blocks are either of interstellar origin or reprocessed condensates from the disk they should be sub-micron in size. Important consequences result if both claims are valid (that particles at the edge of protoplanetary disks appear large from their extinction and that fractal aggregates do not appear large from their extinction), namely that particles at the edge of the disk must be compact. This requires at minimum one additional mechanism besides particle growth by aggregation. Either the growing fractal dust aggregates are compacted in some way or there is an effective transport of dust from the inner parts of the disk [11]. Since time scales are much smaller at the inner portion of the disk, compact aggregates from growth as well as fragmentation during later stages exist concurrently.

It has been shown that fractal growth is important in dense cores of molecular clouds [27], but we ask a less obvious question, whether small, fractal-like aggregates can also be present in significant quantities in the diffuse interstellar medium. If so, it would significantly influence the optical properties obtained from the medium. It is not an outrageous idea to think of interstellar dust particles as aggregates. Already Mathis assumed this in 1986 [28] and models depending on rather porous grains continue to be discussed [29]. There is not yet justification for why interstellar dust particles should be fractal-like aggregates. However, one observation might lead us to this conclusions: the linear dependence between total to selective extinction R_V and wavelength of maximum polarization λ_{max} can be explained by assuming aggregate structure and changing the number of monomers per aggregate slightly. Here, R_V is defined by

$$R_V = \frac{A_V}{A_B - A_V}, \tag{6}$$

with A being the extinction in magnitude with respect to the visible (V) and blue (B). We set this in relation to the wavelength of maximum polarization which for the general diffuse interstellar medium follows a linear dependence as found by Whittet & van Breda 1978 [31]:

$$R_v = (5.6 \pm 0.3)\lambda_{\max}. \tag{7}$$

The results are plotted in Fig. 5. Whether this simple linear relationship showing up in both observations and in modeling results is coincidental has yet to be resolved.

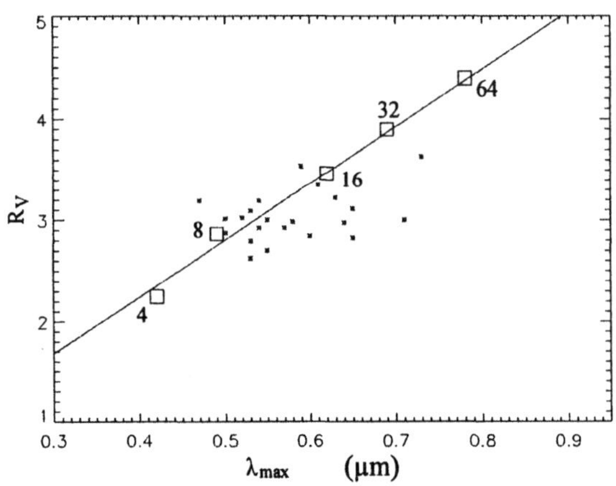

Figure 5: The solid line denotes the linear dependence between total to selective extinction and the wavelength of maximum polarization as found by Whittet and van Breda (1978) for the diffuse interstellar medium [31]. The squares mark the values for aggregates consisting of the printed number of monomers (taken from Wurm and Schnaiter, 2002) [30]. Smaller dots show data taken from Taurus dark clouds [32].

Within the calculations we also find that a very small degree of coagulation is plausible. The data given by Whittet *et al.* 2001 [32] for the Taurus dark clouds (also shown in Fig. 5) show a small increase in R_v with λ_{max}. Within the model this suggests coagulation of approximately 8 to 16 monomers in an aggregate, a doubling in size. In the ballistic cluster-cluster aggregation process, a single cluster size dominates the distribution at a given time [13]. This means that a doubling in size happens on one coagulation timescale τ_c, which is the average time it takes an aggregate to collide with another aggregate. The lifetime of the corresponding parts of the clouds or the time they spend in the given dense form must match the coagulation timescale. In a simple model the coagulation timescale is given by

$$\tau_c = \frac{1}{Nv\sigma}. \tag{8}$$

Here, N is the particle number density, v is the collision velocity, and σ is the collision cross section. Assuming, for example, an H_2 density of $10^4/cm^3$, a dust to gas mass ratio of 0.01, an average particle of 100nm radius, and turbulent collision velocities of a few m/s, τ_c would be on the order of 10^6 years. This matches the timescale on which the Taurus clouds have formed stars [33]. A very small degree of coagulation is thus possible. It has to be noted that the aggregates that would explain interstellar polarization and extinction would only consist of 8 to 16 monomers, which is rather small. Such small aggregates could also already be the result of aggregation shortly after or during the time of condensation where densities might be sufficiently large.

Another feature that might emphasize the need for a more detailed morphological treatment of interstellar dust particles is the 217.5nm extinction bump. This ubiquitous feature of the diffuse interstellar medium is usually interpreted as an electronic absorption band of solid carbonaceous material. Theoretical investigations using Mie theory for spherical particles and optical constants of bulk graphite indeed fit the mean interstellar bump quite well. It is clear that this assignment is a strong simplification of nature since graphite is an anisotropic material and therefore can hardly form spherical nanoparticles. As laboratory astrophysics matured, this interpretation became more questionable, since measurements on realistic carbonaceous dust material (i.e. soot) failed to fit the narrow width of the interstellar 217.5 nm bump. Soot particles are fractal aggregates composed of tiny nanometer-size spherical monomers (see Fig. 6). It has been argued that fractal clustering has a strong influence on the extinction properties of soot particles and might account for the discrepancies between measurements and observations. By measuring the UV-VIS spectral extinction properties of soot particles at different clustering stages it might be shown that the width of the bump is indeed correlated to the degree of clustering [35]. This holds especially for very small aggregates of up to some ten monomers where the electromagnetic interactions between neighboring particles have a significant influence on the soot extinction (see Fig. 7).

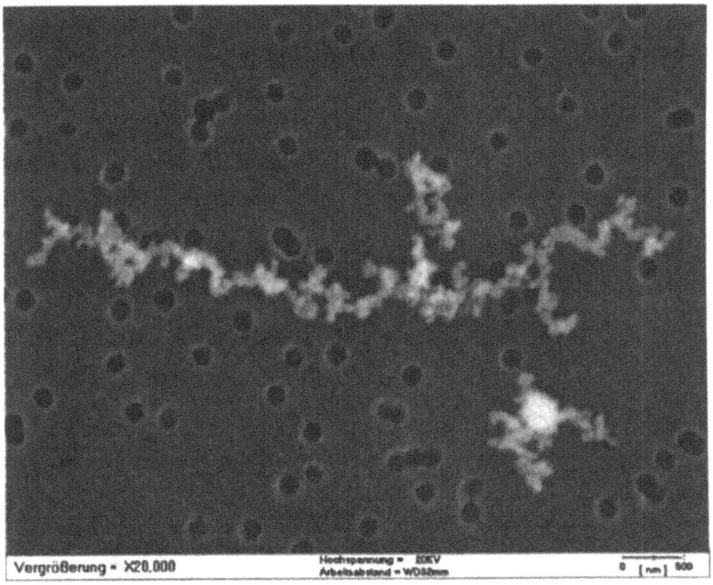

Figure 6: SEM micrograph of a soot aggregate emitted by a commercial Diesel engine [34]. A fractal dimension of 2 was determined for the freshly emitted soot.

Interstellar soot particles are formed in the expanding outflows of carbon-rich asymptotic giant branch stars (AGB). These fractal soot aggregates are altered in the interstellar medium, e.g. by shattering in shock waves or by ice mantle formation within dense molecular clouds. As a result of these processes a specific morphology and composition of the carbonaceous aggregates might occur giving rise for the 217.5 nm extinction bump and its variations with respect to different interstellar environments. Though the problem is not yet fully solved, it is likely that very small carbonaceous fractals will also be present in the diffuse interstellar medium [36].

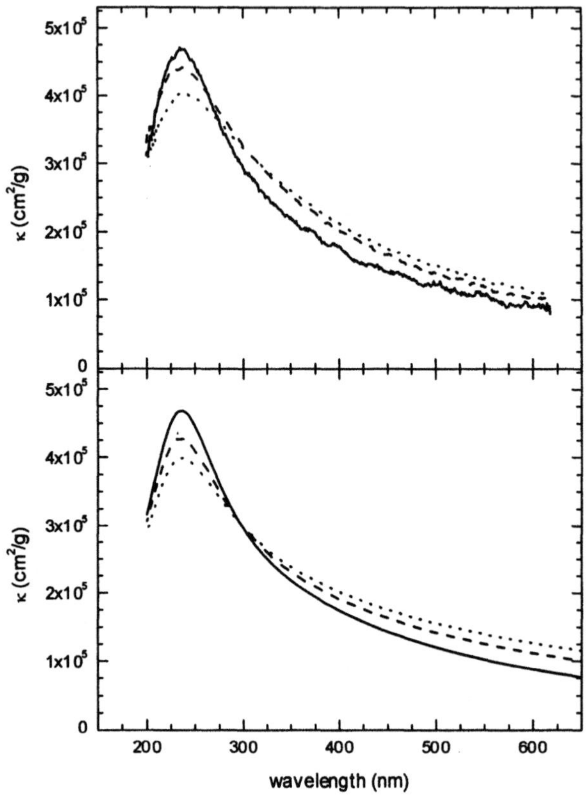

Figure 7: Mass specific UV-VIS extinction spectra of soot particles at different coagulation stages measured by matrix-isolation spectroscopy [35] (top). The feature decreases and broadens with increasing clustering within the sample. Calculations using an extended Mie-code for sphere aggregates by F. Rouleau (bottom) reflect the same trend. The curves represent the result for a sphere (solid line), a 4 member CCA (dashed line), as well as a 32 member CCA (dotted line).

4. Conclusions

It would certainly be improper to treat every cosmic dust population as a collection of fractal dust aggregates. On the other hand there are a number of environments where these types of dust particles play an important role, e.g. in molecular clouds and certainly in protoplanetary disks. In these cases treating aggregates as fractal-like aggregates might offer a more comprehensive insight and simpler explanation for a number of phenomena with only a few parameters. Even though it seems more complicated to assume the particles are fractal-like, in many respects it becomes simpler. It is easier to characterize and generate, for example, a fractal-like aggregate, than many types of particle systems. Most importantly, however, we must consider the observational record and whether the evidence agrees with the simulations.

References

1. E. K. Jessberger, T. Stephan, D. Rost, P. Arndt, M. Maetz, F. J. Stadermann, D. E. Brownlee, J. P. Bradley, and G. Kurat; in: Interplanetary Dust, ed. E. Grün, B. Å. S. Gustafson, S. F. Dermott, and H. Fechtig (Springer), pp 253 (2001).
2. J. M. Greenberg and B. Å. S. Gustafson, A&A, **93**, 35 (1981).
3. E. V. Petrova, K. Jockers, and N. N. Kiselev, Icarus, **148**, 526 (2000).
4. K. Serkowski, in IAU Syposium 52, Interstellar Dust and Related Topics, ed. J. M. Greenberg and H. C. van de Hulst (Kluwer), pp 145 (1973).
5. D. Fabian, Th. Henning, C. Jäger, H. Mutschke, J. Dorschner, and O. Wehrhan, A&A, **378**, 228 (2001).
6. A. Li and J. M. Greenberg, A&A, **323**, 566 (1997).
7. N. V. Voshchinnikov and D. A. Semenov, AstL, **26**, 679 (2000).
8. H. B. Throop, J. B. Bally, L. W. Esposito, and M. J. McCaughrean, Science, **292**, 1686 (2001).
9. G. Wurm, J. Blum, and J. E. Colwell, Icarus, **151**, 318 (2001).
10. G. Wurm, J. Blum, and J. E. Colwell, Phys. Rev. E, **64**, 046301 (2001).
11. W. R. Skinner and J. M. Leenhouts, Metic, **28**, 439 (1993).
12. J. Blum and G. Wurm, Icarus, **143**, 138 (2000)
13. G. Wurm and J. Blum, Icarus, **132**, 125 (1998).
14. M. Schnaiter and G. Wurm, Appl. Opt., **41**, 1175 (2002).
15. S. K. Friedlander, Smoke, Dust, and Haze (Oxford Univ. Press) (2000).
16. P. Meakin, Rev. Geophys., **29**, 317 (1991).
17. G. Wurm and J. Blum, ApJ, **529**, L57 (2000).
18. R. Sablotny, PhD thesis, FSU Jena (1996).
19. H. Kimura, JQSRT, **70**, 581 (2001).
20. S. Kempf, PhD thesis, FSU Jena (1998).
21. J. Blum, G. Wurm, S. Kempf, and Th. Henning, Icarus, **124**, 441 (1996).
22. R. Botet, & R. Jullien, J. Phys. A: Math. Gen., **19**, L907 (1986).
23. J. Blum and 26 colleagues, Phys. Rev. Lett., **85**, 2426 (2001).
24. S. V. W. Beckwith, Th. Henning, and Y Nakagawa, in: Protostars and Planets IV, ed. V. Mannings, A. P. Boss und S. S. Russell (Univ. of Arizona Press), pp 533 (2000).
25. S. J. Weidenschillung and J. N. Cuzzi. in: Protostars and Planets III, ed. E. H. Levy and J. I. Lunine (Univ. of Arizona Press), pp 1031 (1993).
26. T. Kozasa, J. Blum, and T. Mukai, A&A, **263**, 423 (1992).
27. V. Ossenkopf, A&A, **280**, 617 (1993).
28. J. S. Mathis, J. S. ApJ, **308**, 281, 1986.

29. N. V. Voshchinnikov and J. S. Mathis, ApJ, **526**, 257 (1999).
30. G. Wurm and M. Schnaiter, ApJ, **567**, 370 (2002).
31. D. C. B. Whittet and I. G. van Breda, A&A, **66**, 57 (1978).
32. D. C. B. Whittet, P. A. Gerakines, J. H. Hough, and S. S. Shenoy, ApJ, **547**, 872 (2001).
33. S. J. Kenyon and L. Hartmann, ApJS, **101**, 117 (1995).
34. H. Saathoff *et al.*, J. Aerosol Sci., submitted
35. M. Schnaiter, H. Mutschke, J. Dorschner, and Th. Henning, ApJ, **498**, 486 (1998).
36. A. C. Andersen, J. A. Sotelo, V. N. Pustovit, and G. A. Niklasson, in:Electromagnetic and Light Scattering by Nonspherical Particles, ed. B. Å. S Gustafson, L. Kolokolova, and G. Videen (ARL), pp 1 (2002).

SCATTERING OF LIGHT BY A SPHERE WITH AN ARBITRARY RADIALLY VARIABLE REFRACTIVE INDEX

A. Y. PERELMAN,[1] T. V. ZINOV'EVA,[2] and I. G. MOSSEEV[1]
[1]*State Forest Technical Academy, St. Petersburg 194021, Russia*
[2]*Sobolev Astronomical Institute, St. Petersburg University, St. Petersburg-Peterhof 198504, Russia*

Abstract. Based on the piecewise-continuous hyperbolic approximation (PCHA), we have developed a numerically stable and accurate algorithm for computation of the internal and scattered fields, as well as energetic characteristics, of a sphere with an arbitrary radially variable complex refractive index. The algorithm is cast in terms of the power functions, which overcomes a number of problems associated with round-off errors. The method of computation is tested with known solutions relating to the particular cases of the problem. The PCHA is proved to be convergent. The PCHA allows one to solve the scattering problem associated with an arbitrary complex radially variable refractive index in terms of the simplest functions. The PCHA makes it possible to construct the formal refractive index contour reproducing the scattering experimental data considerably more accurately than the Mie theory. This result is of importance in remote sensing problems. Several examples of calculations for the scattering function of cosmic fluffy dust particles are presented.

1. Introduction

Light scattering by spherically symmetric obstacles is of importance when the optical characteristics of various particles (aerosols, hydrosols, cosmic dust, and so on) are investigated [1-3]. These characteristics are used in chemical

G. Videen and M. Kocifaj (eds.), Optics of Cosmic Dust, 103–118.

engineering, environmental science, biological research, wave propagation and remote sensing. There are a great number of publications [4-15] on different aspects of scattering by inhomogeneous spherical particles.

The main aim of our study is to develop an effective method that can be used to solve the scattering problem dealt with by any radially variable refractive indices. The problem is formulated as follows: Let $R = 2\pi a/\lambda$ be the size parameter (a is the radius of a spherical particle, λ is the wavelength in vacuum), $\rho=2\pi r/\lambda$ be the wave distance (r is the distance from the center of a particle), and

$$\mu(\rho) = \begin{cases} M(\rho), & 0 \le \rho < R, \\ M, & \rho > R \end{cases} \tag{1}$$

is the contour of the refractive index. Here, $M(\rho)$ is an arbitrary (real or complex) radially variable refractive index of the particle and M is a real constant. The problem is to determine the internal and scattered electromagnetic fields, and in the end, a certain collection of the optical quantities (scattering function, efficiencies, etc.), which can be evaluated experimentally.

The Mie - Debye method of separating the variables in case of the variable refractive index $M(\rho)$ differs only in the radial equation from the equations for the homogeneous sphere [2,6,10]. Namely, the equation for the radial components is written as [6,16]

$$Y_n''(\rho) - \beta M'(\rho)M^{-1}(\rho)Y_n'(\rho) + \left[M^2(\rho) - \frac{n(n+1)}{\rho^2} \right] Y_n(\rho) = 0, \tag{2}$$

where $\beta = 2$ and $\beta = 0$ correspond to the E-wave and H-wave, respectively. The solution of the set of two differential equations (2) (with $\beta = 2$ and $\beta = 0$) is expressed in terms of the well-known mathematical functions in the exceptional cases [1,17], e.g., if

$$M(\rho) = \gamma \rho^{\alpha}, \tag{3}$$

in which α and γ are constants [1,18]. For this reason, the numerical methods of integration (2) for specific profiles of $M(\rho)$ are usually used. So, the computations are carried out for profiles

$$M(\rho) = m_o(1 + A\rho^2)^{-1}, \qquad M(\rho) = m_o(1 - A\rho^2)^{1/2}$$

describing the generalized lenses of the Maxwell fisheye and Luneberg types, respectively [3,12,15], as well as for some profiles of $M(\rho)$ having oscillating structure [6]. Some papers [1,4,5,7] treat a piecewise constant profile of the refractive index (a stratified sphere). Models of a droplet with radially varying refractive index have also been studied by means of geometric optics in [8,9].

The present method to solve the problem under consideration is based on the fact that there exists the profile

$$M(\rho)=\gamma\rho^{-1}, \tag{4}$$

to which corresponds the linear independent solutions of equation (2) of the form

$$\rho^{-0.5\pm\nu_n} \quad (\beta=2), \qquad\qquad \rho^{0.5\pm\nu_n}\,(\beta=0), \tag{5}$$

where

$$\nu_n=\sqrt{(n+0.5)^2-\gamma^2}\,. \tag{6}$$

As a final result, the simplicity of functions (5) explains the effectiveness of the new method. Its advantage becomes especially apparent for profiles of $M(\rho)$ given by different real or complex functions within subintervals of the interval $0\le\rho<R$. Such a complicated structure of $M(\rho)$ occurs in the model of a non-absorbing sphere with an arbitrarily located highly absorbed spherical inclusion considered in [13].

It has been shown [16,18] that the scattering functions, the efficiencies and other optical characteristics are expressed in terms of the Mie theory formulas, with the scattering coefficients $a_n=A_{1n}$ (electric wave) and $b_n=A_{2n}$ (magnetic wave) dependent on the refractive index profile. The following representations of the extinction Q_{ext} and backscattering Q_b efficiency factors, and the scattering function $i(\mu)$

$$\begin{cases} Q_{ext}=\dfrac{2}{R^2}\mathrm{Re}\displaystyle\sum_1^\infty(2n+1)(a_n+b_n), \\ Q_b=\dfrac{1}{R^2}[\ \displaystyle\sum_1^\infty(2n+1)(-1)^n(a_n+b_n)\ ]^2, \end{cases} \tag{7}$$

$$\begin{cases} i(\mu)=|S_1(\mu)|^2+|S_2(\mu)|^2, \\ S_1(\mu)=\displaystyle\sum_1^\infty\frac{2n+1}{n(n+1)}[a_n\pi_n(\mu)+b_n\tau_n(\mu)]\ , \\ S_2(\mu)=\displaystyle\sum_1^\infty\frac{2n+1}{n(n+1)}[b_n\pi_n(\mu)+a_n\tau_n(\mu)] \end{cases} \tag{8}$$

for a sphere with the size parameter R are valid [19,20]. Here, $\mu=\cos\Theta$, Θ is the scattering angle and

$$\begin{aligned} \pi_n(\mu)&=P_n'(\mu), \\ \tau_n(\mu)&=-\mu P_n'(\mu)+n(n+1)P_n(\mu) \end{aligned} \tag{9}$$

are the angle - dependent functions [2].

2. Piecewise continuous hyperbolic approximation

Let the contour of the refractive index $\mu(\rho)$ be represented in the form

$$\mu(\rho)=\begin{cases} m_o, & 0\le \rho<\rho_o, \\ M(\rho), & \rho_o<\rho<R, \\ M, & \rho>R. \end{cases} \tag{10}$$

We suppose that the refractive index of a spherical particle is specified by any continuous function $M(\rho)$ satisfying the conditions

$$M(\rho_o)=m_o, \qquad M(R)=M\,. \tag{11}$$

As compared with (1), the homogeneous kernel (its size parameter equals to any ρ_o, $0<\rho_o<R$, and m_o is an arbitrary real or complex constant) of the particle in contour (10) being introduced. Let us denote

$$\Delta=R-\rho_o, \qquad \delta=\frac{\Delta}{R}100\% \tag{12}$$

the absolute and relative widths of the coating. If $100-\delta<<1$, the contour (10) is actually coincident with the contour (1), and because of the stability of the Maxwell equations, the solutions of the problem corresponding to these contours are close to each other. In turn, if $\delta<<1$, the solution of the Mie problem

$$\mu(\rho)=\begin{cases} m_o, & 0\le\rho<R, \\ M, & \rho>R \end{cases} \tag{13}$$

and the one corresponding to condition (10) also should be close to each other. This assertion will be proved for the piecewise continuous hyperbolic approximation (PCHA) defined below. It also will be proved that the PCHA works well, independent of the value of δ.

By definition, the PCHA is associated with the contour of the refractive index (10), represented by the following model (Figure 1)

$$\mu(\rho)\approx m(\rho), \tag{14}$$

in which

$$m(\rho)=\begin{cases} m_o, & 0\le\rho<\rho_o, \\ x_iM(x_i)\rho^{-1}, & \rho_{i-1}<\rho<\rho_i\ (i=1,\ldots,k) \\ \mu_k=M, & \rho>R. \end{cases} \tag{15}$$

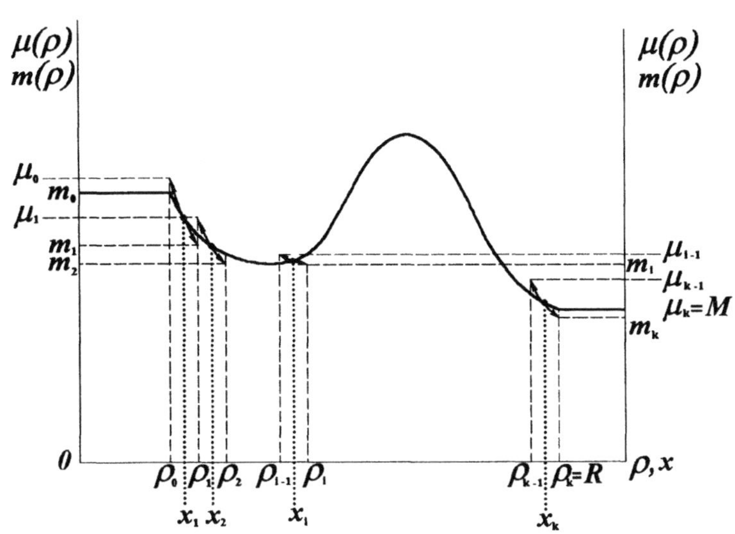

Figure 1: Typical model of the real refractive index contour.

Here, $\rho_1,\dots,\rho_{k-1}$ and $x_1,\dots,x_k$ are arbitrary sets of numbers satisfying the condition

$$\rho_o < x_1 < \rho_1 < x_2 < \rho_2 < \dots < \rho_{k-1} < x_k < \rho_k, \tag{16}$$

and k is any integer (the order of approximation).

The optical characteristics are determined by equations (7), in which the Mie coefficients a_n and b_n must be replaced by the generalized Mie coefficients A_{1n} and A_{2n}, respectively. The coefficients A_{1n} and A_{2n} are computed on the basis of the algorithm elaborated for the PCHA [18,21,22] (see Appendix).

The developed PCHA is analogous to the quadrature formula, which allows any square to be approximated by a collection of simple figures. The PCHA and, for example, the Simpson rule are based on the hyperbolas and parabolas, respectively. The solution of scattering problems for a stratified sphere and the one associated with the PCHA are similar to the straight line and Simpson rules in integration theory. The main advantage of the PCHA consists in utilization of the functions of the simplest analytical nature.

3. Accuracy and stability of the PCHA

The PCHA appears to be the proper tool to determine the electromagnetic field and its energy characteristics within the optical range for a number of contours $\mu(\rho)$ of the form (10) [21-23]. To illustrate this fact, we shall consider two typical models of the refractive index $M(\rho)$ in the coating $\rho_o < \rho < R$ relating to the contour (10), namely,

$$M(\rho) = \alpha_1 + \alpha_2 e^{\rho}, \tag{17}$$

$$M(\rho) = \sum_{n=0}^{3} \alpha_n \rho^{-n}. \tag{18}$$

By virtue of the continuity conditions (11), the parameters α_1 and α_2 involved in model (17) are uniquely found. We have

$$M(\rho) = m_o + (m_o - M)(e^{\Delta} - 1)^{-1}(1 - e^{\rho - \rho_o}), \quad \rho_o \le \rho \le R, \tag{19}$$

where Δ is given by (12). Model (18) has two free parameters, which will be determined either by smoothing the profile of the refractive index

$$M'(\rho_o) = M'(R) = 0, \tag{20}$$

or by means of equations

$$\alpha_o = \alpha_3 = 0. \tag{21}$$

Let us investigate the rate of convergence of the PCHA and its stability. When computing a certain optical quantity, the PCHA is said to be convergent in itself if four significant digits of this quantity are retained as the order of approximation k is increased. Figure 2 shows the variations of the extinction Q_{ext} and backscattering Q_b efficiency factors (the values of $Q_{ext} = Q_{ext}^{\delta}$ and $Q_b = Q_b^{\delta}$ are marked over the corresponding axes), and the order of approximation k (values of k are in square brackets), which guarantee the convergence, as functions of the relative width δ of the coating (values of δ are in parentheses). It is seen that

$$\lim_{\delta \to 0} Q_{ext}^{\delta} = Q_{ext}^{Mie}, \qquad \lim_{\delta \to 0} Q_b^{\delta} = Q_b^{Mie}, \tag{22}$$

i.e., Q_{ext}^{δ} and Q_b^{δ} at $\delta = 0$ do, respectively, equal Q_{ext} and Q_b of the Mie theory. The calculations were carried out for $m_o = 1.5$, $M = 1.0$, λ=0.55 μm and $a = 2\,\mu m$. Similar results are obtained for various collections of input data [21]. Moreover, the limiting relation of the type (22) is fulfilled when the presented solution is tested with the known one for a stratified sphere [18].

Let us now consider the influence of the refractive index profile variations on the scattering characteristics. Typical results of calculations are shown in Figure 3. Curve 1 describes the Mie extinction efficiency (the width of the coating $\Delta = 0$), while curves 2-4 are related to the generalized Mie extinction efficiencies with $\Delta = 0.1$. The curves 2,3 and 4 correspond to the models (18) with condition (21); (18) with condition (20); and (19); respectively. Figure 3

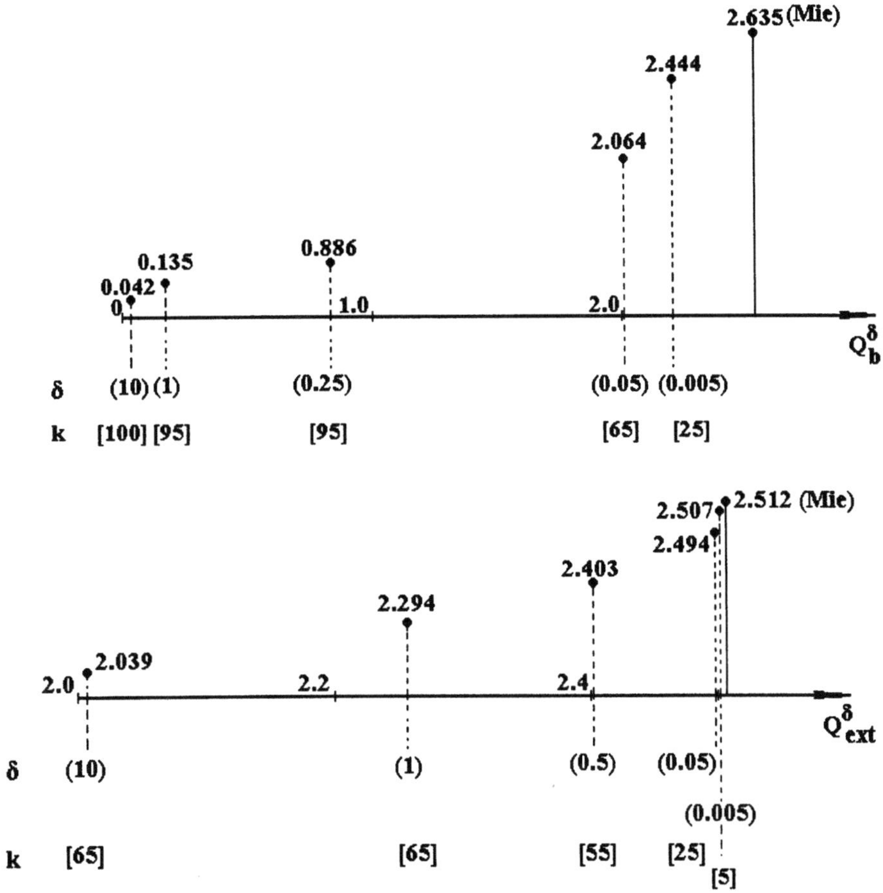

Figure 2: Efficiencies Q_{ext}^{δ} (top) Q_b^{δ} (bottom) as functions of the relative width δ and the corresponding orders of approximation k.

shows that the curves 1-4 are, actually, coincident. Here, the width $\Delta = 0.1$ is small. If Δ is increased then curves 2-4 no longer remain close to each other, but begin to deviate from the Mie curve (1), especially in the region of small size parameters. Detailed calculations show that factor Q_{ext} is weakly dependent on the choice of the model of the profile $M(\rho)$, while for factor Q_b and scattering function $i(\mu)$, the dependence is perceptible. However, the solution of the problem is, in fact, insensitive to the degree of smoothness of the contour $\mu(\rho)$ without stationary points, and in the case of dielectric particles, any monotone refractive index profile $M(\rho)$ can be chosen [22].

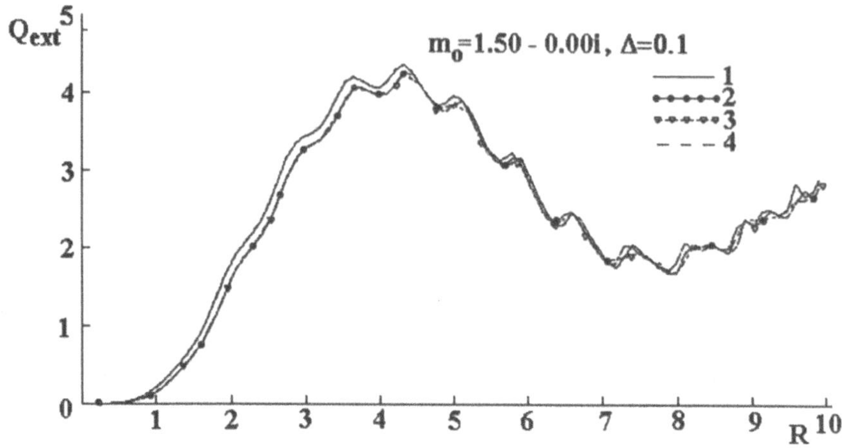

Figure 3: Influence of the refractive index contour type on the extinction efficiency.

We shall use the model of $M(\rho)$ given by Eq. (19). The results of calculations of quantities (7) are presented in Figures 4 and 5 dealing with Q_{ext} and Q_b, respectively. In the case of Q_{ext}, the curves corresponding to $\Delta = 0.05$ for all values of the size parameters ρ and the refractive index $m = n - ik$ are actually coincident with the Mie curve ($\Delta = 0$), while the curves corresponding to $\Delta = 1$ are slightly differ from the Mie curve. These differences are greater in the backscatter, where the factor $Q_b = Q_b(\Delta)$ reveals considerable deviations from the Mie factor $Q_b = Q_b(0)$, and, what is more, these deviations grow appreciably if the imaginary part of the refractive index increases. A variety of the interesting phenomena involve light-scattering resonances [13,14]. Let us introduce the following notations: $Q_{ext} = Q_{ext}(R,\ \Delta)$ is the extinction efficiency dependent on the absolute width Δ of the intermediate layer $(R > \Delta \geq 0)$, $R^{res}(\Delta)$ and $Q_{ext}^{res}(\Delta)$ are the position and the magnitude of the first resonance peak. The results of computations corresponding to the four-parameter profile $M(\rho)$ of the type (18) [on condition (20)] for $m_o = 9.0 - 0.5i$ are presented in Figures 6-8. Figure 6 shows the behaviour of $Q_{ext}(R,\ \Delta)$ for $\Delta = 0$ (Mie scattering), $\Delta =$ 0.001 and $\Delta = 0.1$.

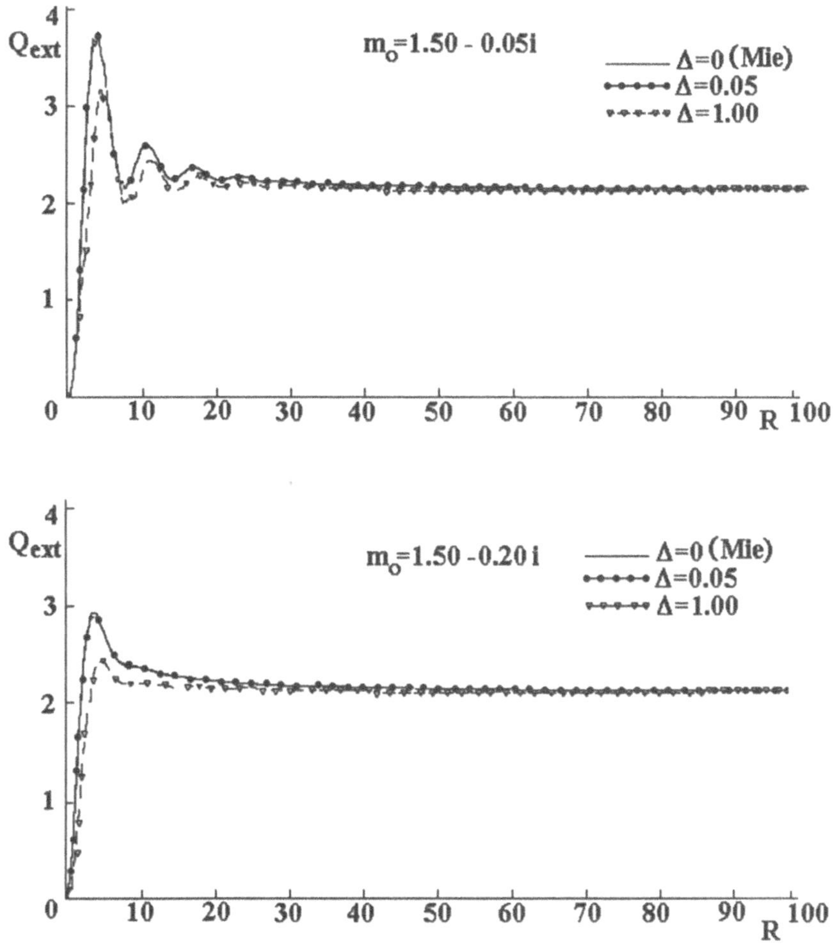

Figure 4: Extinction efficiency as the function of the size parameter R for the absolute widths Δ = 0, 0.05, 1 and the refractive indices $m_o = 1.5 - ik$, (k=0.05, 0.20).

The curves $Q_{ext}(R,\ 0)$ and $Q_{ext}(R,\ 0.001)$ actually coincide. This fact illustrates continuity of the dependence of $Q_{ext}(R,\ \Delta)$ on the width Δ. It is seen that

$$R^{res}(0) = 0.35, \qquad Q_{ext}^{res}(0) = 3.22, \tag{23}$$

$$R^{res}(0.1) = 0.24, \qquad Q_{ext}^{res}(0.1) = 14.0. \tag{24}$$

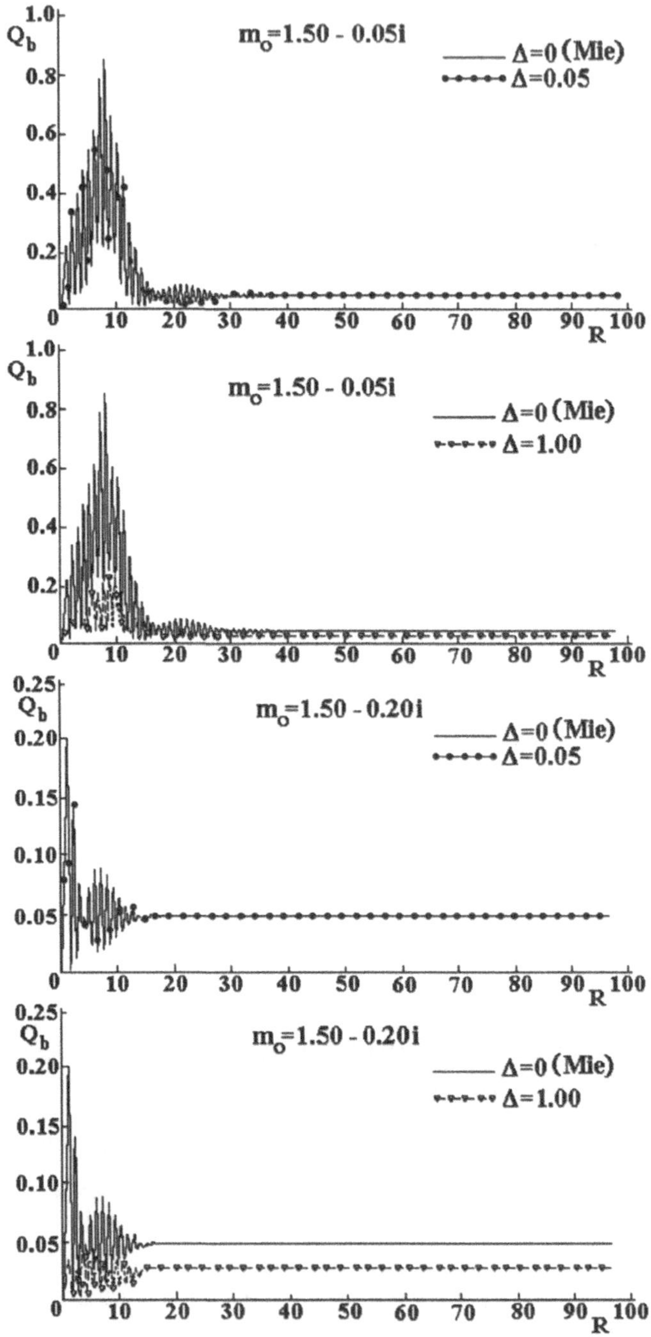

Figure 5: Backscattering efficiency as the function of the size parameter R for the absolute widths $\Delta = 0, 0.05, 1$ and the refractive indices $m_o = 1.5 - ik$ (k=0.05, 0.20).

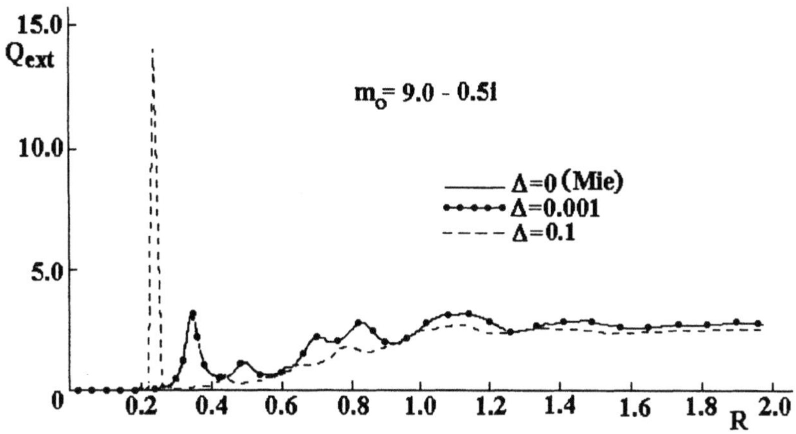

Figure 6: The first resonance peak in the Mie and generalized Mie theories.

According to Eqs. (23) and (24), there exists a profile $M(\rho)$ such that the magnitude of the extinction efficiency at the first resonance peak can grow more than four times, as compared with that corresponding to the Mie scattering. In turn, the width Δ being increased, the function $R^{res}(\Delta)$ decreases monotonically (Figure 7), while the function $Q_{ext}^{res}(\Delta)$ is oscillating, its absolute maximum is attained at Δ=0.10 (Figure 8).

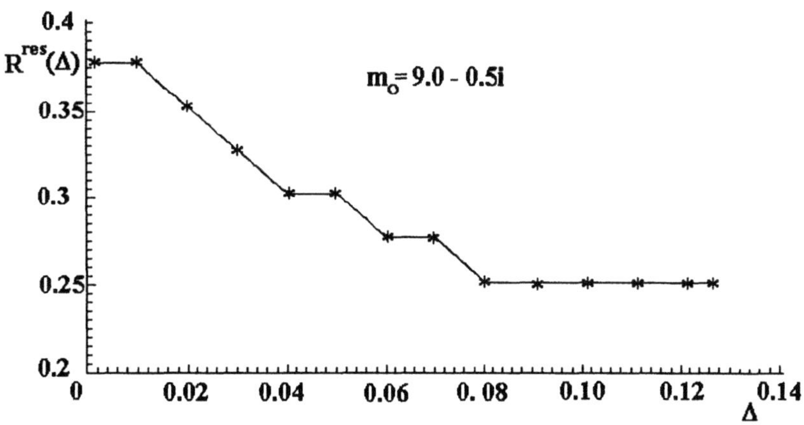

Figure 7: Position of the first resonance peak as the function of the absolute width Δ.

The functions $R^{res}(\Delta)$ and $Q_{ext}^{res}(\Delta)$ depend on the choice of the profile $M(\rho)$. For example, the computations show that for the profile $M(\rho)$ of the form (19) these functions are monotonically increasing and decreasing, respectively. Hence, by varying the refractive index profile within the coating, we are able to demonstrate how the magnitude of the first resonance peak may be enhanced or suppressed, and its position changed.

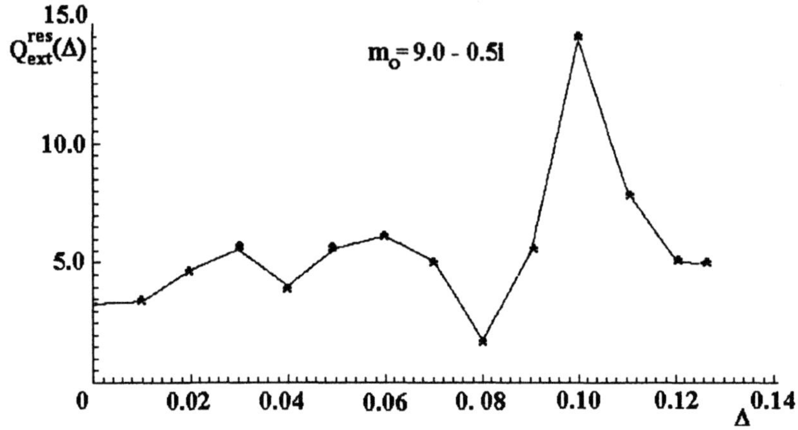

Figure 8: Magnitude of the first resonance peak as the function of the absolute width Δ.

4. Some applications

The particular case of the PCHA associated with the simple model of a two-layer dielectric spherical particle has been discussed in [16]. The model was applied to study the different optical characteristics of an antireflection-coated particle [23].

The presented PCHA extends the set of scattering problems for which it approximates to a fairly high degree of accuracy the experimental data, because of arbitrary profile of $M(\rho)$ can be used with any (real or complex) values of m_o in contour (1) of the refractive index $\mu(\rho)$. The PCHA works well for statistical ensembles of irregular particles randomly distributed through the volume being illuminated. It is true for star-shaped particles, which are encountered among phytoplankton cells in sea water, for dielectric and absorbing fluffy particles of cosmic dust, and so on.

We assume a real polydisperse system that average size parameter $\overline{R}$ describing the ensemble of particles. The contour (1) of the refractive index $\mu(\rho)$ is said to be simple if its profile $M(\rho)$ has no stationary points,

$$M'(\rho) \neq 0 \qquad (\rho_o < \rho < R). \tag{25}$$

As for the scattering function $i = i(\mu)$, its measured values $i_{exp}(\mu)$ can be satisfactorily restored in terms of a simple contour if the function $i_{exp}(\mu)$ is associated with an ensemble of non-absorbing particles. For absorbing particles, the experimental data can be represented on the basis of a simple contour more accurately, as compared with Mie theory, on condition that both Δ and R (i.e., R may differs from $\overline{R}$) are varied. For example, the scattering function measured in [24], $i_{exp}(\mu)$ $(m_o = 1.65 - 0.25i,\ \overline{R} = 31.2)$ is reproduced by means of a simple contour (1) with a profile of the form (19) (at Δ=1.0 and $R = 69.2 \neq \overline{R}$) more accurately than by using the Mie theory (Figure 9). At the same time, Figure 9 shows that, even in the case of $R \neq \overline{R}$, the coincidence between the experimental and theoretical data is not satisfactory. It should be added that the condition $R = \overline{R}$ is of importance when solving remote sensing problems. To provide the condition $R = \overline{R}$ and improve the approximation, the set of contours (1) needs to be expanded. In fact, it is sufficient to assume that the contour (1) can have a stationary point, or, for the case of dielectric particles, the one-modal profiles $M(\rho)$ can be used. Then, the width $\Delta = R - \rho_o$, the stationary point ρ^* $(\rho_o < \rho^* < R)$ and $M^* = M(\rho^*)$ (M^* is a complex number) are three parameters to be varied. We shall employ the profiles of the form (18) in the improved approximation.

Let us return to the situation discussed above discussed (Figure 9). The results of the improved approximation of $i_{exp}(\mu)$ are shown in Figure 10. We can conclude that the approximation with the parameters (m_0=1.65-0.25i, $R = \overline{R} = 31.2$)

$$\Delta = 0.27, \qquad \rho^* = \rho_o + \frac{\Delta}{3}, \quad M^* = 2m_o, \tag{26}$$

in which ρ_0 = 30.93, is acceptable. The corresponding profile (18) is described by the following coefficients

$$\begin{cases} \alpha_0 = (143 + 33.8i) \cdot 10^5, \\ \alpha_1 = -(134 + 31.5i) \cdot 10^7, \\ \alpha_2 = (417 + 98.0i) \cdot 10^7, \\ \alpha_3 = -(432 + 102i) \cdot 10^9. \end{cases} \tag{27}$$

Hence, within the framework of the PCHA, it is possible to choose the refractive index profile so that the constructed scattering function is numerically close to the one obtained by measurements for particles of different shape and structure. It allows the kernel of the integral equation in the inverse problem to be adjusted on the basis of the experimental scattering information. This fact can be used in remote sensing [25].

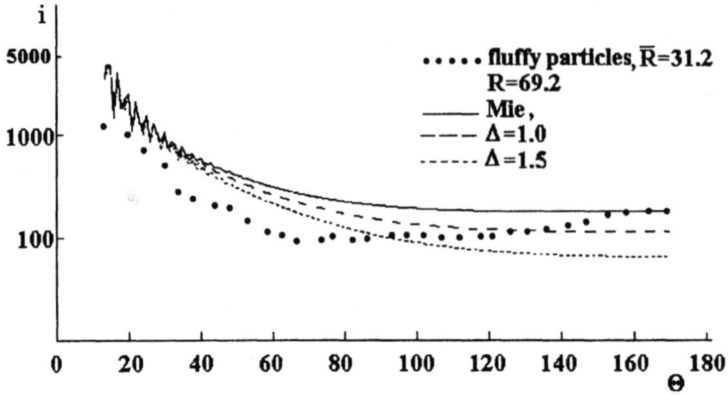

Figure 9: Scattering function for fluffy particles, m_0=1,65-0.25i (the contour of the refractive index without stationary points).

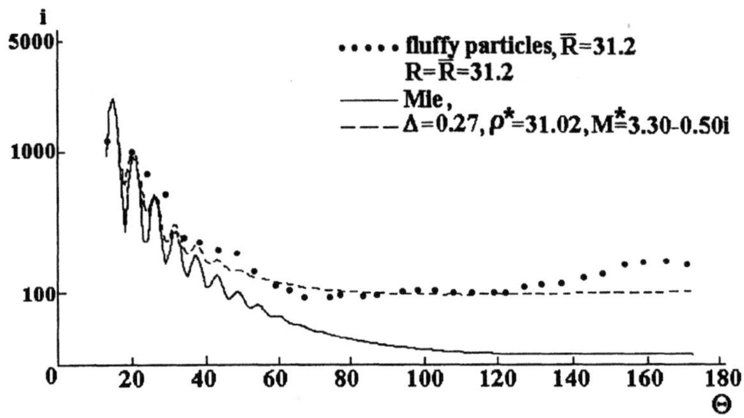

Figure 10: Scattering function for fluffy particles, m_0=1,65-0.25i (the contour of the refractive index with stationary point).

5. Conclusion

We present a methodology whereby the Mie theory can be generalized for an arbitrary complex index profile by the additional calculation of some power functions. Because of the greatest simplicity of the power functions, the problems of computer overflows are overcome, as well as round-off errors and computation times are greatly reduced in comparison with conventional methods of solution. It becomes apparent especially in cases of an absorbing particle, and particles having thick coatings. Equation (1) for the radial components can approximately be solved using the Runge Kutta method [6]. The computations should be carried out for any radial profile, separately. The

PCHA is based on the exact solutions of Eq. (2), which are represented by functions of the simplest analytical form. The functions are used in constructing the algorithm to solve the scattering from systems with an arbitrary radial profile of the refractive index. The PCHA reproduces the experimental data considerably better than Mie theory does, and makes it possible to formally improve the kernel of the integral equations used in remote sensing and establish feasible ranges for different scattering characteristics.

6. Acknowledgments

This work was supported by INTAS grant (open call 99/652).

Appendix A. The PCHA to compute the generalized scattering coefficients A_{1n} and A_{2n}

Given the contour (15) and the numbers ρ_i, x_i satisfying condition (16). For the sake of definiteness, we set

$$\rho_i = \rho_o + i\frac{\rho_k - \rho_o}{k} \ (i = 0,1,...,k), \ \ x_i = \frac{\rho_{i-1} + \rho_i}{2} \ (i = 1,...,k). \quad \text{(A1)}$$

Let us denote

$$m_i = \lim_{\substack{\rho\to\rho_i \\ \rho<\rho_i}} m(\rho), \qquad \mu_i = \lim_{\substack{\rho\to\rho_i \\ \rho>\rho_i}} m(\rho) \qquad (i = 0,1,...,k), \quad \text{(A2)}$$

$$\tau_{1i} = m_i^2 \mu_i^{-2}, \quad \tau_{2i} = 1 \ \ (i = 0,1,...,k), \quad \text{(A3)}$$

$$\nu_{in} = \left[(n+0.5)^2 - \gamma_i^2\right]^{1/2}, \qquad \gamma_i = x_i M(x_i). \quad \text{(A4)}$$

We have ($j = 1,2$)

$$A_{jn} = -B_{jn}(\psi_n)/B_{jn}(\zeta_n), \quad \text{(A5)}$$

where (from now on the subscript n being omitted)

$$B_j(g) = (a_{jk}\Delta_{j+} + c_{jk}\Delta_{j-})\delta_{j-}(g) - (b_{jk}\Delta_{j+} + d_{jk}\Delta_{j-})\delta_{j+}(g), \quad \text{(A6)}$$

in which

$$\Delta_{j\pm} = \begin{vmatrix} \rho_o^{-1.5+j\pm\nu_1} & \psi(m_o\rho_o) \\ (-1.5+j\pm\nu_1)\rho_o^{-2.5+j\pm\nu_1} & \tau_{jo}^{-1} m_o \psi'(m_o\rho_o) \end{vmatrix}, \quad \text{(A7)}$$

$$\delta_{j\pm}(g) = \begin{vmatrix} \rho_k^{-1.5+j\pm\nu_k} & g(\mu_k\rho_k) \\ (-1.5+j\pm\nu_k)\rho_k^{-2.5+j\pm\nu_k} & \tau_{jk}\mu_k g'(\mu_k\rho_k) \end{vmatrix}, \quad \text{(A8)}$$

In turn, the coefficients a_{jk}, b_{jk}, c_{jk}, d_{jk} are found by the scheme as follows. We have

$$a_{j1} = d_{j1} = 1, \ b_{j1} = c_{j1} = 0 \ \ (k = 1) \quad \text{(A9)}$$

and ($k \geq 2$)

$$\begin{bmatrix} a_{jk} & b_{jk} \\ c_{jk} & d_{jk} \end{bmatrix} = \prod_{i=2}^{k} \begin{bmatrix} \alpha_{ji} & \beta_{ji} \\ \gamma_{ji} & \delta_{ji} \end{bmatrix}. \quad \text{(A10)}$$

Here,

$$\begin{cases} \alpha_{ji+1} = \omega_{ji}(\nu_i, \nu_{i+1}), & \beta_{ji+1} = \omega_{ji}(\nu_i, -\nu_{i+1}), \\ \gamma_{ji+1} = \omega_{ji}(-\nu_i, \nu_{i+1}), & \delta_{ji+1} = \omega_{ji}(-\nu_i, -\nu_{i+1}), \end{cases} \tag{A11}$$

where

$$\omega_{ji}(a,b) = 0.5\rho_i^{-a+b}\{1 + a^{-1}[\ 0.5 + \tau_{ji}(-0.5 + b)\ ]\}. \tag{A12}$$

References

1. M. Kerker, "The Scattering of Light and Other Electromagnetic Radiation" (Academic Press, New York, 1969).
2. C.F. Bohren and D.R. Huffman, "Absorption and Scattering of Light by Small Particles" (Wiley, New York, 1983).
3. M. Born and E. Wolf, "Principles of Optics" (Pergamon New York, 1964).
4. A.L. Aden and M. Kerker, J. Appl. Phys, **22**, 1242 (1951).
5. K.S. Shifrin, Izv. Acad. Sc. Geophys. USSR **2**, 15 (1952).
6. D.Q. Chowdhurry, S.C. Hill, and P.W. Barber, J. Opt. Soc. Am. **A8**, 1702 (1991).
7. L.Kai and P. Massoli, Appl. Opt. **33**, 501 (1994).
8. P. Massoli, Appl. Opt. **37**, 3227 (1998).
9. M. Schneider and D. Hirleman, Appl. Opt. **33**, 2379 (1994).
10. P.J. Wyatt, Phys. Rev. **127**, 1837 (1962).
11. R. Bhandari, Appl. Opt. **24**, 1060 (1985).
12. J.I. Mikulski and E.L. Murphy, IEEE Trans. Anten. Propag. **AP-11**, 169 (1963).
13. G. Videen and P. Chylek, Opt. Commun. **158**, 1 (1998).
14. G. Videen and P. Chylek, Opt. Commun. **98**, 313 (1993).
15. B.R. Johnson, Appl. Opt. **35**, 3286 (1996).
16. A.Y. Perelman, Appl. Opt. **18**, 2307 (1979).
17. B.S.Westcott, Electron. Lett. **4**, 572 (1968).
18. A.Y. Perelman. Appl. Opt. **35**, 5452 (1996).
19. H.C. van de Hulst, "Light Scattering by Small Particles" (Dover, New York, 1981).
20. K.S. Shifrin, "Physical Optics of Ocean Water" (Am. Inst. Phys. New York, 1988).
21. A.Y. Perelman and T.V. Zinov'eva, Opt. Spectrosc. **92**, 481 (2002) (in press).
22. A.Y. Perelman and T.V. Zinov'eva, Opt. Atmosph. Ocean **14**, 669 (2001).
23. K.S. Shifrin, A.Y. Perelman, and A.M. Kokorin, Opt. Spectrosc. **59**, 597 (1985).
24. R.H. Giese, K. Weiss, R.H. Zerull, and T. Ono, Astr. Astrophys. **65**, 265 (1978).
25. A.Y. Perelman and I.G. Mosseev, "Method of optical similarity in remote sensing," in Electromagnetic and Light Scattering by Non-Spherical Particles, ed. B. Gustafson, L. Kolokolova, and G. Videen (ARL, Adelphi, 2002) 259.

CALCULATION OF OPTICAL FIELDS INSIDE SPHEROIDAL PARTICLES OF COSMIC DUST: COMPARISON OF DIFFERENT METHODS: GMT, *T* - MATRIX, SVM

V.A. BABENKO AND P.K. PETROV
Stepanov Institute of Physics, Belarus National Academy of Sciences, Skaryna pr. 68, 220072 Minsk, Belarus

Abstract. We use three methods (General Multipole Technique (GMT), Separation of Variables Method (SVM), and *T*-matrix approach) to calculate the optical fields inside spheroidal particles. The parallel use of these significantly different approaches allows us to obtain benchmark results. We compare the codes with respect to computer time, precision and effectiveness for internal field calculations for prolate spheroidal grains of astronomical silicates.

1. Introduction

The thermal effects (heating, evaporation, destruction, etc.) that occur during electromagnetic wave interaction with particles of cosmic dust play an important role in physics of the interstellar medium (ISM). Effective numerical methods for evaluation of internal electromagnetic fields are in great demand in such areas as astrophysics, laser technology and manifold industrial applications where thermo-optical and nonlinear effects in small non-spherical particles need to be investigated. What follows is a list of existing numerical methods that can serve the purpose:

- Discrete Dipole Approximation (DDA);
- Finite Difference Time Domain technique (FDTD);
- Separation of Variables Method (SVM);
- Variants of the General Multipole Technique (GMT).

The first two methods are simple to implement and universal, but very slow and resource demanding. The SVM is very fast and accurate, but its applicability is restricted to spherical and spheroidal particles only. It also involves very complex (and sometimes unreliable) algorithms for spheroidal functions and eigenvalue calculations. These computational difficulties are most pronounced for oblate spheroids.

G. Videen and M. Kocifaj (eds.), Optics of Cosmic Dust, 119–130.

Two modifications of GMT are of special interest. First is the extended boundary condition, T-matrix method proposed by Waterman [1], which includes spherical vector multi-poles centered at the origin. Although this method has proven to be effective in solving the external scattering problems, its efficacy in calculation of internal fields has not been carefully investigated thus far. The *T*-matrix method, widely used for light scattering calculation for spheroidal grains, may be implemented to calculate the internal fields, provided that the Rayleigh hypothesis for internal problems is valid, which hinders the correct computation of fields for spheroids having large aspect ratios, $a/b > \sqrt{5}$ (a,b – the major and minor semiaxes) [2].

The second modification, which we call simply GMT [3], with the lowest order multi-poles distributed along the segment of a coordinate axis proved to be extremely convenient when dealing with axially symmetrical particles. The GMT is based on the idea of alternative choice of complete function systems to approximate the wave fields. Such functions are built on the base of fundamental solution of Maxwell's equations and, as a rule, have quite simple form. It is essential that the function system used can be fitted roughly to the geometry of a scatterer. In particular, the system of lowest order multi-poles distributed along the symmetry axis of the scatterer we used in our calculations has proven to be almost independent on elongation of the body. In contrast, analogous calculations in the frame of the conventional T-matrix approach failed as the body elongation exceeds some limit. Unlike SVM, the GMT is not bound to a specific geometry of a scatterer [3]. Moreover, the possibility to distribute the fictitious sources in the complex space imparts additional flexibility to this method. Our calculations suggest that GMT proves to be nearly as fast as SVM and much more simple in practical realization.

The parallel application of these radically different methods allows us to obtain benchmark results. We develop computer codes for the optical field calculation inside spheroidal particles using these three approaches.

2. GMT-approach

The idea of a General Multipole Technique for satisfying the boundary conditions for partial differential equations appears to be first introduced by Kupradze [4]. Consider the boundary problem for Maxwell's equations. Let three-dimensional space R^3 consist of a closed surface ∂V, its interior V_i and its exterior $R^3 \setminus V_i$. Let a point within V_i be the origin. The mathematical formulation of our boundary problem of scattering of an incident field $(\mathbf{E}_0, \mathbf{H}_0)$ by a homogeneous dielectric object is:

$$\nabla \times \mathbf{E}_t(\mathbf{x}) = i\mu_t \mathbf{H}_t(\mathbf{x}), \quad \nabla \times \mathbf{H}_t(\mathbf{x}) = -i\varepsilon_t \mathbf{E}_t(\mathbf{x}), \quad t = s, i$$

with boundary conditions

$$\mathbf{n} \times (\mathbf{E}_s + \mathbf{E}_0) = \mathbf{n} \times \mathbf{E}_i, \; \mathbf{n} \times (\mathbf{H}_s + \mathbf{H}_0) = \mathbf{n} \times \mathbf{H}_i \tag{1}$$

and radiation conditions

$$\mathbf{H}_s(\mathbf{x})\times\mathbf{x}/|\mathbf{x}|+\mathbf{E}_s(\mathbf{x})=O\left(|\mathbf{x}|^{-1}\right),\ \mathbf{E}_s(\mathbf{x})\times\mathbf{x}/|\mathbf{x}|+\mathbf{H}_s(\mathbf{x})=O\left(|\mathbf{x}|^{-1}\right),\ |\mathbf{x}|\to\infty,$$

holding uniformly over all possible directions.

One way to solve this problem is to transform it into the first order integral equation by means of Huygens principle. One can express the fields outside and inside V_i through the surface currents $\mathbf{j}_t^{e}(\mathbf{x})=-\mathbf{n}(\mathbf{x})\times\mathbf{H}_t(\mathbf{x})$;

$\mathbf{j}_t^{m}(\mathbf{x})=\mathbf{n}(\mathbf{x})\times\mathbf{E}_t(\mathbf{x});\quad \mathbf{x}\in\partial V$:

$$\begin{bmatrix}\mathbf{E}_i(\mathbf{x})\\ \mathbf{H}_i(\mathbf{x})\end{bmatrix}\chi_{V_i}(\mathbf{x})=\int_{\partial V}\widehat{G}^i(\mathbf{x},\mathbf{y})\begin{bmatrix}\mathbf{j}_i^e(\mathbf{y})\\ \mathbf{j}_i^m(\mathbf{y})\end{bmatrix}ds_y, \tag{2}$$

$$\begin{bmatrix}\mathbf{E}_s(\mathbf{x})\\ \mathbf{H}_s(\mathbf{x})\end{bmatrix}\chi_{V_s}(\mathbf{x})=\int_{\partial V}\widehat{G}^s(\mathbf{x},\mathbf{y})\begin{bmatrix}\mathbf{j}_s^e(\mathbf{y})\\ \mathbf{j}_s^m(\mathbf{y})\end{bmatrix}ds_y,\ \chi_{V_s}(\mathbf{x})=\begin{matrix}1, & \mathbf{x}\in V\\ 0, & \mathbf{x}\in R^3\setminus V\end{matrix} \tag{3}$$

where the free space Green's tensor has the components

$$\widehat{G}^t_{\binom{11}{22}}(\mathbf{x},\mathbf{y})=-i\begin{pmatrix}\mu_t\\ \varepsilon_t\end{pmatrix}\left(\hat{I}+\chi_t^{-2}\nabla_x\bullet\nabla_x\right)\frac{\exp(i\chi_t r)}{4\pi r},$$

$$\widehat{G}^t_{\binom{21}{12}}(\mathbf{x},\mathbf{y})=\mp\nabla_x\frac{\exp(i\chi_t r)}{4\pi r}\hat{I},\quad \chi_t=\sqrt{\mu_t\varepsilon_t},\ r=|\mathbf{x}-\mathbf{y}|.$$

Let us suppose that we have a complete system of functions satisfying Maxwell equations in each subregion and approximating the solution. Having such a system at hand, we need to satisfy only the boundary conditions. To obtain the integral equation we can use part of Eq. (3):

$$\mathbf{E}_s(\mathbf{x})\chi_{V_s}(\mathbf{x})=\int_{\partial V}\left[\widehat{G}_{11}{}^s(\mathbf{x},\mathbf{y})\mathbf{j}_s^e(\mathbf{y})+\widehat{G}_{12}{}^s(\mathbf{x},\mathbf{y})\mathbf{j}_s^m(\mathbf{y})\right]ds_y.$$

Having replaced $\mathbf{j}_s^e$, $\mathbf{j}_s^m$ according to the boundary conditions, we obtain the expression

$$0=\int_{\partial V}\left[\widehat{G}_{11}{}^s(\mathbf{x},\mathbf{y})\left(\mathbf{j}_i^e(\mathbf{y})-\mathbf{j}_0^e(\mathbf{y})\right)+\widehat{G}_{12}{}^s(\mathbf{x},\mathbf{y})\left(\mathbf{j}_i^m(\mathbf{y})-\mathbf{j}_0^m(\mathbf{y})\right)\right]ds_y,\ \mathbf{x}\in V_i, \tag{4}$$

which can be considered the variant of the boundary condition (extended or integral boundary condition), for if the boundary condition (1) holds for given surface currents, so does Eq. (4) [5]. Obviously, the analogous expression follows from Eq. (2). Note that unlike Eq. (1) the inner and outer surface currents are now uncoupled. Expression (4) is the first-kind integral equation for unknown surface currents $\mathbf{j}_s^e$, $\mathbf{j}_s^m$.

Two important ideas were suggested to make this equation manageable for numerical calculations.

First, Kupradze [4] has proven *Theorem 1: The system of surface fields*

$$\mathbf{\Phi}^t(\mathbf{x},\mathbf{x}_n,\mathbf{a})=\frac{1}{\chi_t^2}\left[\mathbf{a}\times\nabla\mathbf{g}(\mathbf{x},\mathbf{x}_n,\chi_t)\right],\ \mathbf{\Psi}^t(\mathbf{x},\mathbf{x}_n,\mathbf{a})=\frac{1}{\chi_t}\left[\nabla\times\mathbf{\Phi}^t(\mathbf{x},\mathbf{x}_n,\mathbf{a})\right],\ t=s,i$$

distributed on the closed auxiliary surface inside (outside) ∂V *is linearly independent and complete on the surface* ∂V *in* $L^2(\partial V)$ *norm. Each element of this system satisfies Maxwell's equations inside (outside)* ∂V. Obviously, linear combinations (in general, integrals) of these functions give another complete system. This simple fact leads to very interesting results. One can use different auxiliary surfaces and different linear combinations of initial functions to match the geometry of a particular scattering problem. In fact, the complete system of spherical vector harmonics [6] (the auxiliary surface degenerates to a point) $\{\mathbf{M}_{\sigma mn}(\mathbf{x}),\ \mathbf{N}_{\sigma mn}(\mathbf{x})\}$, $\sigma = odd, even;\ m = 0,1,...; n = 1,...,m;$

lowest order multipoles [3] (auxiliary surface is a segment of a curve) $\{\mathbf{M}_{\sigma mn}(\mathbf{x}-\mathbf{p}_n),\ \mathbf{N}_{\sigma mn}(\mathbf{x}-\mathbf{p}_n)\}$, $\sigma = odd, even;\ m = 0,1,...; l = m (m>0),$ $l = 1\ (m=0);\ n = 1,...,N$;

and multiple multipoles [3] $\{\mathbf{M}_{\sigma mn}(\mathbf{x}-\mathbf{p}_n),\ \mathbf{N}_{\sigma mn}(\mathbf{x}-\mathbf{p}_n)\}$, $\sigma = odd, even;\ m =$ $0,1,...;\ n = 1,...m;$ ($\mathbf{p}_l$ are distributed arbitrarily inside ∂V) can be obtained this way.

Second, Eq. (4) can be replaced with a more convenient equivalent expressions, *Theorem 2* [7,8]: *Let* $\{\mathbf{n}\times\boldsymbol{\Phi}^s_n, \mathbf{n}\times\boldsymbol{\Psi}^s_n\}$ *be the function system of tangential fields on* ∂V. *Each function satisfies Maxwell's equations outside* ∂V. *Then the system of integral equations*

$$\int_{\partial V}\left[(\mathbf{j}^e_i - \mathbf{j}^e_0)\left(\mathbf{n}\times\boldsymbol{\Phi}^s_n\right) + i\sqrt{\frac{\mu_s}{\varepsilon_s}}(\mathbf{j}^m_i - \mathbf{j}^m_0)\left(\mathbf{n}\times\boldsymbol{\Psi}^s_n\right)\right]ds = 0, \tag{5}$$

$$\int_{\partial V}\left[(\mathbf{j}^e_i - \mathbf{j}^e_0)\left(\mathbf{n}\times\boldsymbol{\Psi}^s_n\right) + i\sqrt{\frac{\mu_s}{\varepsilon_s}}(\mathbf{j}^m_i - \mathbf{j}^m_0)\left(\mathbf{n}\times\boldsymbol{\Phi}^s_n\right)\right]ds = 0$$

is equivalent to Eq.(4).

This system is called the extended integral equations. We can replace the unknown surface currents with their finite approximation by another complete system $\{\mathbf{n}\times\boldsymbol{\Phi}^s_n, \mathbf{n}\times\boldsymbol{\Psi}^s_n\}$ for the inner region

$$\mathbf{j}^e_i = \mathbf{n}\times\sum_{n=1}^{N}\left[\alpha_n\boldsymbol{\Phi}^i_n + \beta_n\boldsymbol{\Psi}^i_n\right],\ \mathbf{j}^m_i = i\sqrt{\frac{\varepsilon_i}{\mu_i}}\mathbf{n}\times\sum_{n=1}^{N}\left[\alpha_n\boldsymbol{\Psi}^i_n + \beta_n\boldsymbol{\Phi}^i_n\right] \tag{6}$$

to obtain from Eqs. (5) the system of linear equations for α_n, β_n.

In our calculations, we used the following complete functions systems:

1. Lowest order spherical multipoles of the third kind satisfying the radiation condition at infinity as weight functions:

$$\boldsymbol{\Phi}^s_n(\mathbf{x}) = \mathbf{M}^{(3)}_{e,oms}(\mathbf{x} - p_n\hat{z}) = -\tau^s_m(c_n)\frac{\xi_s(r_n)}{r_n}\begin{pmatrix}\cos\\ \sin\end{pmatrix} m\varphi\,\hat{\varphi} +$$

$$+ m\pi^s_m(c_n)\frac{\xi_s(r_n)}{r_n}\begin{pmatrix}-\sin\\ \cos\end{pmatrix} m\varphi\left[\sin(\theta-\theta_n)\hat{r} + \cos(\theta-\theta_n)\hat{\theta}\right];$$

$$\mathbf{\Psi}_n^s(\mathbf{x}) = \mathbf{N}^{(3)}_{e,oms}(\mathbf{x} - p_n\hat{z}) = m\pi_m^s(c_n)\frac{\xi_s'(r_n)}{r_n}\begin{pmatrix}-\sin \\ \cos\end{pmatrix} m\varphi\,\hat{\varphi} +$$

$$+ s(s+1)\sqrt{1-c_n^2}\,\pi_m^s(c_n)\frac{\xi_s(r_n)}{r_n^2}\begin{pmatrix}\cos \\ \sin\end{pmatrix} m\varphi\left[\cos(\theta-\theta_n)\hat{r} - \sin(\theta-\theta_n)\hat{\theta}\right] +$$

$$+ m\pi_m^s(c_n)\frac{\xi_s(r_n)}{r_n}\begin{pmatrix}\cos \\ \sin\end{pmatrix} m\varphi\left[\sin(\theta-\theta_n)\hat{r} + \cos(\theta-\theta_n)\hat{\theta}\right];$$

2. Lowest order spherical multipoles of the first kind which are regular at the origin to approximate the inner surface fields with

$$\mathbf{\Phi}_l^s(x) = \mathbf{M}^{(1)}_{e,oms}(\chi_i(\mathbf{x} - p_n\hat{z})) = -\tau_m^s(c_l)\frac{\psi_s(\chi_i r_l)}{\chi_i r_l}\begin{pmatrix}\cos \\ \sin\end{pmatrix} m\varphi\,\hat{\varphi} +$$

$$+ m\pi_m^s(c_l)\frac{\psi_s(\chi_i r_l)}{\chi_i r_l}\begin{pmatrix}-\sin \\ \cos\end{pmatrix} m\varphi\left[\sin(\theta-\theta_l)\hat{r} + \cos(\theta-\theta_l)\hat{\theta}\right];$$

$$\mathbf{\Psi}_l^s(\mathbf{x}) = \mathbf{N}^{(1)}_{e,oms}(\chi_i(\mathbf{x} - p_n\hat{z})) = m\pi_m^s(c_n)\frac{\psi_s'(\chi_i r_l)}{\chi_i r_l}\begin{pmatrix}-\sin \\ \cos\end{pmatrix} m\varphi\,\hat{\varphi} +$$

$$+ s(s+1)\sqrt{1-c_l^2}\,\pi_m^s(c_l)\frac{\psi_s(\chi_i r_l)}{(\chi_i r_l)^2}\begin{pmatrix}\cos \\ \sin\end{pmatrix} m\varphi\left[\cos(\theta-\theta_n)\hat{r} - \sin(\theta-\theta_n)\hat{\theta}\right] +$$

$$+ m\pi_m^s(c_l)\frac{\psi_s(\chi_i r_l)}{\chi_i r_l}\begin{pmatrix}\cos \\ \sin\end{pmatrix} m\varphi\left[\sin(\theta-\theta_n)\hat{r} + \cos(\theta-\theta_n)\hat{\theta}\right];$$

$m = 1,\dots M;\ n = 1,\dots,N;\ l = 1,\dots,L;\ s = m(m>0);\ s = 1(m=0)$

Here $p_n(p_l)$ are points distributed along the z-axis, (r,θ,φ) are the spherical coordinates of point $\mathbf{x}$; $(r_{n(l)},\theta_{n(l)},\phi_{n(l)})$ are the spherical coordinates of the same point in the coordinate system shifted on $p_n\,(p_l)$ distance from the origin along z-axis. The relation among these coordinates is

$$r_{n(l)} = \sqrt{r^2 + p_{n(l)}^2 - 2rp_{n(l)}c},\quad c_{n(l)} = \left(rc - p_{n(l)}\right)/r_{n(l)};\quad \varphi_{n(l)} = \varphi;$$
$$c = \cos\theta;\quad c_{n(l)} = \cos(\theta_{n(l)});$$

$\xi_{n(l)}, \psi_{n(l)}$ are the Riccati-Hankel and the Riccati-Bessel functions; $\pi_m^s(c) = P_m^s(c)/\left(1-c^2\right)^{1/2}$, $\tau_m^s(c) = -\left(1-c^2\right)^{1/2} dP_m^s(c)/dc$ where $P_m^s(c)$ are the associated Legendre polynomials. Having substituted these functions into Eqs. (5 and 6), we obtain the linear system for coefficients $\alpha_{e,omsn}, \beta_{e,omsn}$. We take advantage here of the fact that for axially symmetric bodies, the system of equations decouple into a set of independent subsystems for each m. To find the electric field inside the particles, the finite sum

$$\mathbf{E}_i(\mathbf{x}) = \sum_{\sigma=e,o}\sum_{m=0}^{M}\sum_{l=1}^{L}[\alpha_{\sigma msl}\mathbf{M}^{(1)}_{\sigma msl}(\chi_i(\mathbf{x} - p_l\hat{z})) + \beta_{\sigma msl}\mathbf{N}^{(1)}_{\sigma msl}(\chi_i(\mathbf{x} - p_l\hat{z}))]$$

is used. The validity of these series approximations throughout the entire inner domain can be proven with the aid of Green's theorem [6].

3. Separation of variables method (SVM)

Our SVM-code for internal field is based on the approach of Voshchinnikov and Farafonov [9] for external problems, which, from a computational point of view, is considerably more efficient than other SVM-schemes. The characteristic features of this approach are the division of fields into two parts $\mathbf{E} = \mathbf{E}_1 + \mathbf{E}_2$, where the first axisymmetric term does not depend on the azimutal angle, and the averaging of the second non-axisymmetric term over φ gives zero, and a new way to select the expansion basis for the second part.

3.1. Axisymmetric part

The spheroidal axisymmetric components of the internal electric field in point (ξ,η,φ) inside the spheroidal prolate particle with complex refractive index M (the outer medium is vacuum) and semifocal distance d may be expressed in the form of a series over the combinations of prolate radial spheroidal functions of the first kind $R_{1l}^{(1)}(c_2,\xi)$ and the prolate angular spheroidal functions $S_{1l}(c_2,\eta)$ and their derivatives, where $c_1 = k_0 d = (2\pi/\lambda)d,\ c_2 = c_1 M$

$$\textit{TM-wave}\quad \left[E_{1\xi}^{(2)}\right]_{\mathrm{TM}} = r_1 \sum_{l=1}^{\infty} b_l^{(2)} R_{1l}^{(1)}(c_2,\xi)\left[\frac{\eta}{\gamma_\eta} S_{1l}(c_2,\eta) - \gamma_\eta S_{1l}'(c_2,\eta)\right],$$

$$\left[E_{1\eta}^{(2)}\right]_{\mathrm{TM}} = r_1 \sum_{l=1}^{\infty} b_l^{(2)} S_{1l}(c_2,\eta)\left[\frac{\xi}{\gamma_\xi} R_{1l}^{(1)}(c_2,\xi) + \gamma_\xi R_{1l}'^{(1)}(c_2,\xi)\right],\quad \left[E_{1\varphi}^{(2)}\right]_{\mathrm{TM}} = 0.$$

$$\textit{TE-wave}\quad \left[E_{1\varphi}^{(2)}\right]_{\mathrm{TE}} = \sum_{l=1}^{\infty} a_l^{(2)} R_{1l}^{(1)}(c_2,\xi) S_{1l}(c_2,\eta),\quad \left[E_{1\xi}^{(2)}\right]_{\mathrm{TM}} = \left[E_{1\eta}^{(2)}\right]_{\mathrm{TM}} = 0,$$

where $\gamma_\xi = \sqrt{\xi^2 - 1},\ \gamma_\eta = \sqrt{1-\eta^2},\ \gamma_\varphi = \sqrt{\xi^2 - \eta^2},\ r_1 = \dfrac{i}{M^2 c_1 \gamma_\varphi}$.

The internal axisymmetric amplitude coefficients $a_n^{(2)}(\mathrm{TE})$ and $b_n^{(2)}(\mathrm{TM})$ $(n = 1,2,\ldots)$ are expressed through the sum of the amplitude coefficients for incident ($a_l^{(0)}(\mathrm{TE})$ and $b_l^{(0)}(\mathrm{TM})$) and scattered ($a_l^{(1)}(\mathrm{TE})$ and $b_l^{(1)}(\mathrm{TM})$) fields:

$$\begin{matrix}\mathrm{TE}\\ \mathrm{TM}\end{matrix}\ \begin{pmatrix} a_n^{(2)} \\ b_n^{(2)} \end{pmatrix} = \frac{1}{R_{1n}^{(1)}(c_2,\xi_0) N_{1n}(c_2)}$$

$$\sum_{l=1}^{\infty}\left[\begin{pmatrix} a_l^{(0)} \\ b_l^{(0)} \end{pmatrix} R_{1l}^{(1)}(c_1,\xi_0) + \begin{pmatrix} a_l^{(1)} \\ b_l^{(1)} \end{pmatrix} R_{1l}^{(3)}(c_1,\xi_0)\right] N_{1l}(c_1)\delta_{nl}^{(1)}(c_2,c_1),$$

where
TE: $a_l^{(0)} = -2i^l N_{1l}^{-2}(c_1) S_{1l}(c_1,\cos\alpha),\ b_l^{(0)} = 0,$
TM: $a_l^{(0)} = 0,\ b_l^{(0)} = 2i^l N_{1l}^{-2}(c_1) S_{1l}(c_1,\cos\alpha),$ (α-the incidence angle)

and $\delta_{nl}^{(m)}(c_2,c_1) = \displaystyle\int_{-1}^{1} \bar{S}_{mn}(c_2,\eta)\bar{S}_{ml}(c_1,\eta)d\eta,$

$\bar{S}_{mn}(c_i,\eta)$ are the normalized angular spheroidal functions $\bar{S}_{mn}(c_i,\eta)=N_{ml}^{-1}(c_i)S_{mn}(c_i,\eta)$, $N_{ml}(c_i)-$ the normalization coefficients [9].

3.2. Non-axisymmetric part

The non-axisymmetric part of the internal field are represented through the scalar potentials U and V:

$$\left[\mathbf{E}_2^{(2)}\right]_{\mathrm{TE}}=\nabla\times\left(U_{\mathrm{TE}}^{(2)}\mathbf{e}_z+V_{\mathrm{TE}}^{(2)}\mathbf{r}\right),\quad \left[\mathbf{E}_2\right]_{\mathrm{TM}}=\frac{i}{M^2k_0}\nabla\times\nabla\times\left(U_{\mathrm{TM}}^{(2)}\mathbf{e}_z+V_{\mathrm{TM}}^{(2)}\mathbf{r}\right),$$

$$\begin{pmatrix}U^{(2)}\\V^{(2)}\end{pmatrix}=\sum_{m=1}^{\infty}\sum_{l=m}^{\infty}\begin{pmatrix}a_{ml}^{(2)}\\b_{ml}^{(2)}\end{pmatrix}S_{ml}(c_2,\eta)R_{ml}^{(1)}(c_2,\xi)\cos m\varphi.$$

Rather cumbersome but straightforward operations give rise to the following expressions for the spheriodal components of the internal electric field:

TM-wave

$$\left[E_{2\xi}^{(2)}\right]_{\mathrm{TM}}=\frac{i\gamma_\xi}{M^2c_1^2\gamma_\varphi^3}\sum_{m=1}^{\infty}\sum_{l=m}^{\infty}\cos m\varphi\left[\left(-2\eta\frac{\gamma_\xi^2}{\gamma_\varphi^2}\left[A_{ml}\right]_{\mathrm{TM}}+C_{ml}\right)R_{ml}^{(1)\prime}(c_2,\xi)S_{ml}(c_2,\eta)-\right.$$

$$-\left(2\eta\frac{\gamma_\eta^2}{\gamma_\varphi^2}\left[B_{ml}\right]_{\mathrm{TM}}+D_{ml}\right)R_{ml}^{(1)}(c_2,\xi)S_{ml}{}'(c_2,\eta)+\gamma_\eta^2\left[A_{ml}\right]_{\mathrm{TM}}R_{ml}^{(1)\prime}(c_2,\xi)S_{ml}{}'(c_2,\eta)+$$

$$\left.+\left[B_{ml}\right]_{\mathrm{TM}}\left(\lambda_{ml}(c_2)-c_2^2\eta^2+\frac{m^2}{\gamma_\xi^2}\right)R_{ml}^{(1)}(c_2,\xi)S_{ml}(c_2,\eta)\right],$$

$$\left[E_{2\varphi}^{(2)}\right]_{\mathrm{TM}}=\frac{-i}{M^2c_1^2\gamma_\varphi^2\gamma_\xi\gamma_\eta}\sum_{m=1}^{\infty}\sum_{l=m}^{\infty}m\sin m\varphi\left[\gamma_\xi^2\left[B_{ml}\right]_{\mathrm{TM}}R_{ml}^{(1)\prime}(c_2,\xi)S_{ml}(c_2,\eta)+\right.$$

$$\left.+\gamma_\eta^2\left[A_{ml}\right]_{\mathrm{TM}}R_{ml}^{(1)}(c_2,\xi)S_{ml}{}'(c_2,\eta)+G_{ml}R_{ml}^{(1)}(c_2,\xi)S_{ml}(c_2,\eta)\right],$$

$$\left[E_{2\eta}^{(2)}\right]_{\mathrm{TM}}=\frac{i\gamma_\eta}{M^2c_1^2\gamma_\varphi^3}\sum_{m=1}^{\infty}\sum_{l=m}^{\infty}\cos m\varphi\left[\left(2\xi\frac{\gamma_\xi^2}{\gamma_\varphi^2}\left[A_{ml}\right]_{\mathrm{TM}}-F_{ml}\right)R_{ml}^{(1)\prime}(c_2,\xi)S_{ml}(c_2,\eta)+\right.$$

$$+\left(2\xi\frac{\gamma_\eta^2}{\gamma_\varphi^2}\left[B_{ml}\right]_{\mathrm{TM}}+E_{ml}\right)R_{ml}^{(1)}(c_2,\xi)S_{ml}{}'(c_2,\eta)+\gamma_\xi^2\left[B_{ml}\right]_{\mathrm{TM}}R_{ml}^{(1)\prime}(c_2,\xi)S_{ml}{}'(c_2,\eta)+$$

$$\left.+\left[A_{ml}\right]_{\mathrm{TM}}\left(-\lambda_{ml}(c_2)+c_2^2\xi^2+\frac{m^2}{\gamma_\eta^2}\right)R_{ml}^{(1)}(c_2,\xi)S_{ml}(c_2,\eta)\right],$$

where $\lambda_{ml}(c)-$ the eigenvalues of prolate spheriodal functions and the following notations are introduced

$$\left[A_{ml}\right]_{\mathrm{TM}}=\xi k_0\left[a_{ml}^{(2)}\right]_{\mathrm{TM}}+\eta c_1\left[b_{ml}^{(2)}\right]_{\mathrm{TM}},\quad \left[B_{ml}\right]_{\mathrm{TM}}=\eta k_0\left[a_{ml}^{(2)}\right]_{\mathrm{TM}}+\xi c_1\left[b_{ml}^{(2)}\right]_{\mathrm{TM}},$$

$$C_{ml}=\gamma_\eta^2c_1\left[b_{ml}^{(2)}\right]_{\mathrm{TM}},\quad D_{ml}=\gamma_\eta^2k_0\left[a_{ml}^{(2)}\right]_{\mathrm{TM}},\quad E_{ml}=\gamma_\xi^2c_1\left[b_{ml}^{(2)}\right]_{\mathrm{TM}},$$

$$F_{ml}=\gamma_\xi^2k_0\left[a_{ml}^{(2)}\right]_{\mathrm{TM}},\quad G_{ml}=\gamma_\varphi^2c_1\left[b_{ml}^{(2)}\right]_{\mathrm{TM}}.$$

TE-wave

$$\left[E_{2\xi}^{(2)}\right]_{\mathrm{TE}} = \frac{1}{c_1\gamma_\varphi\gamma_\xi}\sum_{m=1}^{\infty}\sum_{l=m}^{\infty}\left[A_{ml}\right]_{\mathrm{TE}} S_{ml}(c_2,\eta)R_{ml}^{(1)}(c_2,\xi)\sin m\varphi,$$

$$\left[E_{2\eta}^{(2)}\right]_{\mathrm{TE}} = -\frac{1}{c_1\gamma_\varphi\gamma_\eta}\sum_{m=1}^{\infty}\sum_{l=m}^{\infty}\left[B_{ml}\right]_{\mathrm{TE}} S_{ml}(c_2,\eta)R_{ml}^{(1)}(c_2,\xi)\sin m\varphi,$$

$$\left[E_{2\varphi}^{(2)}\right]_{\mathrm{TE}} = \frac{\gamma_\xi\gamma_\eta}{c_1\gamma_\varphi^2}\sum_{m=1}^{\infty}\sum_{l=m}^{\infty}\cos m\varphi\Big[\left[A_{ml}\right]_{\mathrm{TE}} S_{ml}(c_2,\eta)R_{ml}^{(1)\prime}(c_2,\xi) -$$

$$-\left[B_{ml}\right]_{\mathrm{TE}} S_{ml}(c_2,\eta)R_{ml}^{(1)\prime}(c_2,\xi)\Big],$$

where

$$\left[A_{ml}\right]_{\mathrm{TE}} = \xi k_0\left[a_{ml}^{(2)}\right]_{\mathrm{TE}} + \eta c_1\left[b_{ml}^{(2)}\right]_{\mathrm{TE}},\ \left[B_{ml}\right]_{\mathrm{TE}} = \eta k_0\left[a_{ml}^{(2)}\right]_{\mathrm{TE}} + \xi c_1\left[b_{ml}^{(2)}\right]_{\mathrm{TE}}.$$

The incident amplitude coefficients $\left(a_{ml}^{(0)}, b_{ml}^{(0)}\right)$ are known

$$\begin{matrix}\mathrm{TE}\\ \mathrm{TM}\end{matrix}\qquad a_{ml}^{(0)} = \mp\frac{4i^{l-1}}{k_0\sin\alpha}\frac{S_{ml}(c_1,\cos\alpha)}{N_{ml}^2(c_1)},\quad b_{ml}^{(0)} = 0$$

and the coefficients for the internal field $\left(a_{ml}^{(2)}, b_{ml}^{(2)}\right)$ are connected with the coefficients for the scattered $\left(a_{ml}^{(1)}, b_{ml}^{(1)}\right)$ and incident fields with the aid of the vector equations. For the TM-mode they have the form

$$\mathbf{X}_1^{(\mathbf{m})} = \widehat{\mathbf{\Delta}}_{21}^{(\mathbf{m})}(\mathbf{Z}_1^{(\mathbf{m})} + \mathbf{F}^{(\mathbf{m})}),\ \ \mathbf{X}_2^{(\mathbf{m})} = \widehat{\mathbf{\Delta}}_{21}^{(\mathbf{m})}\mathbf{Z}_2^{(\mathbf{m})}, \tag{7}$$

where the definitions for vectors

$$\mathbf{X}_1^{(\mathbf{m})} = \left[k_0 a_{ml}^{(2)} R_{ml}^{(1)}(c_2,\xi) N_{ml}(c_2)\right]_{l=m}^{\infty},\ \mathbf{X}_2^{(\mathbf{m})} = \left[c_1 b_{ml}^{(2)} R_{ml}^{(1)}(c_2,\xi) N_{ml}(c_2)\right]_{l=m}^{\infty},$$

$$\mathbf{Z}_1^{(\mathbf{m})} = \left[k_0 a_{ml}^{(1)} R_{ml}^{(3)}(c_1,\xi) N_{ml}(c_1)\right]_{l=m}^{\infty},\ \mathbf{Z}_2^{(\mathbf{m})} = \left[c_1 b_{ml}^{(1)} R_{ml}^{(3)}(c_1,\xi) N_{ml}(c_1)\right]_{l=m}^{\infty}, \tag{8}$$

$$\mathbf{F}^{(\mathbf{m})} = \left[k_0 a_{ml}^{(0)} R_{ml}^{(1)}(c_1,\xi) N_{ml}(c_1)\right]_{l=m}^{\infty},$$

and the square matrix

$$\widehat{\mathbf{\Delta}}_{21}^{(\mathbf{m})} = \left[\delta_{nl}^{(m)}(c_2,c_1)\right]_{n,l=m}^{\infty} \tag{9}$$

are introduced (all quantities in Eqs.(7-9) must be evaluated at $\xi = \xi_0$). For the TE-mode these vector equations are much more complex (the upper subscript (m) and notation $\xi = \xi_0$ are omitted for brevity):

$$\mathbf{X}_1 = \left(\xi_0^2\widehat{\mathbf{I}} - \widehat{\mathbf{\Gamma}}_{22}^2\right)^{-1}\Big[\left(\frac{\xi_0^2}{M^2}\widehat{\mathbf{\Delta}}_{21} - \widehat{\mathbf{\Gamma}}_{22}\widehat{\mathbf{\Gamma}}_{21}\right)(\mathbf{Z}_1 + \mathbf{F}) +$$

$$+\frac{\xi_0}{M^2}\left(\widehat{\mathbf{\Gamma}}_{21} - M^2\widehat{\mathbf{\Gamma}}_{22}\widehat{\mathbf{\Delta}}_{21}\right)\mathbf{Z}_2\Big],$$

$$\mathbf{X}_2 = \left(\xi_0^2\widehat{\mathbf{I}} - \widehat{\mathbf{\Gamma}}_{22}^2\right)^{-1}\left[\left(\xi_0\widehat{\mathbf{\Gamma}}_{21} - \frac{\xi_0}{M^2}\widehat{\mathbf{\Gamma}}_{22}\widehat{\mathbf{\Delta}}_{21}\right)(\mathbf{Z}_1 + \mathbf{F}) + \right.$$

$$\left. + \left(\xi_0^2\widehat{\mathbf{\Delta}}_{21} - \frac{1}{M^2}\widehat{\mathbf{\Gamma}}_{22}\widehat{\mathbf{\Gamma}}_{2121}\right)\mathbf{Z}_2\right],$$

where $\widehat{\mathbf{I}}$ is the unit matrix, and matrices $\mathbf{\Gamma}_{\mathbf{ij}}^{(\mathbf{m})}$ are

$$\widehat{\mathbf{\Gamma}}_{\mathbf{ij}}^{(\mu)} = \left[\int_{-1}^{1}\eta\bar{S}_{ml}(c_i,\eta)\bar{S}_{mn}(c_j,\eta)d\eta\right]_{l,n=m}^{\infty}.$$

One can find the connections between vectors $(\mathbf{Z}_1,\mathbf{Z}_2)$ and $\mathbf{F}$ in [9]. Eventually, the electrical field intensity at point (ξ,η,φ) inside the spheroidal particle is

$$B(\xi,\eta,\varphi) = \left|E_{1\xi}^{(2)} + E_{2\xi}^{(2)}\right|^2 + \left|E_{1\eta}^{(2)} + E_{2\eta}^{(2)}\right|^2 + \left|E_{1\varphi}^{(2)} + E_{2\varphi}^{(2)}\right|^2$$

4. Summary and numerical results.

We compare the coded techniques with respect to computer time, precision and effectiveness for internal field calculations for prolate spheroidal dust grains composed of astronomical silicates. In this section we demonstrate the inherent effectiveness and accuracy of our SVM and GMT codes.

In this work we compare three numerical techniques applied to calculation of internal fields in small spheroidal particles. We conduct the comparative calculations not only for cosmic grains, but for other practical systems, including elongated aerosol particles in the Earth atmosphere. In so doing, the input parameters of particles (the complex refractive index, the maximal size parameter ρ_{max}, the aspect ratio ε) varied over a wide range. Here we only mention the primary conclusions resulting from these calculations:

a) The GMT method has the greatest capabilities compared with SVM and T-matrix with respect to size parameters ($\rho_{max} \approx 60 \div 100$) and aspect ratios (up to 100:1). The T-matrix approach breaks down at relatively small ρ_{max} and at aspect ratios of about 4:1. The SVM approach is intermediate between GMT and T-matrix approaches in these respects. We suspect that the drawbacks of our SVM-calculations (as distinct from T-matrix) are dictated by the shortcomings of our code rather than by the intrinsic properties of the SVM. Efforts to improve the calculation of the spheroidal functions and eigenvalues can drastically enhance the range of effective applicability of SVM.

b) The realization of SVM and GMT codes requires computer times of one order of magnitude greater than that of T-matrix code. In so doing, the GMT is only slightly slower than SVM at relatively small size parameter $\rho_{max} \leq 12$, but the computer time for GMT is abruptly increased with ρ_{max}.

c) In the ranges of parameters where the parallel comparison with GMT and SVM is possible, the results of these methods coincide.

In the case of cosmic dust the situation is simpler due to the relatively small grain size. As an illustration, the results of comparative calculations of the fields (namely, the electrical field intensity $B(\xi,\eta,\varphi)$) inside the prolate (aspect ratio 1:5 and 1:10) spheroidal grains of astronomical silicate with $r_{max} = 0.5\mu$ and 1.0μ at wavelengths $\lambda = 1.06\mu$ and 0.55μ are depicted in Figs.1 and 2.

We use the optical constants of astronomical silicate: $n = 1.5$, $\kappa = 0.0077$ for $\lambda = 1.06\ \mu$; $n = 1.72$, $\kappa = 0.03$ for $\lambda = 0.55\ \mu$.

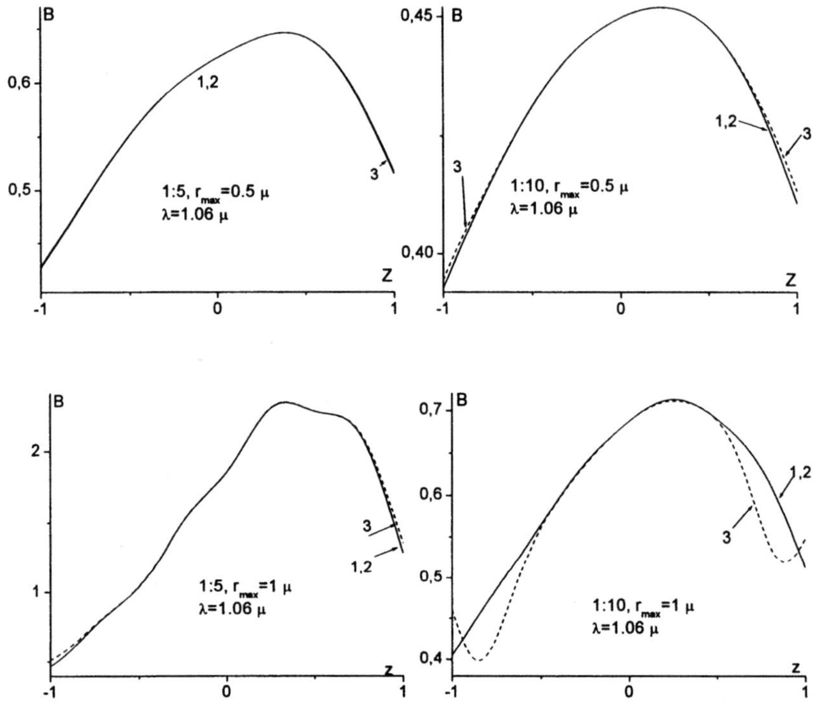

Figure 1. Electrical field intensity along the symmetry axis of the prolate spheroidal grain of astronomical silicate with aspect ratios 1:5 and 1:10 and great semiaxes 1.0 and 0.5 μ. Incident wave with wavelength $\lambda = 1.06\mu$ propagates along the symmetry axis. Calculations are performed by GMT(*1*), SVM(*2*) and T-matrix(*3*).

It should be pointed out that Figs.1 and 2 illustrate the simplest degenerate situation: the incident wave propagates along the symmetry axis of a spheroid (*z*-axis) and the internal fields *B* are calculated along this axis. In this case the field contains the azimutal mode $m = 1$ only and the polarization of the incident wave is of no concern. Nevertheless, such plots are very illustrative because, as our experience suggests, the most remarkable features of the internal field (maximums, resonances, etc.) are located on this axis. In the non-degenerate case, when the incident polarized wave propagates at some angle to the *z*-axis and the internal field must be calculated throughout the grain volume, the presentation of results needs the 3-D graphics and is more cumbersome. In our codes the field are calculated in nodes of the 3-D spatial grid in the spheroidal coordinates, for further solution of heat-conduction equation over the time-spatial grid with sources calculated in the first step.

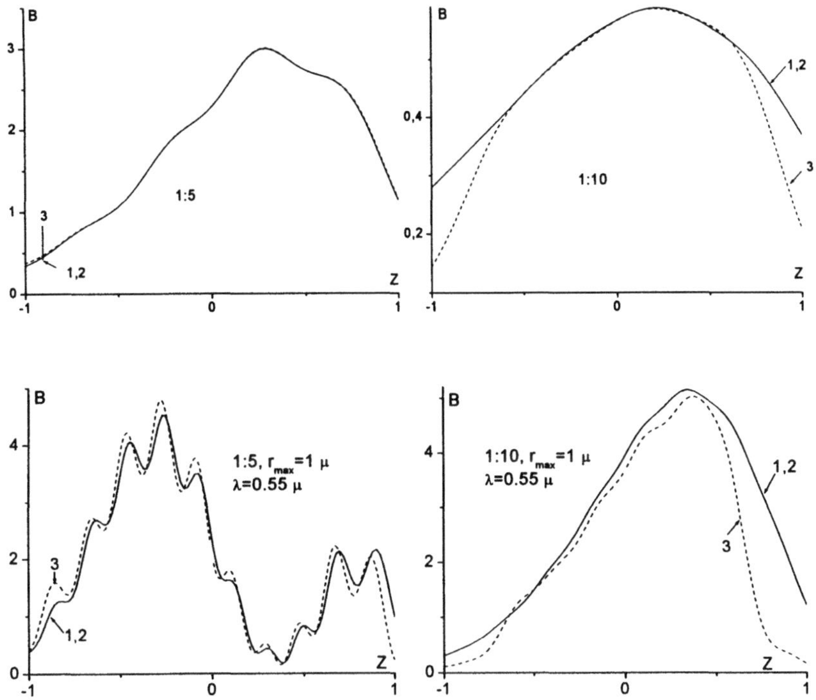

Figure 2. Electrical field intensity along the symmetry axis of the prolate spheroidal grain of astronomical silicate with aspect ratios 1:5 and 1:10 and great semiaxes 1.0 and 0.5 μ . Incident wave with wavelength $\lambda = 0.55\mu$ propagates along the symmetry axis. Calculations are performed by GMT(*1*), SVM(*2*) and T-matrix(*3*).

It is clear from the graphs that the GMT and SVM curves coincide with sufficient accuracy. Actually, the intensity values coincide with one another at least up to four decimal figures in all cases for cosmic dust. The most surprising is that the values calculated via T-matrix approach, which are expected to be correct only within the inscribed sphere, proved to be quite accurate far away from it. However, our calculations show that the T-matrix approach fails outside the inscribed sphere for non-degenerate cases.

The basic conclusion of the work is that the GMT and SVM calculations converge to the same results. These methods are important for the calculation of the optical fields inside spheroidal grains of cosmic dust and following investigation of thermal effects.

5. Acknowledgment

The work was financially supported by the INTAS grant 99/652.

References

1. J.Waterman, Proc. IEEE, **53**, 805 (1985).
2. V.G. Farafonov, Opt. i Spektrosk.(in press), (2002).
3. A.Doicu, Y.Eremin, and Th.Wriedt, "Acoustic and Electromagnetic Scattering Analysis using Discrete Sources" (Academic, San Diego, 2000).
4. V.D. Kupradze, Success Mathematical Sci.,**22**, 58 (1967).
5. M.S.Zhdanov, "The analogs of Cauchy-type integrals in geophysical fields theory", (Moscow Univ. Press ,Moscow, 1984).
6. Y.A.Eremin and A.G.Sveshnikov, "The Discrete Sources Method in Electromagnetic Diffraction Problems" (Moscow Univ. Press ,Moscow, 1992).
7. H.Kersten, Math. Meth. Appl. Sci.,**7**, 40 (1985)
8. Th.Wriedt, A.Doicu},J.Mod.Opt. **45**, 199 (1998).
9. N.V.Voshchinnikov and V.G.Farafonov, Astroph.Space Sci. **204**, 19 (1993)

THERMAL EFFECTS OF RADIATION ON DUST PARTICLES

LIUDMILA G. ASTAFYEVA
National Academy of Sciences of Belarus,
Stepanov Institute of Physics,
Skorina Pr. 68, Minsk 220072, Belarus

Abstract. Cosmic dust grains are an important component of the interstellar medium. They play a central role in the star-formation process, the energy balance of gas clouds, etc. Thermal heating effects are important, for among other things, the infrared properties of dust grains. To calculate the thermal effect of various kinds of radiation on dust particles the absorption and heating of particles must be known over a wide interval of particle composition and sizes. As dust grains may be both uniform and composite collections of particles of distinct materials, including voids agglomerated together. Many such particles can be simulated by homogeneous spheres and by a series of concentric spherical layers, each of which has the optical properties of one material or vacuum [1]. Single-layered grains of amorphous silicate and carbonaceous material are considered. Then composite grains with two materials: amorphous carbon and vacuum or amorphous silicate particles coated with a carbonaceous "mantle" are examined [2]. The mathematical formulation of the problem of heating of multi-layered particles consists of solving the heat-conduction equation in spherical coordinates with corresponding initial and boundary conditions, taking into account the non-uniform heat release inside the particles and temperature dependencies of optical and thermo-physical properties of the grain material.

1. Introduction

Interactions of laser radiation with metallic and metallic oxide particles have been the subject of many theoretical investigations into the heating and destruction of homogeneous and hollow spherical aerosols under the action of intense laser radiation [3]-[5]. The particle sizes were varied from 5 to 40 μm. The particles were irradiated by unpolarized plane-wave CO_2 laser beam of

G. Videen and M. Kocifaj (eds.), Optics of Cosmic Dust, 131–142.

wavelength λ = 10.6 μm, with laser fluxes in the range ~ 10^6 W/cm^2. The heating of lidar-irradiated three-layer spherical particles was analyzed theoretically and numerically in [6].

In this chapter we investigate the heating of one- and two-layered spherical particles by radiation of wavelength λ = 0.55 μm. For the calculations we use optical constants of actual amorphous carbon ("Be 1" tabulated in [7]) and optical constants of amorphous silicate taken from [8]. The refraction index of carbon is $n = 2.08$, the absorption index of carbon is $\chi = 0.801$, for silicate $n = 1.576$, $\chi = 0.000022$, for air $n = 1$, $\chi = 0$.

2. Theory

The expressions of the components of electromagnetic fields inside one-layered spherical particle can be obtained from the theory of diffraction of electromagnetic radiation by a homogeneous sphere. They can be written as follows:

$$E_r = (Nk_0 r)^{-2} \cos\varphi \sum_{l=1}^{\infty} l(l+1) \frac{\psi_l(Nk_0 r)}{\psi_l(Nk_0 R)} \alpha_l Q_l(\theta) \sin\theta, \tag{1}$$

$$E_\theta = (Nk_0 r)^{-1} \cos\varphi \sum_{l=1}^{\infty} \frac{\psi_l(Nk_0 r)}{\psi_l(Nk_0 R)} \left[\alpha_l D_l(Nk_0 r) S_l(\theta) + i\beta_l Q_l(\vartheta)\right], \tag{2}$$

$$E_\varphi = -(Nk_0 r)^{-1} \sin\varphi \sum_{l=1}^{\infty} \left[\alpha_l D_l(Nk_0 r) Q_l(\theta) + i\beta_l S_l(\theta)\right], \tag{3}$$

where r, θ and φ are radius and angular spherical coordinates; R is the spherical particle radius; N is the complex refractive index of the particle material ($N = n - i\chi$); k_0 is the wave number in vacuum ($k_0 = 2\pi/\lambda$), λ is radiation wavelength, $\psi_l(z)$ is the Ricatti-Bessel function and $\psi_l'(z)$ is the derivative of the corresponding function. $D_l(z) = \psi_l'(z)/\psi_l(z)$ is logarithmic derivatives of the Ricatti-Bessel functions. The angular functions $Q_l(\theta)$ and $S_l(\theta)$ are defined with the aid of the Legendre polynomials $P_l(\cos\theta)$ and their derivatives.

The internal field expansion coefficients α_l and β_l are defined from

$$\alpha_l = Ni^l \frac{(2l+1)}{l(l+1)} \varsigma_l(k_0 R)\left[D_l(Nk_0 R) - NG_l(k_0 R)\right]^{-1}, \tag{4}$$

$$\beta_l = -Ni^l \frac{(2l+1)}{l(l+1)} \varsigma_l(k_0 R)\left[G_l(k_0 R) - ND_l(Nk_0 R)\right]^{-1}, \tag{5}$$

where $\zeta_l(z)$ is Ricatti-Hankel function, $\zeta_l'(z)$ is the derivative of the corresponding function. $G_l(z) = \zeta_l'(z)/\zeta_l(z)$ is logarithmic derivatives of the Ricatti-Hankel functions.

The expressions for the components of the electromagnetic fields inside different layers of a two-layered spherical particle can be obtained from the theory of diffraction of electromagnetic radiation from a multi-layered sphere. The components of the electric fields at given point within the core can be written as follows:

$$E_r^{(1)} = (N_1 k_0 r)^{-2} \cos\varphi \sum_{l=1}^{\infty} l(l+1) \frac{\psi_l(N_1 k_0 r)}{\psi_l(N_1 k_0 R_1)} A_l^{(1)} Q_l(\theta) \sin\theta, \tag{6}$$

$$E_\theta^{(1)} = (N k_0 r)^{-1} \cos\varphi \sum_{l=1}^{\infty} \frac{\psi_l(N k_0 r)}{\psi_l(N k_0 R_1)} \left[A_l^{(1)} D_l(N_1 k_0 r) S_l(\theta) + i B_l^{(1)} Q_l(\vartheta) \right], \tag{7}$$

$$E_\varphi^{(1)} = -(N k_0 r)^{-1} \sin\varphi \sum_{l=1}^{\infty} \frac{\psi_l(N k_0 r)}{\psi_l(N k_0 R_1)} \left[A_l^{(1)} D_l(N_1 k_0 r) Q_l(\theta) + i B_l^{(1)} S_l(\theta) \right], \tag{8}$$

The expressions for the components of the electric fields within the outer layer have the form

$$E_r^{(2)} = (N_2 k_0 r)^{-2} \cos\varphi \sum_{l=1}^{\infty} l(l+1) \left[\frac{\psi_l(N_2 k_0 r)}{\psi_l(N_2 k_0 R_2)} A_l^{(2)} + \frac{\varsigma_l(N_2 k_0 r)}{\varsigma_l(N_2 k_0 R_1)} \tilde{A}_l^{(2)} \right] \times \tag{9}$$

$$\times Q_l(\theta) \sin\theta,$$

$$E_\theta^{(2)} = (N_2 k_0 r)^{-1} \cos\varphi \sum_{l=1}^{\infty} \left[\frac{\psi_l(N_2 k_0 r)}{\psi_l(N_2 k_0 R_2)} A_l^{(2)} D_l(N_2 k_0 r) + \frac{\varsigma_l(N_2 k_0 r)}{\varsigma_l(N_2 k_0 R_1)} \tilde{A}_l^{(2)} \right] S_l(\theta) +$$

$$+ i \left[\frac{\psi_l(N_2 k_0 r)}{\psi_l(N_2 k_0 R_2)} B_l^{(2)} + \frac{\varsigma_l(N_2 k_0 r)}{\varsigma_l(N_2 k_0 R_1)} \tilde{B}_l^{(2)} \right] Q_l(\vartheta), \tag{10}$$

$$E_\varphi^{(2)} = -(N_2 k_0 r)^{-1} \sin\varphi \sum_{l=1}^{\infty} \left[\frac{\psi_l(N_2 k_0 r)}{\psi_l(N_2 k_0 R_2)} A_l^{(2)} D_l(N_2 k_0 r) + \frac{\varsigma_l(N_2 k_0 r)}{\varsigma_l(N_2 k_0 R_1)} \tilde{A}_l^{(2)} \right] Q_l(\theta) +$$

$$+ i \left[\frac{\psi_l(N_2 k_0 r)}{\psi_l(N_2 k_0 R_2)} B_l^{(2)} + \frac{\varsigma_l(N_2 k_0 r)}{\varsigma_l(N_2 k_0 R_1)} \tilde{B}_l^{(2)} \right] S_l(\vartheta), \tag{11}$$

In Eqs. (6) - (11), $N_i = n_i - i\chi_i$ are the complex refractive indices within layer i, R_1 is the radius of the core, R_2 is the outer radius of the particle. The coefficients $A_l^{(i)}, \tilde{A}_l^{(i)}, B_l^{(i)}, \tilde{B}_l^{(i)}$ in Eqs. (6) - (11) are

$$A_l^{(1)} = i^{l-1} \frac{2l+1}{l(l+1)} \frac{N_1}{N_s} \frac{1}{\psi_l(N_2 k_0 R_2) \varsigma_l(N_2 k_0 R_1)} \frac{1}{\varsigma_l(N_s k_0 R_2)} \frac{1}{\Delta_1}, \tag{12}$$

$$B_l^{(1)} = i^{l-1} \frac{2l+1}{l(l+1)} \frac{1}{\psi_l(N_2 k_0 R_2) \varsigma_l(N_2 k_0 R_1)} \frac{1}{\varsigma_l(N_s k_0 R_2)} \frac{1}{\Delta_2}, \tag{13}$$

$$A_l^{(2)} = i^l \frac{2l+1}{l(l+1)} \frac{N_1}{N_s} \frac{1}{\varsigma_l(N_s k_0 R_2)} \frac{\tilde{L}}{\Delta_2}, \tag{14}$$

$$\tilde{A}_l^{(2)} = \left(-i^l\right) \frac{2l+1}{l(l+1)} \frac{\psi_l(N_2 k_0 R_1)}{\psi_l(N_2 k_0 R_2)} \frac{1}{\varsigma_l(N_s k_0 R_2)} \frac{\tilde{M}}{\Delta_1}, \tag{15}$$

$$B_l^{(2)} = i^l \frac{2l+1}{l(l+1)} \frac{N_2}{N_s} \frac{1}{\varsigma_l(N_s k_0 R_2)} \frac{L}{\Delta_1}, \tag{16}$$

$$\tilde{B}_l^{(2)} = \left(-i^l\right) \frac{2l+1}{l(l+1)} \frac{N_2}{N_s} \frac{\psi_l(N_2 k_0 R_1)}{\psi_l(N_2 k_0 R_2)} \frac{1}{\varsigma_l(N_s k_0 R_2)} \frac{M}{\Delta_1}, \tag{17}$$

where

$$\Delta_1 = \frac{\psi_l(N_2 k_0 R_1)}{\psi_l(N_2 k_0 R_2)} \frac{\varsigma_l(N_2 k_0 R_2)}{\varsigma_l(N_2 k_0 R_1)} M \cdot H - L \cdot F, \tag{18}$$

$$\Delta_2 = \frac{\psi_l(N_2 k_0 R_1)}{\psi_l(N_2 k_0 R_2)} \frac{\varsigma_l(N_2 k_0 R_2)}{\varsigma_l(N_2 k_0 R_1)} \tilde{M} \cdot \tilde{H} - \tilde{L} \cdot \tilde{F} \tag{19}$$

and

$$F = G_l(N_s k_0 R_2) - N_{23} D_l(N_2 k_0 R_2),$$

$$H = G_l(N_s k_0 R_2) - N_{23} G_l(N_2 k_0 R_2),$$

$$L = G_l(N_2 k_0 R_1) - N_{12} D_l(N_1 k_0 R_1),$$

$$M = D_l(N_2 k_0 R_1) - N_{12} D_l(N_1 k_0 R_1).$$

The expressions for $\tilde{F}, \tilde{H}, \tilde{L}, \tilde{M}$ can be obtained from the corresponding expressions for F, H, L, M by the replacement of N_{ij} for N_{ji}, where $N_{ij} = N_i/N_j$. In studies of the energy distribution inside particles the quantity of interest is the internal intensity calculated from

$$I = \vec{E} \times \vec{E}^{*}. \tag{20}$$

For spherical particles it follows that

$$I = |E_r|^2 + |E_\vartheta|^2 + |E_\varphi|^2. \tag{21}$$

To obtain the temperature distribution inside a particle, we solve the following heat conduction equation in spherical coordinates with the origin at the particle center subject to corresponding initial and boundary conditions:

$$c_i(T_i)\rho_i \frac{\partial T_i}{\partial t} = \frac{1}{r^2}\frac{\partial}{\partial r}\left(\kappa_i(T_i) r^2 \frac{\partial T_i}{\partial r}\right) + \frac{1}{r^2 \sin\theta}\frac{\partial}{\partial \theta}\left(\kappa_i(T_i)\sin\theta \frac{\partial T_i}{\partial \theta}\right) + Q_i(r,\theta,t,T_i) \tag{22}$$

where $0 < r < R_i$, $0 < \theta < \pi$. $T_i(r,\theta,t)$ is the local temperature within the particle, $c_i(T_i)$, ρ_i, and $\kappa_i(T_i)$ are the material heat capacity, density and thermal conductivity, respectively. The temperature within the particle depends only on two coordinates r and θ, because the radiation is not polarized and the third coordinate φ is absent. The power density of heat sources at a given point within the particle $Q_i(T_i)$ is determined by equation

$$Q_i(T_i) = I_0 \cdot I_i \cdot (4\pi n_i \chi_i / m_s \lambda) \tag{23}$$

where I_0 is incident radiation intensity, I_i is defined by Eqs. (20) - (21) in the corresponding layer. At the center, interfaces and on the surface of the particle the following conditions must be satisfied:

$$|T_1(0,\theta,t) < \infty,| \quad 0 \le \theta \le \pi, \quad t > 0, \tag{24}$$

$$T_1(R_1,\theta,t) = T_2(R_1,\theta,t), \qquad T_2(R_2,\theta,t) = T_3(R_2,\theta,t), \tag{25}$$

$$\kappa_1(T_1)\frac{\partial T_1(R_1,\theta,t)}{\partial r} = \kappa_2(T_2)\frac{\partial T_2(R_1,\theta,t)}{\partial r}, \tag{26}$$

$$\kappa_2(T_2)\frac{\partial T_2(R_2,\theta,t)}{\partial r} = \kappa_3(T_3)\frac{\partial T_3(R_2,\theta,t)}{\partial r}, \tag{27}$$

$$-\kappa_3(T_3)\frac{\partial T_3(R_2,\theta,t)}{\partial r} = \sigma\varepsilon_\lambda T^4, \tag{28}$$

$$\left.\frac{\partial T_i(R_i,\theta,t)}{\partial \theta}\right|_{\theta=0} = \left.\frac{\partial T_i(R_i,\theta,t)}{\partial \theta}\right|_{\theta=\pi} = 0, \quad i=1,2,3, \tag{29}$$

where σ is the Stefan-Boltzmann constants, ε_λ is the spectral coefficient of blackness.

Finding a solution to the system (22) with the conditions (24) - (29) can only be done numerically. For this purpose, the local one-dimensional iterative scheme was created. The iteration cycle is continued up to the achievement of the required accuracy. The temperature dependencies of thermophysical properties of carbon and silicate are taken into account in the calculations. Using approximate formulas obtained from the data presented in (9) - (12), the values of thermal conductivity $\kappa(T)$ and heat capacity $c(T)$ are corrected at any time according to the temperature at every mesh point.

3. Results

Numerical results are presented for cases of one-layered (homogeneous) spheres of silicate and carbon, and two-layered spheres of air-carbon and silicate-carbon irradiated by radiation with wavelength λ = 0.55 μm with fluxes 10^3 W/cm^2 assuming the initial temperature T_0 = 210 K.

The intensity distribution inside one-layer or two-layer particles has a substantially inhomogeneous character. The maximum is located near the particle surface in the illuminated or shadow hemisphere and depends on the particle structure and size. In the case of a homogeneous silicate particle (Figure 1), the region of maximum intensity lies in the shadow hemisphere near the particle surface that is the result of the focusing of radiation by the silicate sphere.

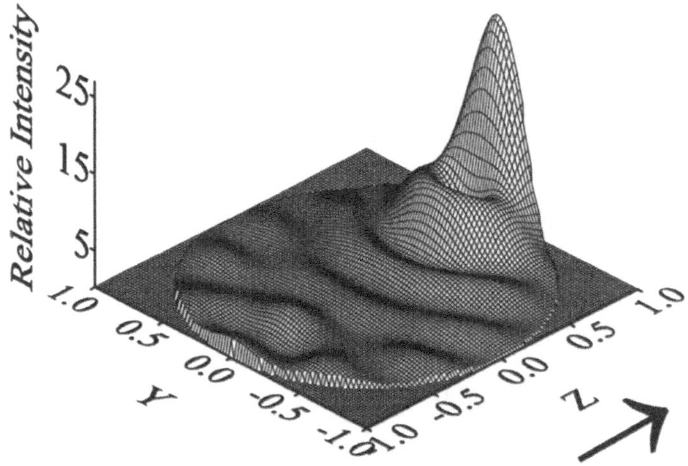

Figure 1: Irradiance distribution inside the silicate grain, $R = 0.5$ μm, $\lambda = 0.55$ μm. The arrow shows the direction of the propagation of incident radiation.

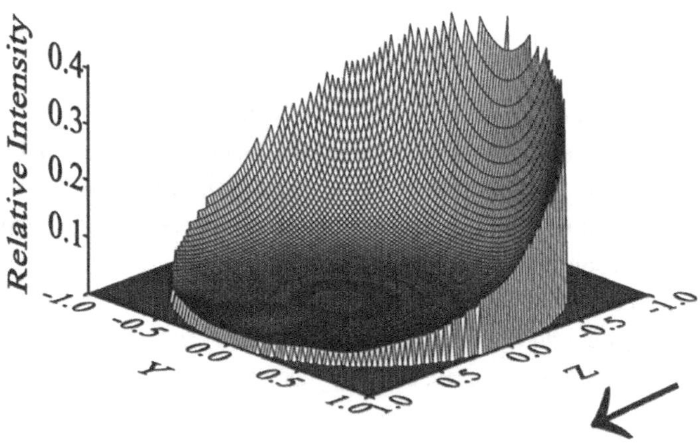

Figure 2: Irradiance distribution inside the carbon grain, $R = 0.5$ μm, $\lambda = 0.55$ μm. The arrow shows the direction of the propagation of incident radiation.

If the homogeneous particle is carbonaceous, the intensity distribution is completely changed (see Figure 2). The maximum intensity occurs within the thin layer near the particle surface located in the illuminated hemisphere. If the particle is two-layered (silicate-carbon), the intensity distribution resembles something middle between silicate and carbon particles (Figure 3).

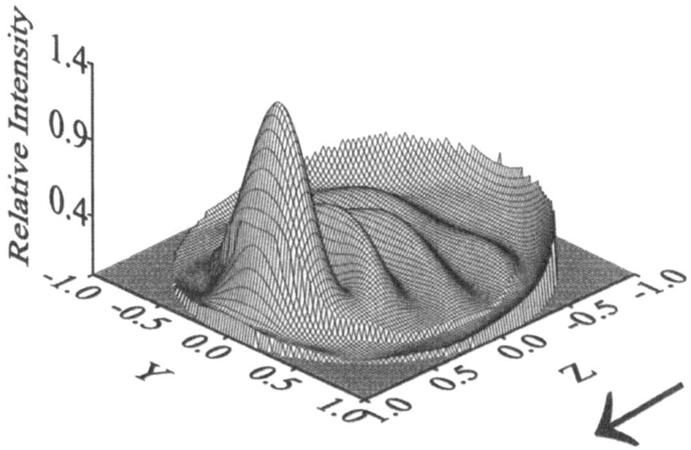

Figure 3: Irradiance distribution inside the silicate-carbon grain, $R_2 = 0.5$ μm, $V_1/V = V_2/V = 0.5$ (V is the volume of two-layered particle, V_1 is the volume of the core, V_2 is the volume of the shell), $\lambda = 0.55$ μm. The arrow shows the direction of propagation of incident radiation.

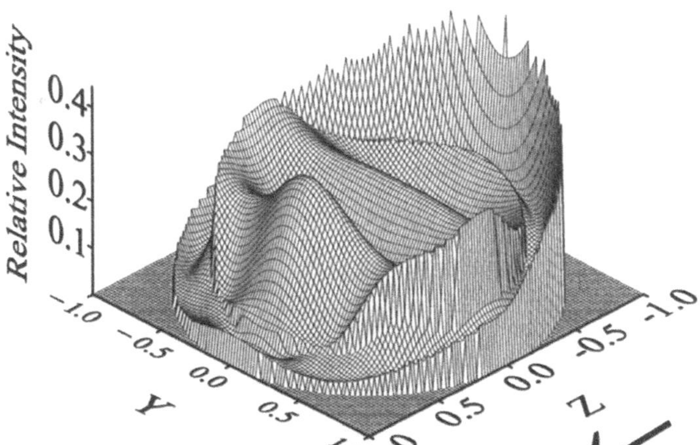

Figure 4: Irradiance distribution inside the hollow carbon grain, $R_2 = 0.5$ μm, $V_1/V = V_2/V = 0.5$, $\lambda = 0.55$ μm. The arrow shows the direction of the propagation of incident radiation.

The maximum intensity region is localized in the shadow hemisphere near the particle surface, as inside silicate particle, in the illuminated hemisphere there is a secondary maximum, as inside the carbon particle. For hollow carbon particles (Figure 4), the maximum intensity region is localized in the illuminated hemisphere as in carbon particles. There is some energy density distribution in the particle core with lower values of the internal intensity than in the illuminated part of the carbon shell and with higher values of the internal intensity than in the shadow part of the carbon shell. The highest intensity maximum occurs within homogeneous silicate particles. The maximum is more than twenty times lower inside two-layered silicate-carbon particles and more than sixty times lower inside carbon and hollow carbon particles.

The magnitude and location of maximum internal intensity plays an essential role during the heating process. Figure 5 shows maximum internal intensity I_{max} as a function of outer particle radius R_2 for the case of silicate, silicate-carbon and carbon particles. It follows from Figure 5 that the dependencies have rather complex character. The values of I_{max} increase, going from carbon to silicate-carbon and then to silicate particles. For carbon particles, the maximum internal intensities remain fairly constant from approximately $R_2 \approx 0.25 \div 0.3$ μm. For silicate particles they grow and begin oscillating with increased outer particle size. For two-layered silicate-carbon grains, the maximum intensity increases when the depth of the carbon shell is decreased, but the behavior as a function of R_2 is much more complicated depending on the thickness of the shell.

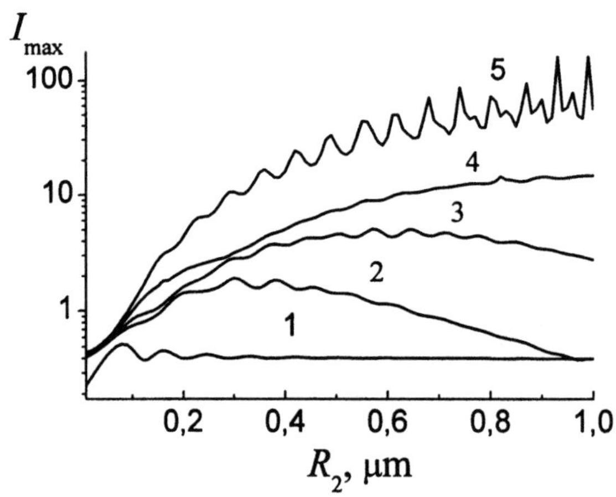

Figure 5. Maximum irradiance inside grains as a function of outer particle radius. 1 – homogeneous carbon grain; 2 – silicate-carbon grain with $V_1/V = V_2/V = 0.5$; 3 – silicate-carbon grain with $V_2/V = 1/3$; 4- silicate-carbon grain with $V_2/V = 1/6$; 5 – homogeneous silicate grain.

We can compare the dependencies of $I_{max}(R_2)$ with the dependencies of absorption efficiencies of such particles on R_2 (Figure 6). We see that when decreasing the thickness of the carbon particle shell, the absorption of the silicate-carbon particles also decreases. The absorption of carbon and silicate-carbon grains at first grows with increasing particle size. Beginning from some value of particle sizes (~ 0.3 μm), the efficiencies remain fairly constant. The absorption of carbon and silicate-carbon grains differs insignificantly. The absorption of silicate grains is the smallest and increases in the considered interval of grain sizes.

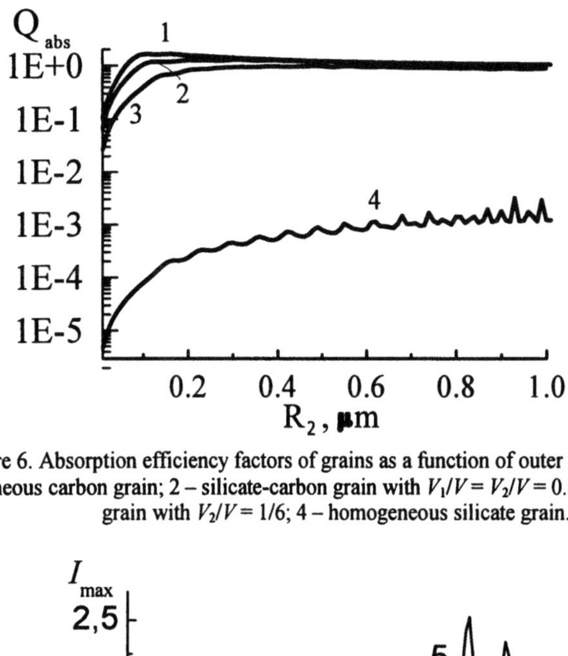

Figure 6. Absorption efficiency factors of grains as a function of outer particle radii. 1 – homogeneous carbon grain; 2 – silicate-carbon grain with $V_1/V = V_2/V = 0.5$; 3 – silicate-carbon grain with $V_2/V = 1/6$; 4 – homogeneous silicate grain.

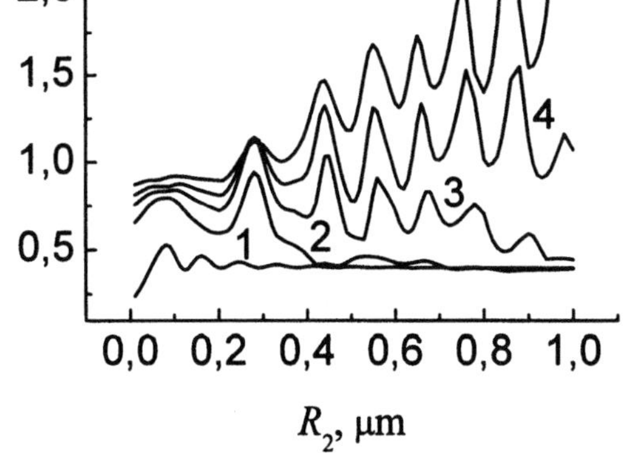

Figure 7. Maximum irradiance inside hollow carbon grains as a function of outer particle radius. 1 – homogeneous carbon grains; 2 – hollow carbon grains with $V_1/V = V_2/V = 0.5$; 3 – hollow carbon grains with $V_2/V = 1/3$; 4- hollow carbon grains with $V_2/V = 1/4$; 5 – hollow carbon grains with $V_2/V = 1/6$.

It is interesting to compare optical properties of silicate-carbon grains with those of hollow carbon particles. Figure 7 gives the dependence of the maximum internal intensity in hollow carbon particles on the outer particle radius. We see that the maxima of the internal intensity in hollow carbon grains grow when the thicknesses of the carbon shells are decreased as in the case of silicate-carbon particles. However, the growth of the maximum internal intensity for hollow carbon particles is substantially less than for silicate-carbon grains. The increase is largest for large grains (R2 ~ 0.8 ÷ 1 μm). For hollow carbon particles maximum irradiance increases approximately six times in this interval of particle sizes, but for silicate-carbon particles this increase is 25 times. The dependencies of I_{max}(R2) for hollow grains have a greater tendency toward oscillation. The character of the dependencies of the absorption efficiency factors on carbon shell thickness is the same for silicate-carbon and hollow carbon grains. Figure 8 shows absorption of hollow carbon particles as a function of grain size. Increasing the particle radius, the absorption rises attaining some maximum value, before decreasing slightly. The thinner the carbon shell thickness, the smaller the growth of absorption of the hollow particles.

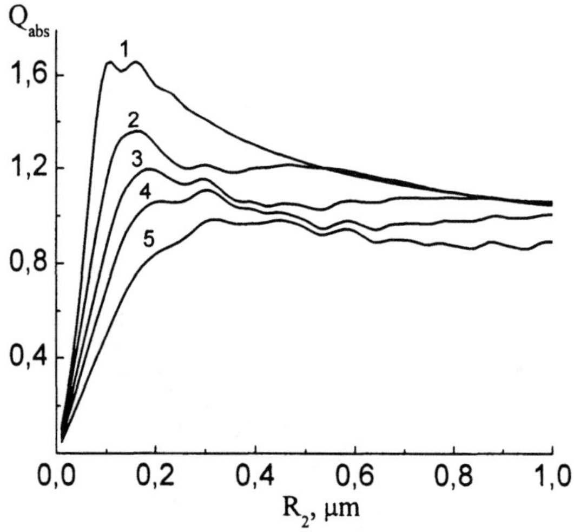

Figure 8. Absorption efficiency factors of hollow carbon grains as a function of outer particle radii. 1 – homogeneous carbon grains; 2 – hollow carbon grains with $V_1/V = V_2/V = 0.5$; 3 – hollow carbon grains with $V_2/V = 1/3$; hollow carbon grains with $V_2/V = 1/4$; 5 – hollow carbon grains with $V_2/V = 1/6$.

The results of heating of different kinds of two-layered grains by radiation with wavelength λ = 0.55 μm are shown in Figure 9. Heating times for temperatures to increase from T = 210 K to T = 260 K grows with outer particle radius for all kinds of grains. The curve describing the heating time of a homogeneous carbon grain lies between those for the two-layered particles. The heating time of silicate-carbon grains is more than for carbon particles. In contrast, the heating time of hollow carbon grains is less than for homogeneous carbon particles.

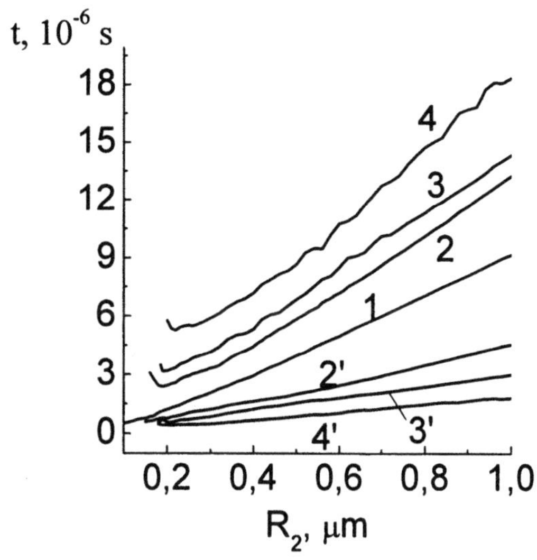

Figure 9. The heating time of two-layered grains to increase from T = 210 K to T = 260 K as a function of the outer particle radius. The effect of variation of carbon fraction is illustrated. 1 – homogeneous carbon particle; 2, 2′ - $V_1/V = V_2/V = 0.5$ for silicate-carbon and hollow carbon grains, accordingly; 3, 3′ - $V_2/V = 1/3$ for silicate-carbon and hollow carbon grains, accordingly; 4, 4′ - $V_2/V = 1/6$ for silicate-carbon and hollow carbon grains, accordingly.

For the thinnest carbon shell, the growth of the heating time for a silicate-carbon grain in comparison with a homogeneous carbon particle is 2.2 times when R_2 = 0.5 μm (curve 3). The decrease of the heating time for hollow carbon grains is 4.3 times (curve 3′). Such dependencies are determined for silicate-carbon grains by the change of the absorption and for hollow carbon grains by the change of the maximum internal intensity. So, the heating of different kinds of two-layered grains differs because of the their structure and can be determined either by the absorption or the internal intensity distribution in the grains.

4. Conclusions

The complex structure of cosmic dust grains has a substantial influence on their internal intensity distributions and, as a result, on their heating. First, the reduction of a strongly absorbing outer carbon layer on a two-layer grain surface leads to a decrease of the heating time of hollow particles. And second, the heating time of silicate-carbon grains rises in comparison to that of carbon grains. In the first case it is determined by the internal intensity distribution in hollow carbon grains. In the second case the main factor defining the heating process is the absorption of silicate-carbon grains.

5. Acknowledgments

The research described in this paper was made possible in part by grant No 99/652 from the INTAS.

References

1. N. V. Voshchinnikov, and J. S. Mathis, The Astronomical Journal, **526**, 257 (1999).
2. B. T. Draine, "Interstellar Grains" in Encyclopedia of Astronomy and Astrophysics (IOP and MacMillan, 2000).
3. A. P. Prishivalko, L. G. Astafyeva, and S. T. Leiko, Appl. Optics, **35**, 965 (1996).
4. L. G. Astafyeva, A. P. Prishivalko, and S. T. Leiko, J. Opt. Soc. Am., **B14**, 432 (1997).
5. L. G. Astafyeva, and A. P. Prishivalko, Int. J. Heat Mass Transfer, **41**, 489 (1998).
6. L. G. Astafyeva, N. V. Voshchinnikov, and L. B. F. M. Waters, Appl. Optics, in press.
7. F. Rouleau, and P. G. Martin, The Astrophysical journal, **377**, 526 (1991).
8. J. Dorschuer, B. Begemann, Th. Henning, C. Jaeger, and H. Mutschke, Astron. Astrophys., **300**, 503 (1995).
9. V. S. Chirkin, "Thermophysical Properties of Material of Nuclear Technics" (Atomizdat, Moscow, 1968).
10. "Thermodynamical Properties of Individual Matter", Handbook (Nauka, Moscow, 1979).
11. N. P. Bansal, R. H. Doremus, "Handbook of Glass Properties" (Academic Press, Orlando, 1986).
12. Landott-Boernstein, "Numerical Data and Functional Relationship in Science and Technology" (Springer, Berlin, 1982).

SIZE DISTRIBUTIONS OF PARTICLES OBTAINED BY INVERSION OF SPECTRAL EXTINCTION AND SCATTERING MEASUREMENTS.

H. HORVATH[1], F.J. OLMO[2], L. ALADOS ARBOLEDAS[2], O. JOVANOVIC[1], M. GANGL[1], W. KALLER[1], C. SANCHEZ[2], H. SAUERZOPF[1] AND S. SEIDL[1]

[1] *Institute of Experimental Physics, University of Vienna, Boltzmanngasse 5, A-1090 Vienna, Austria*

[2] *Departamento de Física Aplicada, Facultad de Ciencias, Universidad de Granada, Fuentnueva s/n SP-18071 Granada, Spain*

Abstract. Suspended particles attenuate and scatter light. Both the attenuation and the scattering is influenced by the particle's size and optical constants of the material. The angular scattering and the wavelength dependent extinction provides a means of obtaining information on the size distribution of the particles. The relation between optical properties and size is given by integral equations which are not easy to solve, since no unique solution exists.

It is possible to predict the accuracy of any inversion procedure through mathematical operations. The information parameter is close to one if the inverted size distribution data have little ambiguity and goes to zero if the inverted size distribution is meaningless. For spectral extinction measurements (in the visible) the range of meaningful size distributions obtained through inversions is between particle radii of 0.1 to 1 μm. For inversion of angular scattering data, the range is between 0.05 and 1.5 μm if scattering angles between 20 and 160° are used. The upper size limit increases if the scattering is also measured at smaller angles.

Simultaneous measurements of the horizontal attenuation, the sky radiation and the size distribution of the aerosol with an impactor has shown that inversions give meaningful results within the range, where the information parameter is high. The usual features of the atmospheric aerosol are also detected by the inversion of the angular scattering obtained from sky radiation measurements.

G. Videen and M. Kocifaj (eds.), Optics of Cosmic Dust, 143–158.

1. Introduction

Optical properties are a powerful tool to obtain information on objects, gases, particles, etc. at a location which can not be accessed or where access is difficult. Examples of this are the determination of chemical elements in the atmosphere of stars, molecules or particles in interstellar space, the velocity due to the red shift of light or images of the earth by satellites. Information on a substance/system can only be obtained, if the optical properties are specific for the substance/system. The wavelength of spectral lines or bands is a good example of the recognition of chemical elements of molecules. Here a unique relation between the spectrum and the substance exists, a mixture of different substances usually does not pose a serious problem for the detection of each of them. Unfortunately, not all optical properties permit a unique relation between a substance and, e.g., a spectrum. One example for this is the scattering and attenuation of light by small particles. For particles much smaller than the wavelength of light, the scattering is symmetric and the attenuation is highest in the blue. For particles much larger than the wavelength of light, the scattering is very pronounced in the forward region, and the light attenuation is independent of wavelength. For particles comparable to the wavelength, the attenuation is mostly higher in the blue and there is a less pronounced forward scatter. A mixture of different particles might have the same properties, thus an estimation of the number of particles and their sizes is difficult. The optical properties are a sum (integral) of the optical properties of the individual particles. The inversion of this sum (integral) frequently is an unstable procedure and may lead to unrealistic results. In the following we treat some basics on inversions and show some results of experimental work.

2. Definitions

The attenuation from the initial flux Φ_o to $\Phi = \Phi_o \cdot \exp(-\sigma_e \cdot x)$ is described by the well known Lambert or Lambert-Beer or Bouguer or Bouguer-Lambert Law. The constant σ_e is called the *extinction coefficient* or *attenuation coefficient*. It is defined as the fraction of radiant flux lost from a collimated beam per unit thickness of aerosol and is given in units of reciprocal length, e.g. in m^{-1}, km^{-1} or Mm^{-1}.

Since the attenuation of radiation can be due to both scattering and absorption, the extinction coefficient can be considered as the sum of the *scattering coefficient* σ_s and the *absorption coefficient* σ_a, i.e. $\sigma_e = \sigma_s + \sigma_a$. The scattering coefficient thereby is defined as the fraction of radiant flux lost due to scattering from a collimated beam per unit thickness of aerosol and is given in units of reciprocal length, e.g. in m^{-1}, km^{-1} or Mm^{-1}. Similarly, the absorption coefficient is defined as the fraction of radiant flux lost from a collimated beam per unit thickness of aerosol due to absorption and is also given in units of reciprocal length, e.g. in m^{-1}, km^{-1} or Mm^{-1}. It is a good habit to add a second subscript g or p to the symbol of the scattering and absorption coefficient, indicating whether it is for the gas molecules or the particles. E.g. $\sigma_{a,p}$ would be the absorption coefficient by particles, $\sigma_{s,g}$ would be the scattering coefficient of the gas molecules.

The angular scattering properties of the aerosol are described by the volume scattering function. Let us assume for this purpose unpolarized light illuminating a volume element dV of aerosol with an irradiance E. The particles and gas molecules scatter part of the illuminating radiation. A certain flux $d\Phi$ is scattered into the solid angle $d\omega$ in the direction given by the scattering angle ϕ_s. The flux $d\Phi$ scattered by the volume element dV into the solid angle $d\omega$ is given by $d\Phi = E \cdot \gamma(\phi_s) \cdot dV \cdot d\Omega$, with $\gamma(\phi_s)$ the *volume scattering function* of the aerosol. It has a dimension of $m^{-1}sr^{-1}$. Obviously, when integrating the volume scattering function over the whole solid angle, the scattering coefficient is obtained: $\sigma_s = \int_{4\pi} \gamma(\phi_s)\, d\Omega$.

For an aerosol having a *number size distribution* $n(r)=(\frac{dN}{dr})$, the number dN of particles within the size interval $[r, r+dr]$ is $dN = n(r) \cdot dr = (\frac{dN}{dr}) \cdot dr$. The *volume size distribution* $v(r)$ gives the volume dV of particles per m^3 in the size interval $[r, r+dr]$ by $dV = v(r) \cdot dr = (\frac{dV}{dr}) \cdot dr$. Similarly, the *mass size distribution* $m(r)=(\frac{dM}{dr})$ gives the mass dM of particles per m^3 in the size interval $[r, r+dr]$ by $dM = m(r) \cdot dr = (\frac{dM}{dr}) \cdot dr$. Mass size distributions can, e.g., be measured by impactors.

The aerosol particles span a size range of four orders of magnitude, therefore it has become a habit to give the diameter interval $[r_1, r_2]$ as logarithmic interval $d \ln r = \ln \frac{r_2}{r_1}$, with $d \ln r = \frac{1}{r} dr$. Then the size distribution for logarithmic diameter intervals is given as: $\frac{dV}{\ln r} = \frac{1}{r}\frac{dV}{dr}$

For spherical particles with density ρ the volume and mass size distributions can be obtained by $(\frac{dV}{dr}) = \frac{4\pi}{3} r^3 \cdot (\frac{dN}{dr})$ and $(\frac{dM}{dr}) = \frac{4\pi}{3} r^3 \cdot \rho \cdot (\frac{dN}{dr})$.

3. The Basic Problem of Inversions

The extinction coefficient of a suspension of N spherical particles per unit volume of radius r is given by

$$\sigma_e(\lambda) = N \cdot r^2 \pi \cdot Q_e(\frac{2r\pi}{\lambda}, m). \tag{1}$$

Q_e is called the extinction efficiency. It is the factor telling which fraction of the light incident on the geometric cross section of the particle interacts with the particle and is removed or deflected from the incident beam. It depends on the refractive index (i.e. the material the particle is made of), the size of the particle and the wavelength of light. For spherical particles it can be obtained by applying one of the many Mie [12] computer codes available in the literature (e.g. [3] [2]). Since Q_e depends on wavelength, the extinction coefficient is a function of wavelength. Normally the particles do not only have one single size, but various sizes and $n(r) \cdot dr$ gives the number of particles occurring in a size interval dr. For comparison with measurements, it is more practical to use the volume size distribution $v(r)$, i.e. the volume of particles in the size interval $[r, r + dr]$, for spherical particles yielding the number of particles per size interval as $dN = \frac{3 \cdot v(r)}{4\pi \cdot r^3} dr$. The extinction coefficient is obtained by integration of all sizes $[r_1, r_2]$ of particles, yielding

$$\sigma_e(\lambda) = \int_{r1}^{r2} \frac{3v(r) \cdot Q_e(\frac{2r\pi}{\lambda}, m)}{4r} \cdot dr. \tag{2}$$

In a similar manner, the volume scattering function for unpolarized light of a suspension of N spherical particles of radius r is given by

$$\gamma(\phi_s, \lambda) = N \frac{1}{k^2} \frac{i_1(\phi_s, \frac{2r\pi}{\lambda}, m) + i_2(\phi_s, \frac{2r\pi}{\lambda}, m)}{2}, \tag{3}$$

with k the wave number $\frac{2\pi}{\lambda}$. The volume scattering function of a suspension of particles having a volume size distribution $v(r)$ is obtained by

$$\gamma(\phi_s, \lambda) = \frac{1}{k^2} \int_{r1}^{r2} \frac{[i_1(\phi_s, \frac{2r\pi}{\lambda}, m) + i_2(\phi_s, \frac{2r\pi}{\lambda}, m)] \cdot 3v(r)}{8\pi r^3} dr. \tag{4}$$

Equations (2) and (4) give a relation between the size distribution of the particles and the optical properties. The optical properties can be measured by remote sensing, and it is desirable to obtain the size distribution of the particles by inverting the integral equation (2) or (4).

The integral equations (2) and (4) can be written in a more generalized way, with different symbols:

$$g(y) = \int_{xmin}^{xmax} K(y, x) f(x) dx. \tag{5}$$

Here $g(y)$ is the measured function (optical property), $f(x)$ is the function to be determined (size distribution function) and $K(y, x)$ is called the kernel of the integral equation. For the inversion of the spectral extinction coefficient we have

$$\begin{matrix} r \longrightarrow x \\ \lambda \longrightarrow y \end{matrix}, \quad \begin{matrix} \sigma_e(\lambda) \longrightarrow g(y) \\ v(r) \longrightarrow f(x) \end{matrix}, \quad \text{and } K(x, y) = \frac{3Q_e(\frac{2x\pi}{y}, m)}{4x}, \tag{6}$$

for the inversion of the angular scattering we obtain for a constant wavelength

$$\begin{matrix} r \longrightarrow x & \gamma(\phi_s) \longrightarrow g(y) \\ \lambda \longrightarrow y' & v(r) \longrightarrow f(x) \end{matrix}, \text{ and } K(x,y) = \frac{3 \cdot [i_1(y, \frac{2x\pi}{\lambda}, m) + i_2(y, \frac{2x\pi}{\lambda}, m)]}{8\pi x^3}. \tag{7}$$

The equation (5) is a Fredholm integral equation of first kind. It must be solved whenever a measurement is made, and depending on the kernel $K(x,y)$, the solution may be simple and unanimous or difficult and ambiguous.

Experimental measurements do not give a continuous function g(y), but values for $g(y)$ at discrete values of y are measured, we call it y_j. Furthermore the measurements are associated with a certain accuracy which we call ϵ_j. Thus equation (5) must be written as

$$g(y_j) = \int_{xmin}^{xmax} K(y_j, x) f(x) dx + \epsilon_j. \tag{8}$$

Using the measured data $g(y_i)$ we must determine solutions $f(x)$ which fulfill (8).

The inversion poses no problem, as long as the kernels for the integral equation $K(y_j, x)$, which we denote as $K_j(x)$, do not have much overlap. This is the case, e.g., when the spectral radiance $R(\lambda)$ of an object is determined by performing the light measurements R_j with a set of narrow band filters having a transmission of $T_j(\lambda)$. The current I_j of a photo sensor having the spectral sensitivity $S(\lambda)$ for filter j is

$$I_j = \int_{\lambda min}^{\lambda max} R(\lambda) \cdot S(\lambda) \cdot T_j(\lambda) \cdot d\lambda. \tag{9}$$

In this case $K_j(x) = S(\lambda) \cdot T_j(\lambda)$. A plot of $K_j(x)$ for interference filters with a bandwidth of $10nm$ and a tri-alkali photomultiplier is shown in Figure 1. There is barely any overlap.

The inversion is impossible if the kernel functions $K_j(x)$ are linearly dependent. An example for this would be the determination of the size of particles by spectral transmission at wavelengths much larger than the size of the particles. In this case, a λ^{-4} dependence of the extinction coefficient always is measured, since the particles are in the Rayleigh scattering regime. An example for the kernels is shown in Figure 1b: In the logarithmic representation, the curves for the different kernels are parallel, which means that they are proportional to each other and thus linearly dependent.

For the two procedures discussed below the kernels are in between the two extremes just considered: For the spectral extinction between 400 and $700nm$ and sizes between $r = 0.01$ and $10\ \mu m$, the kernels are shown in

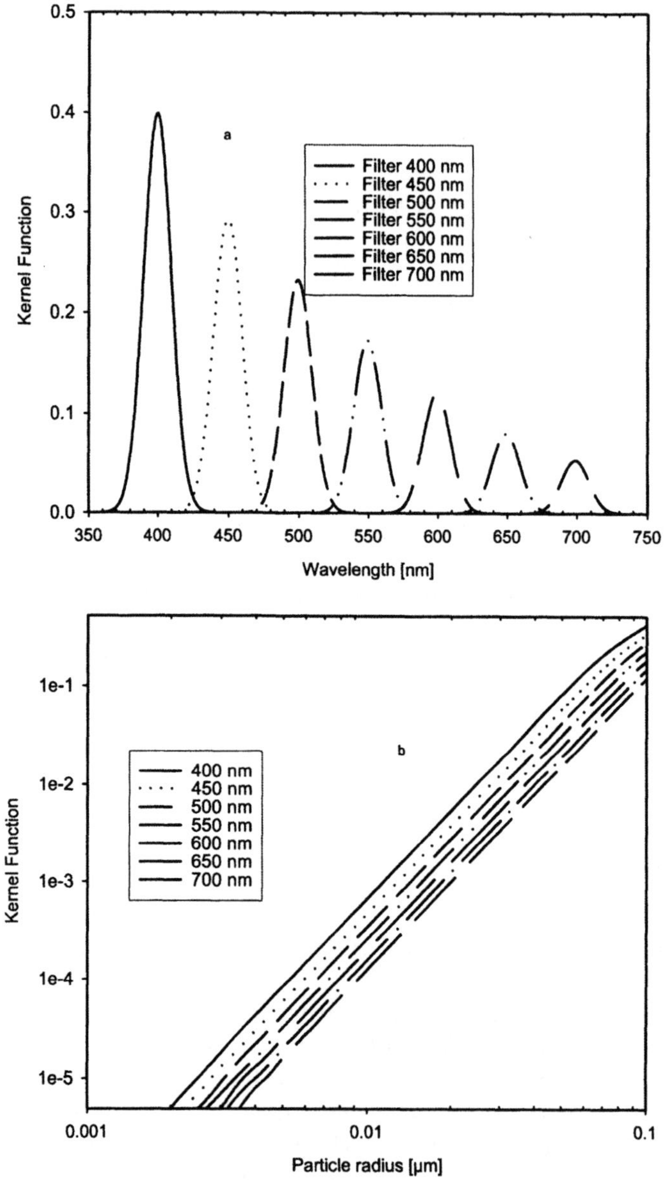

Figure 1. a: Kernels for a well posed problem: determining the spectral radiance with a set of interference filters. b: Kernels for a case where inversion is impossible: Rayleigh scattering.

Figure 2a: There is some overlap but for radii between 0.05 and 1 μm the kernel functions are different. For angular scattering measurements between $\Theta = 2^o$ and 40°, the kernels can be seen in Figure 2b: Again the kernels do overlap, but are distinctly different for particle radii between 0.05 and 10 μm. The scattering at small angles is especially important, since large particles predominantly scatter in the near forward region.

The inversion of optical data to the size distribution is an ill-posed problem. It is characterized by two facts: (1) For one set of measured optical data g_j, there exist several solution functions $f(x)$ which fulfill eq. (8) within the error margin. (2) Small changes in the measured value may lead to considerable changes in the size distribution function. Some simple considerations given below will demonstrate this easily:

It can be shown that for the kernel functions $K_j(x)$, the integral $\int_{xmin}^{xmax} K^2(x)dx$ is finite, thus they can be considered as elements of a Hilbert space L_2 (see e.g. [6]). We can define a scalar product $\langle f|g\rangle = \int_{xmin}^{xmax} f(x)g(x)dx$. If it is zero, the functions f and g are considered orthogonal, denoted by $f \perp g$.

Using this nomenclature, equation (8) can be written as

$$g_j = \langle K_j|f\rangle + \epsilon_j \tag{10}$$

with $g_j = g(y_j)$. If we have $j = 1 \cdots m$ measurements g_j, we use m different kernel functions K_j for the inversion. The kernels K_j span a subspace of L_2 by linear combination which we call $H_k = \text{span}\{K_1, \dots K_m\}$. If the kernels are linearly independent, it has the dimension m and it is possible to determine m orthogonal functions Φ_k with $\langle \Phi_k|\Phi l\rangle = \delta_{k,l}$ and all functions $s \in H_k$ can be written as a linear combination $s = \sum_{j=1}^{j=m} s_j \cdot \Phi_j$. We call s_j the Fourier coefficients. All functions $w \in L_2$ which are perpendicular to elements of H_k form the orthogonal complement $H_k^\perp$, thus if $w \in H_k^\perp$, then $w \perp \Phi_i$ and also $w \perp K_j$ for all i. Since each element of a Hilbert space can be written as the sum of one element of a subset and one element of its orthogonal complement, the solution function of the inversion can be written as $f = s + w$ with $s \in H_k$ and $w \in H_k^\perp$. This can be seen immediately by inserting in (10) yielding: $g = \langle K_j|s\rangle + \langle K_j|w\rangle + \epsilon_j$. Since $\langle K_j|w\rangle = 0$, any function contained in the orthogonal complement can be added to s and $s + w$ is a solution of the inversion problem. Thus s is the mathematically relevant solution function and w is an infinite number of functions which can be added to s, causing the ambiguity of the solution. This does not mean that the function w is unnecessary. It is that s, although the mathematically relevant solution, may be unrealistic for the specific problem (e.g. having spikes, negative values, etc.). By adding w a realistic solution function for the inversion problem can be obtained.

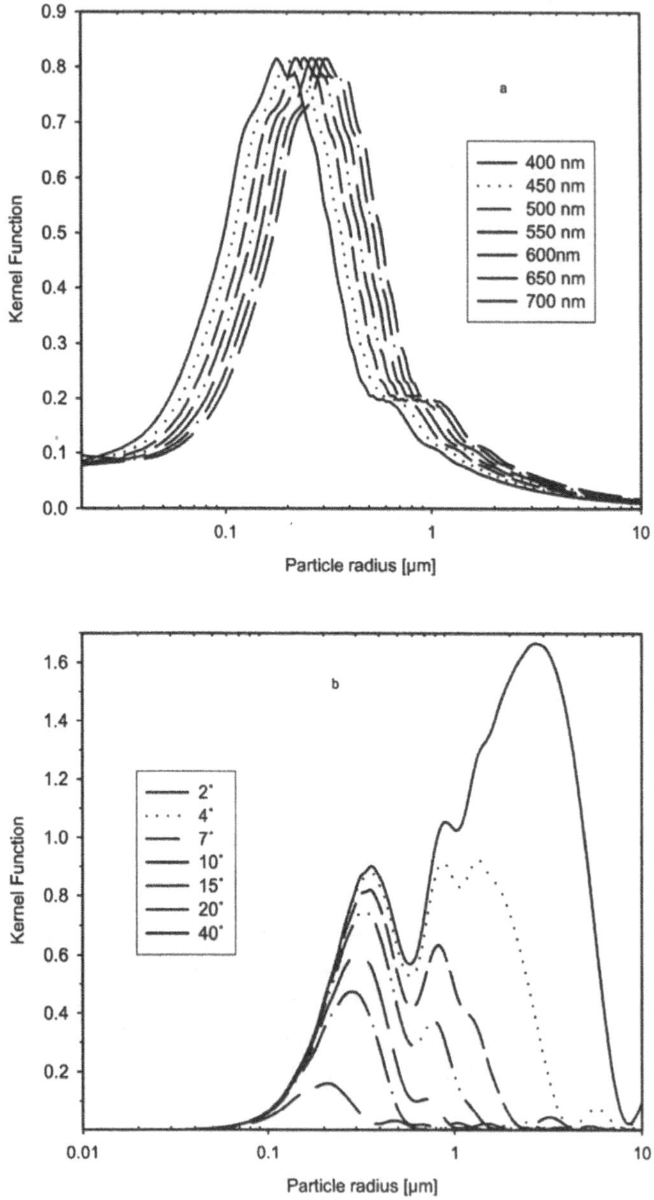

Figure 2. Kernels for inversion of spectral extinction measurements (left) and angular scattering (right).

The fact that an infinite number of functions $w \in H_k^{\perp}$ can be added to the solution function raises the question whether a realistic solution function $f(x)$ of (8) is unique and this question definitely can be answered: "no." Regarding the uncertainty ϵ_j of the measurement, this is also obvious. But it would be desirable to have a set of possible solution functions, where the difference between the solutions is not too large. This means that the contribution of the functions w is small compared to s. The ratio $\kappa = \frac{\|s\|}{\|f\|}$ with $\|f\| = \sqrt{\langle f|f\rangle}$ is a measure for the importance of the invisible contribution w and is called the information parameter [5]. If the solution of the inversion problem is unanimous then $\|w\| = 0$ and $\kappa = 1$, the smaller κ, the larger the contribution of the invisible part of the solution function.

For a given size distribution function f, the value of κ can be determined in the following way: For the kernel functions K_j the orthonormal basis Φ_i is determined. Since $\langle f|\Phi_j\rangle = \langle s{+}w|\Phi_j\rangle = \langle s|\Phi_j\rangle$ the Fourier coefficients r_i for the function s can be determined by $s_i = \langle f|\Phi_j\rangle$, and $\|s\|^2 = \langle s|s\rangle = \sum s_i^2$.

The information content obtained from inversions of optical parameters can best be seen by taking likely size distributions and calculating κ using the above method. We have taken lognormal size distributions characterized by

$$\frac{dV}{d\ln r} = \mathrm{LND}(V, r_g, \sigma_g) = V \cdot \frac{1}{\sqrt{2\pi} \cdot \ln \sigma_g} \exp[-\frac{(\ln r - \ln r_g)^2}{2\ln^2 \sigma_g}] \qquad (11)$$

with V the total volume of the particles, r_g is the geometric mean diameter and σ_g the geometric standard deviation. The geometric volume mean radius determines the location of the maximum, and the standard deviation determines the width of the size distribution curve. Very frequently the size distribution of the atmospheric aerosol is characterized by one lognormal distribution or a sum of two or three ones.

The information parameter is calculated as a function of the mean radius and results are shown in Figure 3, both for inversions of spectral extinction coefficient measurements ($\lambda = 400(25)700nm$) as well as angular scattering data ($\phi_s = 15(5)160^o$). In all cases the geometric standard deviation of the size distribution is assumed 1.8 (this is an average value for the modes of the atmospheric aerosol, see [16]. Results for other widths differ little.

One can immediately see, that the information parameter is larger than 0.8 for particle radii between 0.1 and 1 μm for the inversion of spectral extinction data. For radii < 0.1 μm, the Rayleigh scattering dominates and is $\propto \lambda^{-4}$ independent of size, thus no information can be obtained and the information parameter converges towards zero. For sizes > 1 μm, the extinction approaches the geometric cross section of the particles, which is independent of wavelength.

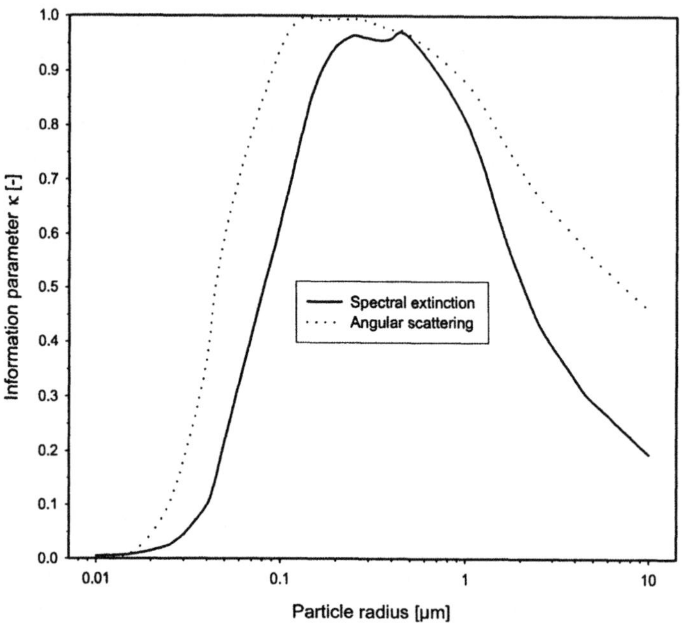

Figure 3. Information parameter for size distribution inversions as a function of the mean particles size of the lognormal size distribution. Curve a: Inversion of spectral extinction measurements in the wavelength range of $\lambda 400(25)700$ nm. Curve b: Angular scattering measurements in the angle range of $\phi_s = 15(5)160^o$.

For the inversion using angular scattering as the data set, the information parameter is similar, except that the range of higher information parameters extends to larger radii. This has its reason in the forward scattering, which strongly increases with increasing size. If still smaller scattering angles are used in the data set for inversion, as is done below, the upper limit of the size range where κ is above 0.8 can further be expanded.

The information parameter tells us which accuracy can be expected when an inversion is performed, but it does not give the solution function of the inverse problem. We already know an infinite number of functions $f = r + w$ that fulfill the integral equation, but only a few functions might be the solution of the specific inverse problem. Obviously the size distribution must be non-negative and should be smooth. The function r does not necessarily has this property. In order to find likely solution functions, a variety of mathematical techniques have been developed to obtain a size distribution function which holds up to the expectations (see e.g. [10], [17] [1] [4], [15] [9] [7] [5] [11] [13]). All methods use procedures to avoid non-negative values and spikes. Usually these methods have been developed to solve a specific problem. If the problem is different the method may not work well or might give solutions similar to the problem the method was trimmed to work for.

The inversion procedures use different methods:

- Matrix inversion: The integral equation is written as a set of linear equations and the solution is found by inverting the matrix. Since the determinant of the set of equations is close to zero, small fluctuations in the measured data can cause a very noisy solution function. Therefore constraints have to be applied to the solution (see e.g. [14])
- Iterative procedures: A first guess solution is assumed and the optical properties for the solution function are calculated. Depending on the deviation between measured and calculated data, some parts of the size distribution function are changed, and again the optical properties are calculated. Again, changes in certain ranges of the size distribution function are performed and so on. This procedure is repeated and if it converges, a solution function is obtained (see. e.g. [15] or [7] or [13])
- Analytical procedures: A large set of possible solution functions is established and their optical parameters are calculated. These functions, or a linear combination, is tested whether the optical parameters fit the measured ones. When agreement with the measurement is achieved, a solution function is found. Normally several solution functions are obtained demonstrating the ambiguity of the solution of the inverse problem.

4. Comparison with Experiments

In the following we compare with some experimental results. The spectral extinction coefficients have been obtained by the University of Vienna Telephotometer [8]. This instrument is a telescope with an attached photometer and measures the radiance of a distant target and the horizon just behind it. The contrast between the two depends on the extinction coefficient, which can be obtained by simple mathematics. The vertical optical depth of the atmosphere is given by $\delta = \int_{bottom}^{top} \sigma_e(h) \cdot dh$, with σ_e the height dependent extinction coefficient. It can be obtained by measuring the incoming direct solar radiation, which is attenuated by the atmosphere. We have used a solar photometer which automatically tracks the sun and obtains spectral solar flux density. The calibration is performed using the Langley method. The same instrument can be used to measure the radiance of the sky. Attenuation is caused by single and multiple scattered light by air molecules and particles. It is possible to separate the multiple scattered light and obtain the scattering phase function.

For the inversion of optical data to size distribution the following methods have been used:

- Analytical procedure. A large set of realistic size distributions are needed. For the atmosphere, the size distribution can well be represented by a sum of lognormal distributions [16]. Therefore we have used as a reservoir of possible size distribution functions the sum of two lognormal functions given by

$$(\frac{dV}{d\ln r}) = \mathrm{LND}(V_1, r_{g1}, \sigma_{g1}) + \mathrm{LND}(V_2, r_{g2}, \sigma_{g2}). \qquad (12)$$

  By varying $0.02 < r_{g1} < 0.8\ \mu m$, $0.2 < r_{g1} < 12\ \mu m$, $1.2 < \sigma_{g1}, \sigma_{g2} < 2.4$ and $0.05 < \frac{V_2}{V_1} < 20$ in steps between a factor of 1.05 to 1.10 a total of $28 \cdot 10^6$ model size distributions is obtained. These are tested, whether the extinction coefficient calculated with these size distributions matches the measured one within an error margin of ϵ (5%). If the test is positive, the model size distribution is considered a solution of the inversion problem. Usually about one hundred thousand to million size distributions fulfill the test. This is an impressive demonstration of the ambiguity of solutions of the inverse problem.
- Iterative procedure. For the inversion of solar and sky radiation measurements the SKYRAD package was used [13]: Several steps have to be performed in series. The direct radiation measurements give the vertical optical depth. The angular scattering function $\gamma(\phi_s)$ is initially set to the measured almucantar sky radiance and a size distribution is estimated. Iteratively the value for $\gamma(\phi_s)$ and the volume size distribution is improved until an agreement between the calculated and measured data is achieved within the error margin. The sky radiation measurements are possible for scattering angles as small as 2° thus information on the size distribution function up to several micrometers are possible.

Figure 4 shows the result of many inversions of a spectral extinction measurement in the wavelength range of 400 to 650 nm. The size distribution given by the solid line is the average of 2 million possible solution functions. The error bars are the standard deviation calculated for the 2 million different solution functions. It is immediately evident that the standard deviation is small only for radii between 0.1 and 1 μm, and this is exactly the range where the information parameter is high (see Figure 3).

The ultimate test for the suitability of an inversion is the comparison with a direct measurement of the size distribution of the aerosol particles.

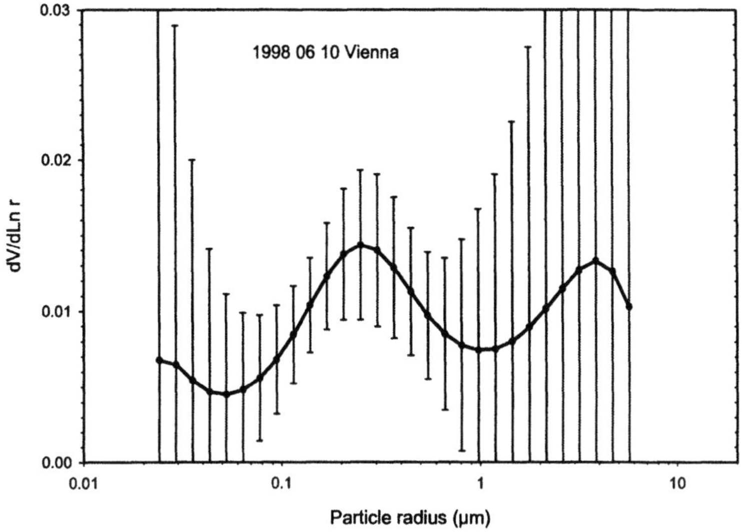

Figure 4. Result of the analytic inversion of a measured spectral extinction coefficient. The solid line is the average of all solution functions. The error bars give the standard deviation.

The size distribution has been measured with a cascade impactor. This instrument passes the aerosol through a series of narrow orifices with high velocity, after which the air is deflected sharply. Particles which have little inertia can follow the gas stream; whereas, particles with high inertia continue almost straight and are deposited on an impaction plate. By increasing the velocity of the gas stream in subsequent nozzles smaller and smaller particles can be deposited. The mass of the particles is weighed and from this, the mass size distribution can be determined. It is converted to a volume size distribution by dividing by the density of the particles. Results of one intercomparison experiment are shown in Figure 5. Also in Figure 5 the size distribution obtained by the inversion of the solar and sky radiation measurement is shown. The sky radiation consists of the light scattered by all particles that are contained in the atmosphere, i.e. the whole vertical column. Therefore the size distribution is given for the particles contained in a column having one square meter base and extending to the top of the atmosphere. The unit of the columnar size distribution thus is volume of particles per square meter, thus m^{-1}. Figure 3 also shows that the size distribution curves are similar within the range where the information parameter is high. This proves that the inversion of optical data is a useful

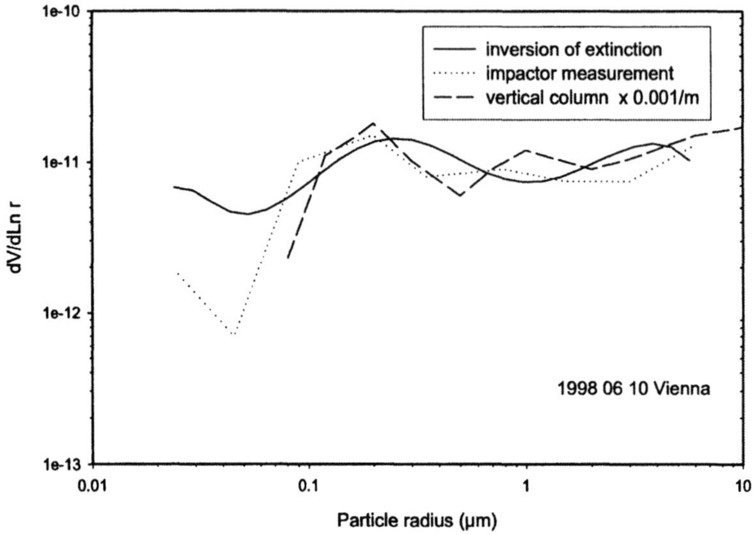

Figure 5. Size distribution of the aerosol in Vienna determined by inversion of horizontal spectral extinction measurements, solar and sky radiation measurement and direct determination of the size distribution by a cascade impactor.

tool within the limits of the inversion, which can be determined by the information parameter. The inversion of the spectral extinction only is good between 0.1 and 1 μm radius since here the information parameter is high. The angular scattering measurements give a good agreement between the measured and the inverted size distribution up to several micrometers, since the forward scattering at a few degrees is dominated by super-micrometer particles.

For illustration the inversion of a few solar and sky radiation measurements is shown in Figure 6. The reliable range extends most likely to 3 to 5 μm radius. The structure of the aerosol consisting of two or more modes is evident. This is a normal feature of the atmospheric aerosol. Whether the sharp minimum at 0.8 μm is realistic needs to be questioned.

5. Conclusions

The size distribution of an aerosol can be obtained by inverting the measured spectral extinction or the angular scattering. By knowing the range of data available (wavelength, scattering angle), the range of reliable data obtained by the inversion can be calculated. It is best characterized by the

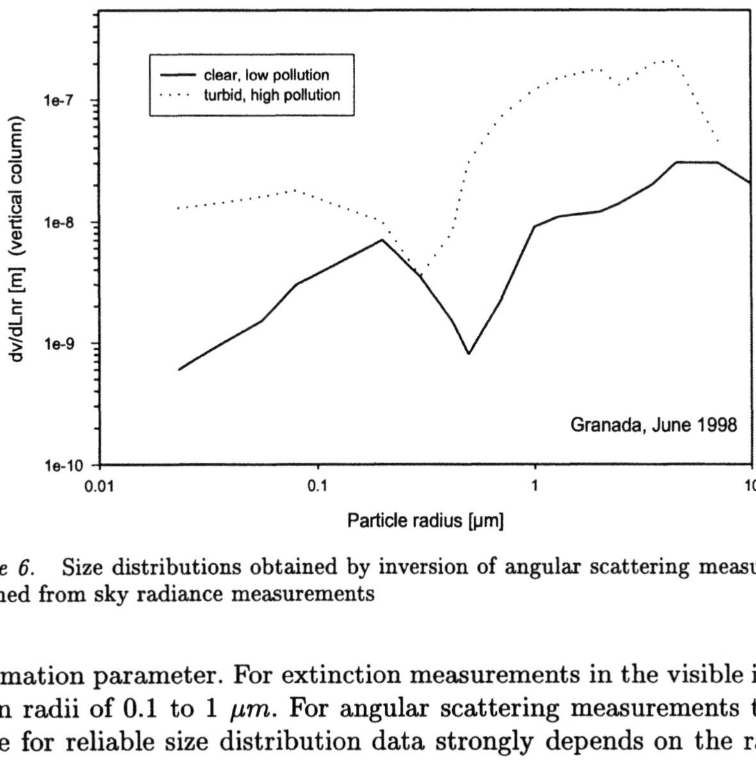

Figure 6. Size distributions obtained by inversion of angular scattering measurements obtained from sky radiance measurements

information parameter. For extinction measurements in the visible it is between radii of 0.1 to 1 μm. For angular scattering measurements the size range for reliable size distribution data strongly depends on the range of scattering angles covered. For angles between 15 and 160° size distributions between .05 and 1.5 micrometer radius can be recovered. If smaller scattering angles are included, the reliable size range can be extended up to 5 μm. The inverted size distribution agree with size distribution measured by an independent method.

6. Acknowledgments

This work was supported in part by the Fonds zur Förderung der wissenschaftlichen Forschung in Österrereich, grant number 12386 and by the Acciones Integradas Spain - Austria, Project 8/98 and la Dirección General da Ciencia y Tecnologia from the Education and Research Spanish

Ministry, project number CLI 98-0957 and REN2000-0903-CO8-CLI. We thank professor T. Nakajima for the software code and his help.

References

1. G.E. Backus, and J.F. Gilbert, Uniqueness of the inversion of gross earth data, *Philos. Trans. R. Soc. London* A bf 266, 123-192, (1970).
2. P. W. Barber, S. C. Hill, *Light scattering by particles: Computational methods.* World Scientific, Singapore (1990).
3. C. F. Bohren, and D. R. Huffman, *Absorption and scattering of light by small particles.* Wiley Interscience, New York (1983).
4. M. A. Box, and S. Lo, Approximate determination of aerosol size distributions. *J. Appl. Meteor.* **15**, 1068-1976, (1976)
5. C. D. Capps, R. L. Henning, and G. M. Hess, Analytic inversion of remote sensing data. *Appl. Opt.* **34**, 3581-87, (1982).
6. C. Dellago, and H. Horvath, On the accuracy of size distribution information obtained from light extinction and scattering measurements - I. basic considerations and models. *J. Aerosol Sci.* **24**, 129-141, (1993).
7. J. Heintzenberg, H. Müller, H. Quenzel, and E. Thomalla, Information content of optical data with respect to aerosol properties: numerical studies with a randomized minimization-search-technique inversion algorithm. *Appl. Optics* **20**, 1308-1315, (1981).
8. H. Horvath, The University of Vienna Telephotometer, *Atmospheric Environment* **15**, 2537-2546, (1981).
9. M. D. King, D. M. Byrne, B. M. Herman, J. A. Reagan, Aerosol size distributions obtained by inversion of spectral optical depth measurements. *J. Atmos. Sci.* 2153-2167, (1978).
10. L. Landweber, An iterative formula for Fredholm integral eqiations of the first kind *Am. J. Math.* **73**, 615-624, (1951).
11. J. M. Livingstone, and P. B. Russel, Retrieval of aerosol size distribution moments from multiwavlenegth particulate extinction measurements. *J. Geophysical Research* **94**, 8425-8433, (1989).
12. G. Mie, Beiträge zur Optik trüber Medien, speziell kolloidaler Goldlösungen. *Ann. Phys.* **25**, 377-455, (1908).
13. T. Nakajima, G. Tonna, R. Rao, P. Boi, Y. Kaufman, and B. Holben, Use of sky brightness measurements from ground for remote sensing of particulate dispersion. *Appl. Opt.* **35**, 2672-2686. (1996).
14. G. E. Shaw, Inversion of optical scattering and spectral extinction measurements to recover aerosol size spectra. *Applied Optics* **18**, 988-993, (1979).
15. Twomey S, *Introduction to the mathematics of inversions in remote sensing and indirect measurements.* Elsevier, New York. (1977)
16. K. T. Whitby, The physical characteristics of Sulfur aerosol. *Atmospheric Environment* **12**, 135-159, (1978).
17. Yamamoto G., and M. Tanaka, Determination of the aerosol size distribution from spectral attenuation measurements. *Appl. Optics* bf 8 447 - 453, (1969)

SIMPLIFIED SOLUTION OF THE INVERSE PROBLEM FOR INSTANTANEOUS COMETARY DUST SIZE DISTRIBUTION

M. KOCIFAJ,[1,2] J. KLAČKA,[3] F. KUNDRACÍK,[4] and G. VIDEEN[5]
[1]*Astronomical Institute, Slovak Academy of Science, Dúbravská 9, 842 28 Bratislava, Slovak Republic*
[2]*SWH a Siemens Company, Dúbravská 9, 842 37 Bratislava, Slovak Republic*
[3]*Astronomical Institute, Faculty of Mathematics, Physics and Informatics, Comenius University, Mlynská dolina, 842 48 Bratislava, Slovak Republic*
[4]*Department of Radiophysics, Faculty of Mathematics, Physics and Informatics, Comenius University, Mlynská dolina, 842 48 Bratislava, Slovak Republic*
[5]*Center for Atmospheric Sciences, Hampton University, Hampton, USA*

Abstract. Available optical measurements indicate that the modal radius r_m of a cometary dust population is in the submicron range and that the dust refractive index m changes slightly in the visible region of the spectrum. A realistic instantaneous particle size distribution $f(r)$ may be determined by processing the measured intensity of continuum at several wavelengths. The solution of the inverse problem for particle size distribution is based on rigorous Mie theory. Additionally, an application of the Rayleigh-Gans approximation enables to construct an accelerated solution scheme since the total intensity of the scattered radiation can then be expressed in an analytical form. However, the range of validity of the approximation is strongly limited to very small submicron-sized particles. The numerical simulations of light scattering by Mie cometary dust particles are performed for two model size distributions – power function $f(r) \propto r^{-n}$ and modified gamma function $f(r) \propto r^a e^{-br}$, which are commonly used to represent real dust populations. It is shown that the cometary dust size distribution may easily be reproduced analysing the spectral behaviour of measured intensity of the scattered radiation $I_\vartheta(\lambda)$. The more rapid increasing of continuum with the wavelength of incident radiation the larger particles are contained in cometary dusty environment.

G. Videen and M. Kocifaj (eds.), Optics of Cosmic Dust, 159–170.

1. Introduction

Information on physical properties of cometary dust, especially on its size distribution, is quite important when considering the physics of comets and their contribution to the Zodiacal Cloud [1]. It is known that the dynamics of such particles is sensitive to particle size and shape [2]. Some particles are definitely lost in space, but certain portions of the population may survive a relatively long time in the Solar System, and may, in such a case, represent relicts from the time of the formation of the Solar System [3]. On the other hand, a presence of various dust populations in coma influences also the analysis of optical measurements of the comets. All these facts result in the necessity to gather information on particle size distributions. The retrieval techniques for size distribution of the cometary dust are usually based on measurements of infrared (IR) thermal emission [4,5,6]. IR observations are efficient mainly for large micron-sized particles. However, the presence of submicron particles was deduced when analysing polarized light of the continuum in visible. Particles of various sizes, shapes and optical properties are intermixed in the dusty coma, so e.g. the large grains or flakes are situated in a mist of small particles [7]. Polarization of the scattered radiation is an important physical quantity that can offer information on the nature and size of these particles [8]. However, applicability of polarization measurements is restricted as they need to be taken for a few scattering angles. Generally, the larger the angular interval, the more valuable the information content. Such data must be collected during a relatively large time period, generally at different comet-Sun distances. As the cometary activity may significantly change with time, the inconsistencies in the data become apparent.

2. Visible continuum and particle size distribution

Experimental optical data obtained for several comets indicates the presence of small submicron particles in the dusty coma. For instance, the spectral energy distribution in the coma of comet Hale-Bopp was dominated by scattering and thermal emission from unusually small submicron sized dust grains [9]. The results show that the optically important grains dominating the visual scattering and near-IR emission from coma had an average radius of $r \leq 0.4$ μm. Such small particles may be successfully characterized by measuring the continuum in visible.

The measurement of absorption and scattering of the solar radiation in coma depends significantly on dust to gas ratio. Absorption of shortwave radiation in the gas environment is caused mainly by electronic transition of molecules. However, the scattered light (in visible) from comets is a superposition from both gas and dust. The strength of the continuum varies from comet to comet and with heliocentric distance. It is not easy to identify the contribution of dust to the scattered light, since the cometary spectrum in the visible region is dominated by emission features from the molecules [8]. Although the emission is a selective process, the composition of gas in coma environment is a variable attribute. Therefore it is difficult to choose adequate wavelengths that will filter out the emissions from the continuum. The IAU commission has defined some filter sets for observation of several comets (e.g. Halley, Hale-Bopp, or Wirtanen), including continuum windows [10].

Measurements of the scattered light in the visible spectral region appear to be a good source of information on cometary dust. However, a dust tail analysis, e.g. a realistic instantaneous size distribution, is an ill-posed problem which fails to fulfill at least the existence of a solution, the uniqueness of the solution, and continuity of the solution on the data function. The solution is affected by intrinsic errors that may be quite large. In addition, the determination of the particle size distribution in coma is also a problem, since appropriate physical theories must take into account particle morphology and composition. The particle asphericity accounts for one of the major complexities when solving the inverse problem of light scattering. Although the scattering properties of the individual non-spherical particles vary significantly with the particle orientation, the volume scattering characteristics of the cometary atmosphere (consisting of polydisperse populations of the particles) can be assumed to be stable. The polydisperse ensembles of small spheroids exhibit essentially the same scattering behaviour as their equivalent-volume spheres [11]. We can therefore apply Mie theory to solve the inverse problem, which significantly simplifies the mathematical formulae utilized in the inversion procedures. The dust tail environment is usually an optically thin environment so we neglect the effect of multiple scattering.

In order to estimate the changes in particle size distribution the model functions are often considered, such as a power function

$$f(r) = Cr^{-n}, \tag{1}$$

where n varies from 3 to 4 for most of comets [10, 18]. However this distribution function overestimates significantly the number of small particles in the dust population. A better choice to fit measured continuum profile is the modified gamma

$$f(r) = Cr^{a}\, e^{-br}. \tag{2}$$

The reddening of the continuum intensity (with respect to that of the Sun) is a general trend observed in cometary dust. As it will be shown, such behaviour is caused by large submicron particles. Many observations were made at small Sun-comet distances, from some tenths of AU to a few AU, at which predominantly large submicron- and micron-sized particles are released from the comets. The production probability of large particles may decrease with increased Sun-comet distance. When smaller particles dominate in the dusty coma, the scattering efficiency of blue light is more efficient. Because of the change in size distribution it is likely that the spectral behaviour of continuum intensity will occur in time.

3. Solution of inverse problem for cometary dust

The inverse problem of extracting the particle size distribution using multispectral measurements of the scattered light requires a solution of the Fredholm integral equation of the first kind

$$I(\lambda)=I_0(\lambda)\left(\frac{\lambda^2}{2\pi}\right)^2\int_{r_1}^{r_2}K(r,\lambda)f(r)dr, \tag{3}$$

where I(λ) is the measured light intensity at wavelength λ, $I_0(\lambda)$ is the intensity of the incident radiation coming from Sun, *K(r, λ)* is dimensionless kernel of the integral equation and *f(r)* is the particle size distribution function. It must be pointed out that an inadequate definition of the particle size distribution function *f(r)* is often a source of inadequate results. One can see that the dimensions of *f(r)* used in equation (3) are [μm^{-1} m^{-2}]. We should not be concerned by the standard volume-dependent size distribution dimensions [μm^{-1} m^{-3}], since the volume changes of *f(r)* in the coma are omitted. The resulting *f(r)* characterizes the columnar size distribution - a dust column density along the trajectory between the observer and source of the measured radiation. This must a be taken into account when comparing the calculated results with those published in the literature [9, 24-27]. Equation (3) represents a mapping in Hilbert space from $L^2[\lambda_1, \lambda_2]$ to $L^2[r_1, r_2]$ constructed over the set of quadratically integrable and continuous functions. Accepting that the measured radiation data cover the visible spectrum $[\lambda_1, \lambda_2]$, the size distribution can successfully be retrieved in submicron range $[r_1, r_2]$. Several methods were developed to solve Fredholm integral equations of this type (e.g. the singular-function theory or eigenfunction theory [12, 13]), although no general rules may be formulated. The significant simplification of the solution of equation (3) is based on the Mellin transform [14] of the kernel. This method is applicable when the kernel of the integral equation belongs to the class of product-type kernels $K(r,\lambda) = K(r/\lambda)$ [15]. The solution of the integral equation (3) may then be in a form

$$f(r)=\int_{\lambda_1}^{\lambda_2}H\left(\frac{r}{\lambda}\right)f(r)d\lambda, \tag{4}$$

when using a two-dimensional Laplace transform. The Mellin transform H of the kernel K is expressed as $H(x)=2K(x)/\pi$ for any argument x. An analytical solution of the inverse problem can be found for small optically soft particles (see Eqs. 8-10), for which the following condition is to be fulfilled

$$\frac{2\pi r}{\lambda}|m-1|<1\,, \tag{5}$$

where m is the particle refractive index. Condition (5) situates particle's size into submicron range if the scattered radiation is measured in visible. It is known that size of particles in the dust tail is about 1 μm [8], and submicron particles were dominated in dusty coma [9, 16, 28] in many cases – e.g., measurements of the dusty coma of comet P/Halley have shown a drastic reduction of the large particles, and a mode position of the particle size distribution was detected near to 0.1 μm [17].

4. Mie theory and approximations

Mie theory is a powerful tool to find an approximate solution to the inverse problem for dust size distribution as it is applicable to particles of any size. A disadvantage is that the particles are assumed to be spherical and homogeneous. Naturally, this fact has a direct impact on the accuracy of the obtained results and lack of the information content. However, there are another sources of significant errors, which are related to intrinsic instability of the solution of integral equation of the first kind (3). Dellago and Horvath [23] have shown that the determination of a mean diameter of monomodal size is possible within ± 35% (using the extinction measurements between 400 and 700 nm), if particle sizes are between 0.4 and 1 μm (which is typical for cometary dust tails). The particle size determination depends also on the refractive index, which may modify the profile of the kernel of the integral equation. We assume here that particle refractive index m changes slightly with wavelength in visible spectrum, so $m(\lambda)=\tilde{m}$. Although this assumption seems to be a weak point of the presented solution, the analysis of the dusty coma of comet Halley has confirmed such behaviour of particle refractive index in visible [19].

Let us assume a particle size distribution in the form of the modified gamma function (Eq. 2) with variable particle modal radius r_m. A modal radius r_m of the size distribution is defined as follows

$$\left[\frac{d}{dr}\left(r^N e^{-br}\right)\right]_{r_m}=0 \tag{6}$$

i.e.

$$r_m = \frac{N}{b} \quad (7)$$

Such a function may fit distributions that are not too narrow, as is typical for dusty environment of cometary coma.

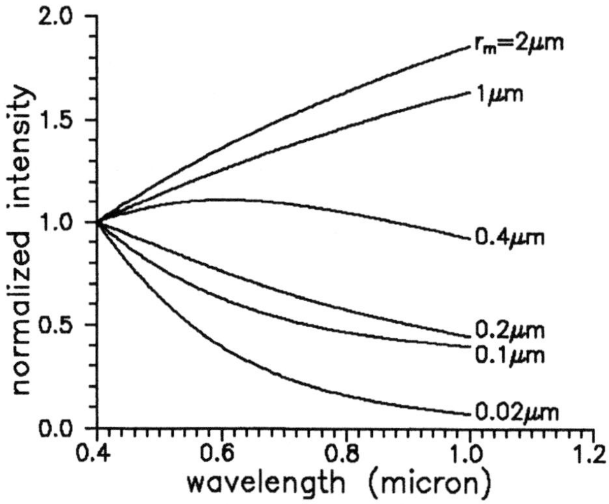

Fig 1: Intensity of the scattered radiation in dusty coma (consisting of dirty ice particles) normalized to the energy distribution of the solar spectrum. Model particle size distribution corresponds to modified gamma function (Eq. 2) with parameters a=2, b=a/r_m [μm^{-1}]. The scattering angle ϑ is 135°.

The presence of small or large particles in dusty coma can be detected immediately when analyzing the spectral behaviour of the intensity of the scattered light. The advantage of this method is that we do not need the absolute values of the intensity of the scattered radiation, and there is no necessity to observe the comet for some days to obtain the intensity of the scattered light at different scattering angles - as it is expected for polarimetric measurements. The computational results for the spectral behaviour of intensity of the scattered radiation obtained using Mie theory are drawn in Fig. 1. In accordance with an example of the visible continuum observation of comet Bennet (Fig. 4) we have calculated the normalized scattered intensity at scattering angle ϑ=135°. The computed intensity is normalized to the energy distribution of the solar spectrum and all values are compared with the intensity at reference wavelength 400 nm.

The presence of submicron particles may be identified by the monotonous decrease of the continuum in visible. The slope of the spectral dependence of normalized intensity is directly referred to the value of modal radius of dust size distribution, as it is clearly shown in Fig. 1. Particles of different size contribute to the scattered radiation in different ways (Fig. 2a-2d).

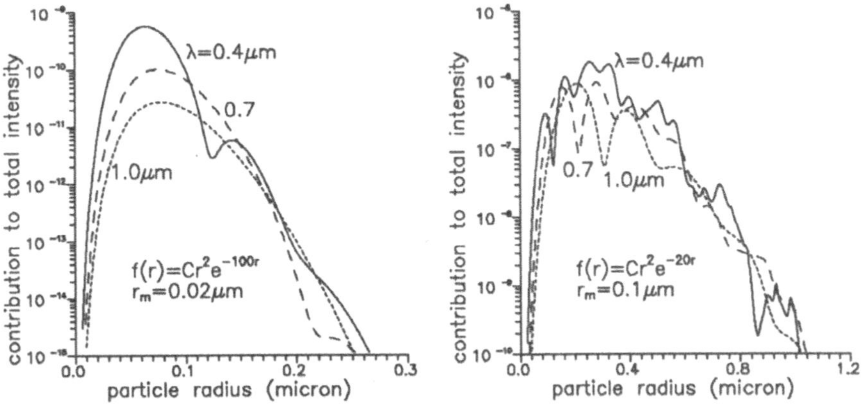

Fig 2a,b: Contribution of different-size particles to the total intensity of the scattered light (at scattering angle 135°; icy particles). Particle size distribution is simulated by modified gamma function (a) with modal radius $r_m = 0.02$ μm and (b) with modal radius $r_m = 0.1$μm.

In general, particles of size close to the modal radius r_m have a dominant contribution in the total intensity of the scattered light at all monitored wavelengths (i.e. λ=400nm, λ=700nm, and λ=1000nm). Although the presence of large micron-sized particles may be identified when analysing the spectral behaviour of the scattered light (Fig.1), it is difficult to estimate the modal radius – because there is no way to distinguish the effect of particles of different size at different wavelengths (Fig. 2d). The differences at the given wavelengths are lost for $r_m > 0.5$ μm (Fig. 2c). Size distributions with the modal radii comparable to the wavelength of the observation can more accurately be determined (Fig. 2b) when measuring the intensity of the scattered light at several wavelengths. The shape of the curves for larger particles presented in Figs. 2a-d more closely resemble those of the particles in the dusty coma.

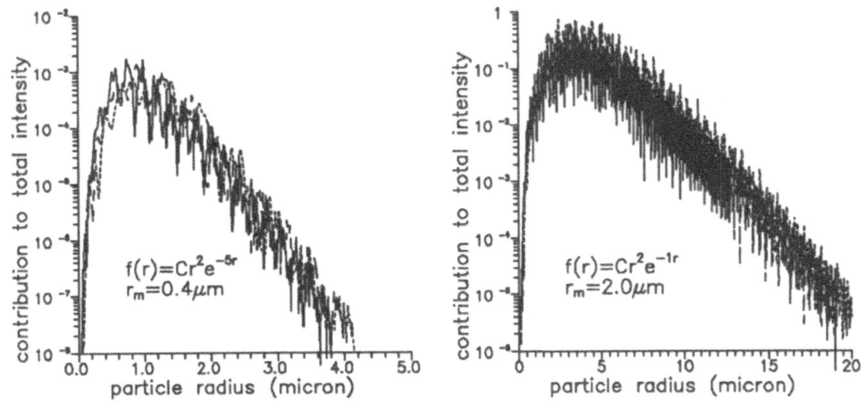

Fig 2c,d: Contribution of the particles of different size to the total intensity of the scattered light (at scattering angle 135°; icy particles). Particle size distribution is simulated by modified gamma function (c) with modal radius $r_m = 0.4$ μm and (d) with modal radius $r_m = 2.0$μm.

The results in Fig.1 were obtained by numerical integration using Mie theory. Significant acceleration of the solution of Eq. 3 may be reached when the normalised intensity decreases rapidly with wavelength; i.e., very small submicron particles are present in the coma and condition (5) is fulfilled. Using this assumption we are able to express the integral (Eq. 3) in analytical form

$$I_\vartheta(\lambda) = C\, I_0\, F(\vartheta) \left(\frac{8\pi}{\lambda} \sin \frac{\vartheta}{2} \right)^{-5} g_2(z), \tag{8}$$

applying Rayleigh-Gans (R-G) approximation for modified gamma distribution (Eq. 2) with parameter a=2. The quantity of C is a normalization constant and function F depends only on the scattering angle ϑ. The value of z depends on parameter b of the size distribution, wavelength, and scattering angle as follows

$$z = \frac{b\lambda}{8\pi} \frac{1}{\sin\left(\frac{\vartheta}{2}\right)} \tag{9}$$

and the function $g_2(z)$ has the following form

$$g_2(z) = \frac{4}{z^3} - 4z\frac{z^2-3}{\left(z^2+1\right)^3} + \frac{12}{z^5} + 12z\frac{z^4-10z^2+5}{\left(z^2+1\right)^5} - 48z\frac{z^2-1}{\left(z^2+1\right)^4}. \tag{10}$$

Values of the function $g_2(z)$ decrease rapidly with increasing value of the parameter z. The derived analytical expression for total intensity of the scattered radiation is quite powerful when analysing the behaviour and changes of the size distribution function for the smallest fraction of cometary dust population. However, the range of validity is restricted as it is documented in Fig. 3. The comparison of the values of phase function S_{11} produced by Mie theory and R-G results in the conclusion that R-G theory can bring good results if the dimensionless size parameter $x = 2\pi r/\lambda$ is less than 1 and particle refractive index is smaller than 2. The theory may be accepted up to $x < 1.5$ in the case the refractive index is less than 1.3 (dirty water ice particles have $\widetilde{m} \approx 1.38$ in the whole visible). On the other hand a presence of dirty ice grains in the dust tail is questionable as these particles are extremely short-lived in comets inside 2 AU from the Sun [29]. Lumme *et al.* [30] performed calculations for slightly absorbing particles (such as ice or silicate with average refractive indices 1.29 + 0.008i and 1.57 + 0.012i) and reconstructed the negative polarization realistically.

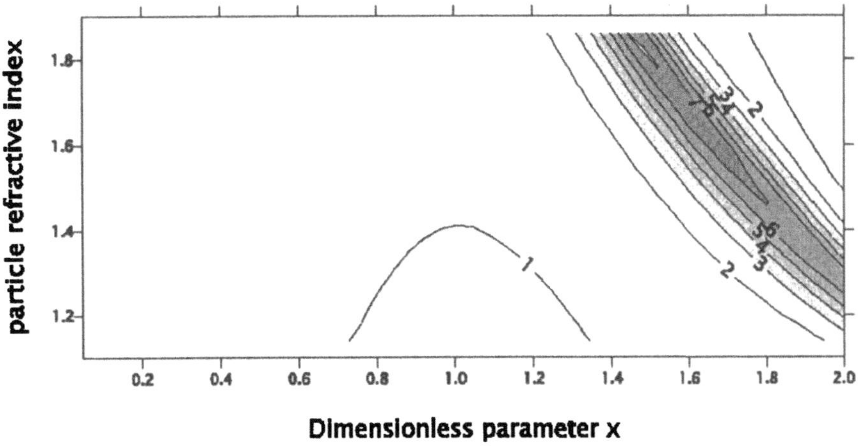

Fig 3: The ratio of the value of phase function S_{11} calculated using Rayleigh-Gans theory to the value of S_{11} obtained by Mie theory for different particle refractive index and size-parameter $x=2\pi r/\lambda$ (scattering angle $\vartheta=135°$).

5. Exemplar application

Comets are usually far from the Sun at small comet phase angles α. Moreover, the heliocentric distance changes only slightly when the comet passes through opposition ($\alpha = 0$). Storrs *et al.* [20] and A'Hearn *et al.* [21] noticed that the dust/gas ratio increases with increasing heliocentric distance. The results of limited observational data on various short-period and long-period comets appear to indicate that the dust/gas ratio does not seem to be related either to the molecular production rates or to the dynamical age of comets. The dust/gas ratio changes from comet to comet. Nevertheless, the measurements of the scattered short-wave radiation by cometary dust at large comet-Sun distances ($\alpha > 90°$) may give enough information on submicron particles. Radiation flux ratio comet/Sun usually increases with wavelength in visible (Fig. 4). The systematic computations using Mie theory have shown that such behaviour cannot be explained if the real particle size distribution is approximated by a power function (Eq. 1). The final scattered light intensity $I_\vartheta(\lambda)$ is namely for such a particle size distribution nearly proportional to λ^{3-n}. An application of the power function cannot produce the observed reddening of the visible continuum (Fig. 4). This is the reason why some models use a summation of power-law distributions with different parameters n. The modified gamma function $f_N(r) \sim r^N e^{-br}$ provides a better fit to the measured continuum profile (Figs. 1 and 4).

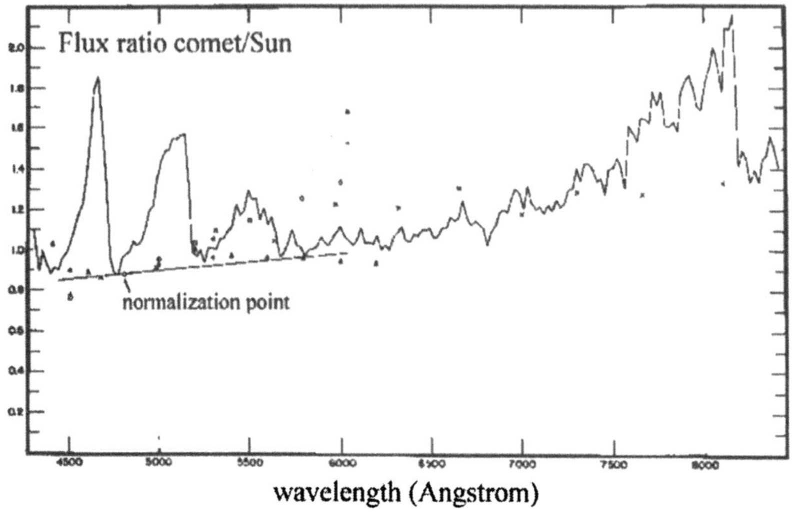

Fig 4: The scattered radiation (scattering angle 137°) as a function of wavelength λ for comet Bennet [22].

Particle modal radius $r_m \sim 1$ μm may be deduced for comet Bennet when comparing the computational simulation of intensity of the scattered radiation by using Mie theory (Fig. 1) with the results of measurement (Fig. 4). Utilization of R-G approximation will result in important underestimation of modal radius of dust size distribution when analysing the continuum such as it is presented in Fig. 4.

6. Conclusion

The simple method for the determination of the instantaneous particle size distribution of cometary dust is presented. The solution of the ill-posed inverse problem for the cometary dust size distribution $f(r)$ is based on processing the measured intensity of the scattered radiation $I_\vartheta(\lambda)$ in visible at several wavelengths λ, and at a fixed scattering angle $\vartheta = \pi-\alpha$, where α is a phase angle of the comet. The angle α defines the position of the comet relatively to the Sun and Earth. The resulting function $f(r)$ is retrieved using the actual data; i.e. it is not necessary to measure the cometary continuum at several phase angles α. Relative (normalised) values of the intensity $I_\vartheta(\lambda)$ construct the input data function. The computed intensity is normalised to the energy distribution of the solar spectrum and all values are compared with the intensity at reference wavelength 400 nm. The reddening of the continuum intensity with respect to that of the Sun is a general trend observed for dusty comets. We have shown that such behaviour is caused by large submicron particles. The slope of the spectral dependence of the normalized intensity can be directly referred to the value of modal radius of the dust size distribution. Although the simple power function is usually used to substitute the realistic size distribution, it is shown that such a substitution cannot explain the observed reddening. Moreover, this function causes problems due to badly defined boundary conditions. On the other hand, the modified gamma distribution appears to be a good approximation for the dust distribution. Such a function may fit successfully the distributions, which are not too narrow (as it is typical for dusty environment of cometary coma). The model gamma function is successfully applied to solve the inverse problem for the dust distribution using the available data for continuum of comet Bennet. It is shown that the slope of the normalised continuum can be obtained when a particle modal radius is close to 1 μm. This result is in good agreement with the information on average particle size in the cometary dust tail. Analytical solution of the inverse problem can be found only for distributions with modal radius less than 0.05 μm. However, the existence of an analytical formulae makes it possible to render the nature (and fundamental character) of changes of $I_9(\lambda)$. Using Mie theory, we are able to construct the graphical representations (such as Fig. 1) for each comet phase angle, which may easily be utilized for in-time determination of the modal radius of particle size distribution.

7. Acknowledgments

The paper was supported by the Sci. Grant Agency VEGA (grants Nos. 1/7067/20, 2/7151/20, and 1/4253/97), by NATO (PST.ARW.977678) and by ARL (N68171-01-M-5712).

References

1. G. J. Flynn, Planet. Space Sci. **42,** 1151 (1994).
2. J. Klačka and M. Kocifaj, J. Quant. Spectrosc. Radiat. Transfer **70**, 595 (2001).
3. D. Jewitt and K. J. Meech , Astrophys. J. **310**, 937 (1986).
4. M. S. Hanner, "In: ESA The Comet Halley Dust and Gas Environment" (N82-23095 13-91, 1981, pp. 67-75).
5. H. Campins and M. S. Hanner, "In: Comets" (A83-13376 03-90, Tucson, AZ, University of Arizona Press, 1982, pp. 341-356).
6. M. S. Hanner (Ed.), "Infrared Observations of Comets Halley and Wilson and Properties of the Grains" (NASA Conference Publication 3004, Washington, D.C., 1988).
7. A. Dollfus, Astron. Astrophys. **213**, 469 (1989).
8. K. S. K. Swamy, "Physics of Comets" (World Scientific Publishing Co Pte Ltd, Singapore, 1986).
9. D. M. Williams, C. G. Mason, R. D. Gehrz, T. J. Jones, C. E. Woodward, D. E. Harker, M. S. Hanner, D. H. Wooden, F. C. Witteborn, H. M. Butner, Astrophys. J. **489**, L91 (1997).
10. K. Jockers, Earth, Moon and Planets **79**, 221 (1997).
11. N. A. Krotkov, D. E. Flittner, A. J. Krueger, A. Kostinski, C. Riley, J. Quant. Spectrosc. Radiat. Transfer **63**, 613 (1999).
12. G. P. Box, K. M. Sealey, M. A. Box, J Atmosph. Sci. **49**, 2074 (1992).

13. G. V. Box, M. A. Box, Appl. Optics **24**, 4525 (1985).
14. K. S. Shifrin, "In: Theoretical problems of light scattering and their applications" (Nauka I Technika, Minsk, 1971; in Russian).
15. W. Schmeidler, "Integralgleichungen mit Anwendungen in Physik und Technik" (AV Geest & Portig, K.-G., Leipzig, 1955).
16. M. Hanner, R. Newburn, Astron. J. **97**, 254 (1989).
17. N. Thomas, H. U. Keller, Astron. Astrophys. **249**, 258 (1991).
18. L. Kolokolova, K. Jockers, Planet. Space Sci. **45**, 1543 (1997).
19. A. K. Sen, M. R. Deshpande, U. C. Joshi, N. K. Rao, A. V. Raveendran, Astron. Astrophys. **242**, 496 (1991).
20. A. D. Storrs, A. L. Cochran, E. S. Barker, Icarus **98**, 163 (1992).
21. M. F. A'Hearn, R. L. Millis, D. G. Schleicher, D. J. Osip, P. V. Birch, Icarus **118**, 223, (1995).
22. G. M. Stokes, Astrophys. J. **177**, 829 (1972).
23. C. Dellago, H. Horvath, J. Aerosol Sci. **24**, 129 (1993).
24. M. Fulle, C. Barbieri, G. Cremonese, Astron. Astrophys. **201**, 362 (1988).
25. D. J. Lien, Astrophys. J. **355**, 680 (1990).
26. R. L. Newburn, H. Spinrad, Astron. J. **90**, 2591 (1985).
27. N. Eaton, J. C. Zarnecki, Mon. Not. R. astr. Soc. **217**, 659 (1985).
28. L. M. Shul'man, "The cometary nuclei " (Nauka, Moscow, 1987, in Russian).
29. M. S. Hanner, Icarus **47**, 342 (1981).
30. K. Lumme, J. Rahola, J. W. Hovenier, Icarus **126**, 455 (1997).

MONTE CARLO MODELING OF COMETARY ATMOSPHERES INCLUDING POLARIZATION

F. MORENO[1], O. MUÑOZ,[1], AND A. MOLINA[1,2]
[1]*Instituto de Astrofísica de Andalucía, CSIC, PO Box 3004, 18008 Granada, Spain*
[2]*Departamento de Física Aplicada, Universidad de Granada, Spain*

Abstract. We present a Monte Carlo model for the transfer of radiation by dust in spherical-shell cometary atmospheres. Our model, which includes the effects of polarization, constitutes an improved version of that by Salo [32]. The code has been designed to compute both the input radiation on the nucleus surface and the output radiation. This has specific applications regarding the interpretation of near-nucleus photometry, polarimetry, and imaging such as planned for the near future space probes. After showing the appropriate validations of the code with other modeling results, we address here the effect of including the full 4×4 scattering matrix into the calculations of the radiative flux impinging on cometary nuclei. As input of the code we use realistic single scattering phase matrices derived by fitting the observed behavior of the linear polarization versus phase angle in cometary atmospheres. The observed single scattering linear polarization phase curves of comets are found to be fairly well represented by a mixture of Mg-rich olivine particles and small carbonaceous particles. The input matrix of the code is thus given by the phase matrix for olivine as obtained in the laboratory [28] plus a variable scattering fraction phase matrix for absorbing carbonaceous particles. Particularly, these fractions are found to be 3.5% for comet Halley and 6% for comet Hale-Bopp, for which the greatest percentage of all the observed comets was found. The total input fluxes computed by ignoring the polarization effects are found to show differences as large as 10% with respect to the fluxes obtained when the full scattering matrix is included in the calculations.

G. Videen and M. Kocifaj (eds.), Optics of Cosmic Dust, 171–182.

1. Introduction

The knowledge of the total energy input on cometary nucleus surfaces is essential because it influences the amount of gas and dust released from those surfaces. The visual presence of a dust coma surrounding the comet nucleus implies the existence of scattering and absorption of light by particles. In the circumnuclear coma region, the multiple scattering effects have been proven to be important [11, 21, 29]. The Monte Carlo method has proven to be useful in computing the radiative energy input into the nucleus in those conditions [29]. The treatment by Salo [29] is the most complete to date, as it incorporates the calculation of the direct, multiple scattering and reradiated infrared fluxes into the cometary surfaces. The method is rather general, being applicable to any phase function, optical thicknesses or single scattering albedo of the coma particles. In the code presented here, we incorporate the full four-element Stokes vector into the calculations, and we use recently derived single scattering matrices for cometary analogues in the laboratory. The code, which in principle is made for spherical nuclei surrounded by spherical-shell dusty atmospheres in which the optical thickness and the spatial distances are related by simple analytical expressions, is intended to be applied to non-spherical surfaces and numerical dust density distributions in the near future. Our final goal is to compute fluxes in real comets, having arbitrary nucleus shapes, non-uniform particle density distributions, and particle size distributions varying with location in the coma. For that objective, *in situ* measurements or computed dust density and size distribution functions will be used as input of the code, and real nucleus shapes will be considered. The dust density distributions could be the output of gas dynamic simulations [7, 8], being the nucleus shapes derived from *in situ* imaging, such as the images given by Giotto flyby of the nucleus of comet Halley, or as those planned for the near future Rosetta Mission to comet Wirtanen. The details of the Monte Carlo code can be found in Moreno *et al.* [24].

2. The Monte Carlo code: description and validation

We consider a spherical nucleus centered at the origin of a cartesian coordinate system surrounded by a dusty coma whose particle scattering properties are described below. To initialize the Monte-Carlo process, a certain number of photons N are released from a plane perpendicular to the x-axis and located at a distance from the origin far enough so that the particle number density is zero. The incoming solar radiation is assumed to be unpolarized. The accuracy in the computed fluxes depends on N, as we discuss below on a few test cases. Following Salo [29], we use a group sampling method by which the photons are released from that plane from a series of concentric rings around the x-axis having increasing area for larger impact parameter b. The outer diameter of the largest annulus, b_{max}, was set to $45R_N$. For impact parameters larger than b_{max}, we found that the contribution to the computed fluxes is negligible.

To maximize the statistical accuracy of both the emergent and incident radiation on the nucleus surface we use the technique of photon weights (see e.g. [27]). Following this technique, a weight $W = 1$ is associated initially to each photon, which is appropriately reduced at each interaction. If the photon extended path intersects the nucleus, the optical depth to the surface, τ_c, is computed, and the energy carried to the nucleus is proportional to $W \exp(-\tau_c)$. Then, the weight is reduced by a factor $1 - \exp(-\tau_c)$. The actual position of the photon is calculated after sampling the optical thickness. Following Collins *et al.* [8], we use two methods to sample the optical thickness. If the photon extended path intersects the cometary surface, the optical depth is sampled from $\tau = -\ln(1-u)$ where u is a random number in the $(0, 1)$ range. If the optical depth sampled from the previous equation is greater than the real optical depth to the surface, a reflection is forced at the corresponding point of intersection of the segment with the surface, the weight W being correspondingly reduced by the albedo A_s of the ground. Otherwise, the photon is allowed to travel to the point at which the real optical depth equals the sampled optical depth. The photon weight W is then multiplied by the single scattering albedo of the particle. The surface is assumed here to be a Lambertian reflector. If the photon extended path does not intersect with the nucleus surface, an interaction is forced at the spatial location corresponding to $\tau = -\ln\{1 - u[1 - \exp(-\tau_{max})]\}$, where τ_{max} is the optical depth from the considered point to infinity. This corresponds to using the truncated probability density distribution as given by $f(\tau) = \frac{\exp(-\tau)}{1-\exp(-\tau_{max})}$ [27]. In this case, the weight is multiplied by $1 - \exp(-\tau_{max})$, and the contribution to the output flux at the corresponding angular direction is then given by $W \exp(-\tau_{max})$. The process ends when the weight reaches a value of $W \leq 10^{-5}$. A new photon is then released from the source.

In order to find the successive geometrical positions of the photon, we need to translate optical distances to geometric distances. This is performed by the same approach as Salo [29]. For these tests, we considered that the particle number density is given by $n(r) \propto 1/r^2$, for which there exists an analytical expression relating optical distances to geometrical distances and vice-versa.

In our formulation, the full four-component Stokes vector is computed at each interaction. To do this, we follow [3]. In a cartesian coordinate system (Fig. 1), the Stokes vector after a scattering event (I, Q, U, V) is computed from that before the event (I', Q', U', V') at each scattering angle as:

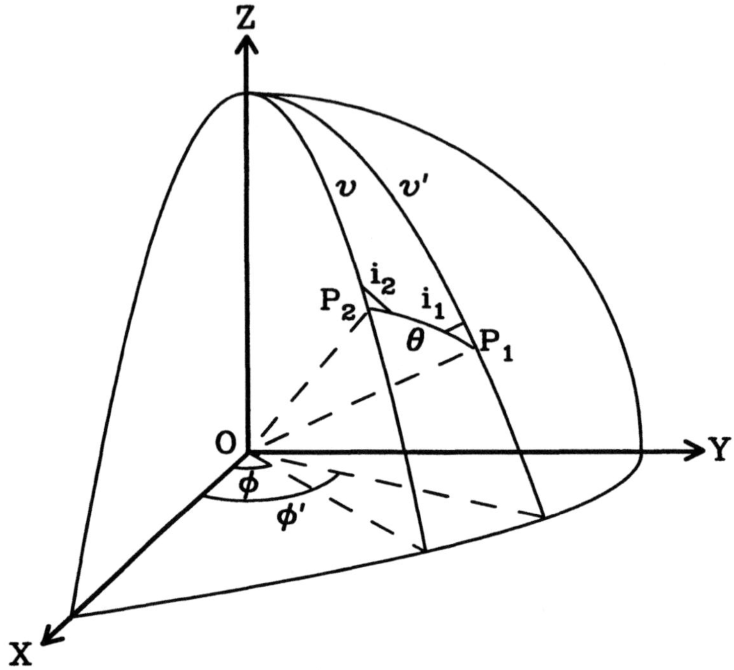

Figure 1. The scattering geometry. The points P_1 and P_2 are given by their polar angles as P_1 ($= (\nu',\phi')$), and P_2 ($=(\nu,\phi)$). The angle θ denotes the scattering angle, i_1 denotes the angle between the meridian plane OP_1Z and the plane of scattering OP_1P_2, and i_2 denotes the angle between the plane OP_2Z and the plane of scattering.

$$\begin{pmatrix} I \\ Q \\ U \\ V \end{pmatrix} = \begin{pmatrix} 1 & 0 & 0 & 0 \\ 0 & \cos 2i_2 & -\sin 2i_2 & 0 \\ 0 & \sin 2i_2 & \cos 2i_2 & 0 \\ 0 & 0 & 0 & 1 \end{pmatrix} \times \begin{pmatrix} Z_{11} & Z_{12} & 0 & 0 \\ Z_{21} & Z_{22} & 0 & 0 \\ 0 & 0 & Z_{33} & Z_{34} \\ 0 & 0 & -Z_{43} & Z_{44} \end{pmatrix} \times$$

$$\begin{pmatrix} 1 & 0 & 0 & 0 \\ 0 & \cos 2i_1 & -\sin 2i_1 & 0 \\ 0 & \sin 2i_1 & \cos 2i_1 & 0 \\ 0 & 0 & 0 & 1 \end{pmatrix} \times \begin{pmatrix} I' \\ Q' \\ U' \\ V' \end{pmatrix}$$ where i_1 and i_2

are the angles between the scattering plane and the planes defined by the Z-axis and the vectors OP_1 and OP_2, respectively. At each interaction, the scattering angle θ is sampled from the normalized phase function by solving the equation $u = \int_0^\theta Z11(\cos\theta')\sin\theta'\,d\theta'$, while u is a random number in the $(0,1)$ interval. For a Henyey-Greenstein phase function, $\cos\theta$ can be derived analytically from the previous formula, but for Mie or numerical phase functions, we used an interpolation algorithm to get θ for a given u. The scattering azimuth is computed by a rejection method as described by [20]. The angles i_1 and i_2 can be obtained by the spherical trigonometry formulas, as given, e.g., by [20]. The scattering matrix is, in the present notation, block-diagonal, which corresponds to the case in which the scattering sample consists of randomly oriented particles with equal amounts of particles and their mirror particles [31].

The thermal reradiation is computed by following the same procedure as described by Salo [29]. It is assumed that the dust is transparent to the thermal infrared radiation, so that no scattering processes take place at those wavelengths. This implies that the dust grains are much smaller than the wavelength at which they radiate, which constitutes an approximation for real comets. At each interaction, the weight of an infrared photon is $W(1 - \varpi)$, and the thermal flux is given by multiplying the weight by $\Omega/4\pi$, where Ω is the solid angle given by:

$$\Omega = 2\pi\{1 - \sqrt{1 - (R_N/r)^2}\}.$$

In order to confirm the ability of the Monte Carlo code to reproduce previous results, we made several validation tests through comparison to previous studies concerning both energy input to cometary surfaces and emitted radiance and polarization from dusty atmospheres. The details of those validations tests can be found in Moreno *et al.* [24].

3. Discussion

We address the effect of neglecting polarization on the energy input on the cometary nucleus by considering realistic phase matrices. Muñoz *et al.* [25, 26] obtained phase matrices for magnesium-rich crystalline olivine as cometary analogue in the laboratory. The crystalline olivine has been clearly identified as a component of the cometary dust from cometary thermal infrared spectra [1, 2]. The single-scattering matrix elements for the sample labeled as "Olivine M" ([25], their Table 2), at the wavelength of 0.633 μm, are shown in Fig. 2 (see the chapter by Muñoz *et al.*[26], their figure 8 for a wavelength of 633 nm). This sample was characterized by an effective radius of $r_{eff} = 2.6\ \mu m$, and an effective variance of $v_{eff} = 5.0$. It is interesting to note that other olivine samples with exactly the same composition and shape of the particles but having different size distributions showed curves similar to that of Fig. 2, specifically regarding the single scattering polarization curve. All of them show a negative branch at small phase angles and a maximum at side scattering angles. This seems to be a general behavior for irregularly shaped mineral particles [32].

Due to physical limitations in the laboratory, the range of scattering angles is approximately 5 to 173 degrees, and no measurements could be made at or near the exact forward and backward directions. Since for our modeling purposes the input matrix should be defined at every angle in the full 0 to 180 degrees range, an extrapolation must be made at near

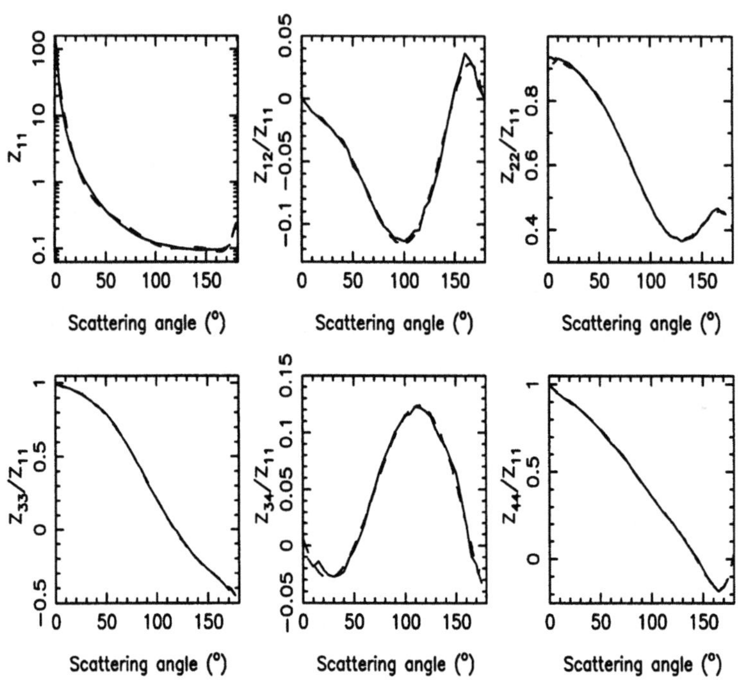

Figure 2. The elements of the single scattering matrix for the olivine sample by Muñoz *et al.* [26], their figure 8 at 633 nm (solid lines), along with the polynomial fits to the data (dashed lines).

both forward and backward directions. The extrapolation was made by fitting the forward peak by a Mie phase function having the same particle size distribution as the corresponding olivine sample. The other matrix elements were fitted using a least-squares polynomial of a variable degree. The resulting fits are shown in Fig. 2 along with the data. We have verified that the fitted scattering matrix elements all satisfy the Cloude coherence matrix tests [5] as described by Hovenier and van der Mee [12].

We assume the dust density distribution function to have a $1/r^2$ dependence, as assumed by Salo [29], corresponding to a constant outflow velocity for the particles ejected from the spherical coma. It is well known that this distribution is unrealistic, due to the aspherical nucleus shape and the large inhomogeneities in dust density distribution across the coma [7, 8]. As described in the introduction, a far more adequate description of both the particle density distribution based on gas dynamic simulations and the surface asphericity is under way. However, the spherical description is adequate enough for our purposes of investigating the effects of polarization on the input radiation.

The linear polarization phase curve observed in comets show a negative branch at phase angles less than about 20 degrees, and then an increase toward a maximum at about 90 degrees. Fig. 3 shows the observed polarization of some comets in the red region of the spectrum, restricted to published narrow-band filter measurements at 684 nm, except for comet Halley at phase angle less than 18 degrees, for which a wide band red filter (centered at 685.5 nm, and having a width of 108 nm) was used. For comets P/Halley 1986 III, Bradfield 1987 XXIX, Liller 1988 V, and Levy 1990 XX, the data by Chernova *et al.* [4] were considered, while for comet Hale-Bopp 1995 O1 the data were taken from Manset and Bastien [19]. For comet Hyakutake 1996 B2, two sources of data are shown, those by Kiselev and Velichko [16], and those published by Manset and Bastien [19]. The differences between both set of measurements might be due to different aperture sizes, and different geocentric and heliocentric distances at the time of the observations [19]. Fig. 3 also shows the fitted single scattering linear polarization curve corresponding to the magnesium-rich olivine.

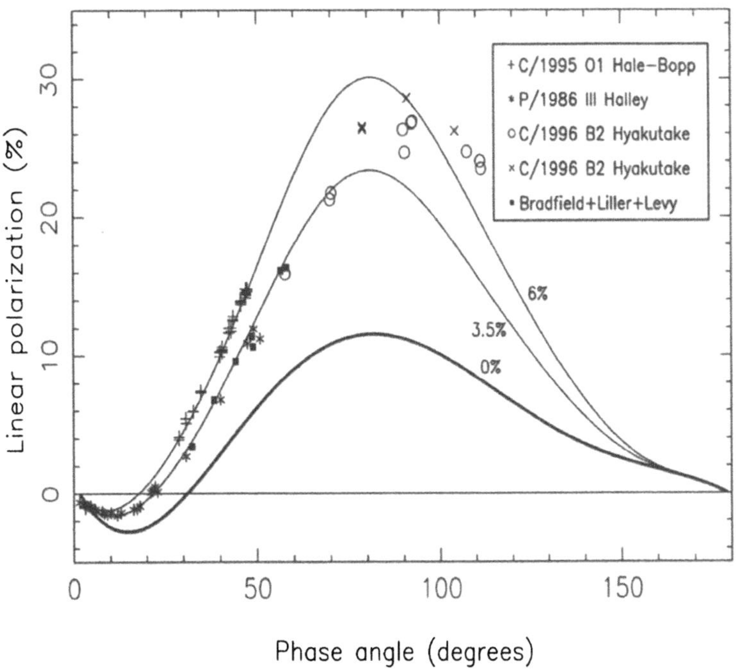

Figure 3. The degree of linear polarization versus phase angle. The symbols represent different comets, as indicated. The observations of comets Bradfield 1987 XXIX, Liller 1988 V, and Levy 1990 XX, are grouped together and represented by solid squares. See text to see the sources of the different datasets. The thick solid line represents the polynomial fit to the single scattering polarization laboratory data from the olivine sample (see figure 2). The two thin solid lines represent the single scattering polarization curves having a small absorbing particle single scattering fraction of 3.5% and 6%, as indicated (see text).

As seen in Fig. 2, all the cometary observations show a larger percentage of linear polarized light than that produced by the olivine sample. In this respect, it is important to note that we have restricted ourselves to published measurements concerning narrow-band data in the red region only (except for Halley at phase angle less than 18 degrees), because in all the wide band measurements, the depolarizing influence of gas emission should be present [15]. In order to fit the observed polarization data, we follow the suggestion by Kissel *et al.* [17] that the comet dust is a mixture of silicate and absorbing material with high carbon content. The combined scattering matrix can be written as $fR_{ij}+(1-f)Z_{ij}$, where Z_{ij} and R_{ij} are the matrix elements of the olivine and carbonaceous particles, respectively, and f is the carbonaceous particles scattering fraction. As a first approximation, we treat those carbonaceous particles as being spheres, and apply the Mie theory to compute the linear polarization. The carbon particles are assumed to have refractive indices of $n_r = 1.88$, and $n_i = 0.71$ at 0.6 μm [9]. We consider those carbon particles to be distributed following a log-normal size distribution characterized by a modal radius r_m, and width σ, that we set to $\sigma = 1.5$. Under these assumptions, the only way to fit the observations is to restrict the size of the carbon spheres to $r_m \lesssim 0.1\mu$m. Particularly, for comets Halle-Bopp and Halley, we found excellent fits by setting $r_m = 0.05$ μm, and scattering fractions of $f = 0.06$ and $f = 0.035$, respectively, as shown in Fig. 3. In the case of Halley, the curve fits the polarization found in other comets as well. For comet Hyakutake, there are significant differences between the two datasets shown in Fig. 3, although most of the points fall somewhere within those two curves (3.5% and 6%).

It should be noted that the specific composition of the small grains would not alter the calculations in principle, as the polarization curve is essentially the same for particles much smaller than the wavelength regardless of refractive index. It is also interesting to note that the higher small-particle scattering fraction of comet Hale Bopp to comet Halley agrees with the larger numbers of smaller particles in comet Hale-Bopp than in comet Halley [18, 13, 22]. The large small-grain population in comet Hale-Bopp is supported by a variety of techniques related to various physical parameters, such as superheat, 10 μm silicate excess, and phase angle dependence of the albedo [22]. It should also be stated that the model results are not appreciably modified if the particle radii are within $\lesssim 0.1$ μm, provided the corresponding scattering fraction is increased. A distribution of spherical absorbing particles having r_m larger than 0.1 μm would not produce the desired polarization-phase angle curve, as they do not show a bell-shaped polarization phase curve. It should be emphasized, however, that all of these arguments assume sphericity for the carbonaceous particles, and that non-spherical particles of much larger dimensions may also have polarization curves similar to that exhibited by small spherical particles (see, e.g., the Discrete-Dipole-Approximation calculations by Yanamandra-Fisher and Hanner [33] on non-spherical carbon particles in the 2.5-5 size parameter range).

For purposes of testing the influence of polarization on the input energy that reaches the nucleus, and in view of the results shown in Fig. 3, we made computations for single scattering matrices having a representative value of 5% for small, spherical, carbonaceous particle scattering fractions, and also, as an upper limit, having a 10% fraction. Fig. 4 shows the fractional differences between polarized and unpolarized total fluxes (direct + multiply scattered + re-radiated), as a function of the solar zenith angle for three values of the normal optical thickness. When polarization is neglected, the radiation fluxes are overestimated in the day-side hemisphere and underestimated in the night-side, reaching maximum differences near the antisolar point. This means that polarization effects should be included in the radiative transfer calculations when a precision of a few percent is required, mainly in the night-side of cometary nuclei.

In our treatment of the Monte Carlo process, all photons make contributions either to the input energy on the nucleus, or to the output flux. The four-element Stokes vector is recorded in our code as a function of the output angles. These calculations are not needed when trying to explain the observed intensity or linear polarization from comets observed from the Earth because most of the light we see is singly scattered light, the circumnuclear coma being hidden by a large amount of particles. Nevertheless, they are useful for close-up photometry or polarimetry, and questions related to nucleus visibility, as the multiple scattering effects show up in the near-nucleus region. These effects could be present even in some ground-based imaging polarimetry, such as reported by Tanga *et al.* [30], or Hadamcik *et al.* [10], or in the sometimes reported non-zero circular polarization in some comets [23], although most authors relate this phenomenon to the presence of aligned particles [28].

As an additional test of the code, we also report some computations concerning the variation of the degree of linear polarization of the output radiation of the spherical-shell coma versus the phase angle. In these computations, the number of initial photons is increased to $N = 5 \times 10^6$, as we need a better resolution in the output angular bins to represent adequately all the features of the single scattering polarization phase angle curve, particularly the negative branch at small phase angles. We consider a spherical nucleus surrounded by a spherical-shell coma having a variable optical depth, and a Halley-like surface albedo ($A_s = 0.04$, [14]). The particles are assumed to have a single scattering albedo of $\varpi = 0.75$, the single scattering matrix being that of the magnesium-rich crystalline olivine sam-

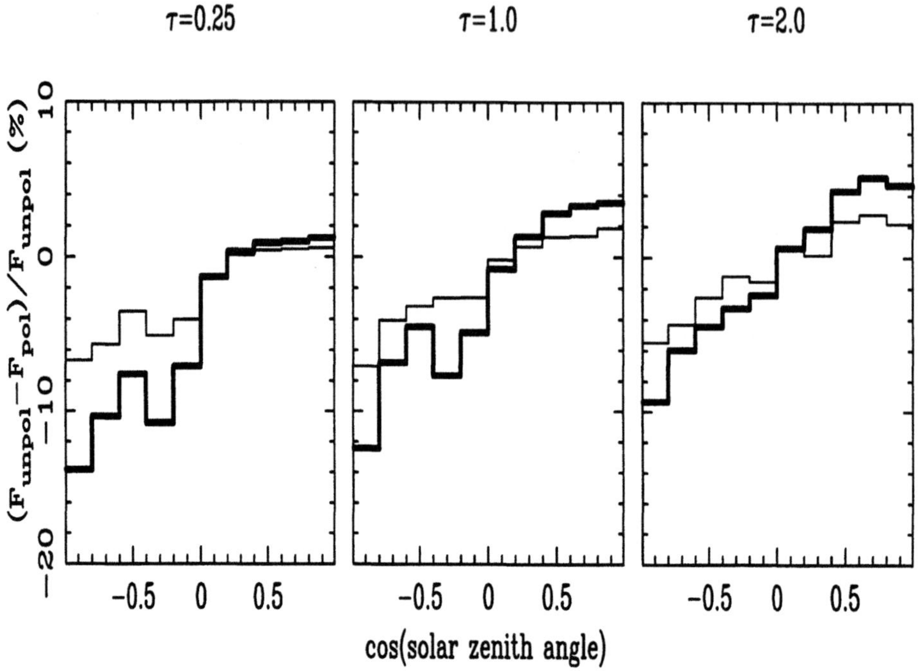

Figure 4. Relative ratios of the total (direct+multiply scattered+re-radiated) flux including polarization (F_{pol}) computed under the assumption of 5% (thin solid lines) and 10% (thick solid lines) scattering fraction for small particles to those fluxes obtained by excluding polarization (F_{unpol}), i.e., all the matrix elements set to zero except the phase function, Z_{11}. The results are given at three different optical depths (τ) as indicated. For those runs, $\varpi = 0.75$ and $A_s = 0$.

ple as described above. We considered three cases, corresponding to three different optical thicknesses, $\tau = 0.25$, $\tau = 1.0$, and $\tau = 2.0$.

Fig. 5 shows the results obtained, along with the fit to the linear polarization data of the olivine sample. In all cases, a decrease in the linear polarization with respect to the single scattering polarization curve is found, although there are some interesting differences regarding the two regions of phase angle $\alpha < 90°$ and $\alpha > 90°$. At phase angles $\alpha < 90°$, the degree of the linear polarization is influenced not only by the scattering by the particles in the coma, but also by the depolarizing influence of a non-zero surface albedo, mainly at small optical depths. This explains why the polarization at $\tau = 0.25$ is slightly smaller than that at $\tau = 1$ or $\tau = 2$ up to about 65° phase angle. As the phase angle approaches 90°, and for phase angles greater than that, the effect of the surface is less and less important. At those phase angles ($\alpha \gtrsim 80$), the degree of linear polarization decreases owing to the increase of the multiple scattering effects as the optical depth becomes larger. The region of the largest differences between the curves cor-

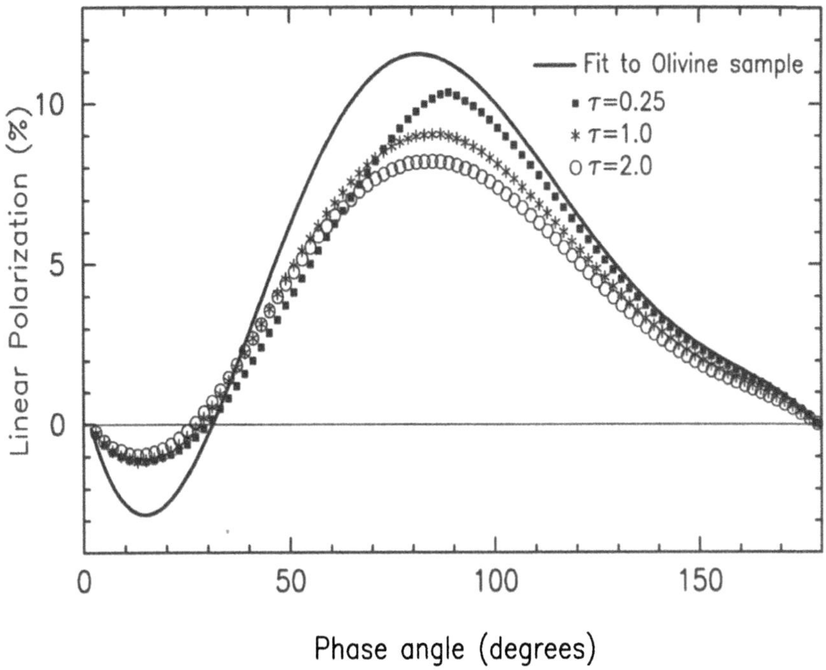

Figure 5. The fitted single scattering linear polarization obtained in the laboratory for olivine (see Muñoz et al. [25, 26], along with the linear polarization of light emerging from a cometary spherical-shell coma obtained by the Monte Carlo code, at three optical thicknesses, as labeled.

responding to the different optical depths is near 90° phase angle, and with that exception, all three curves display very similar values. Thus, within the framework of the model presented here, comets having a wide range in dust optical depths give almost the same degree of linear polarization as a function of the phase angle when observed at distances relatively close to the nucleus, and when Halley-like surface albedo conditions are applicable. Note that the cases of very small optical depth are not treated here: when $\tau \lesssim 0.1$, and independently of the surface albedo (assumed to be non-zero), the degree of linear polarization at phase angles $\alpha < 90°$ would be negligible, as virtually all the photons in the output flux have suffered the depolarizing influence of the surface. Also, for surface albedoes higher than that obtained for Halley, the depolarizing influence of the surface would be larger, so that the output flux would be less and less polarized at phase angles $\alpha < 90°$ as the surface albedo increases, mainly for small optical depth conditions.

4. Acknowledgments

This work was supported by Plan Nacional del Espacio (contract PNE-001/2000-C-01) and Programa Nacional de Astronomía y Astrofísica (contract AYA2001-1177).

References

1. J.R. Brucato, L. Colangeli, V. Menella, P. Palumbo, and E. Bussoletti, Planet. Space Sci., **47**, 773 (1999).
2. H. Campins, and E.V. Ryan, Astrophys. J., **341**, 1059, (1989).
3. Chandrasekhar, S. "Radiative Transfer", Dover, New York (1960).
4. G.P. Chernova, N.N. Kiselev, and K. Jockers, Icarus, **103**, 144 (1993).
5. S.R. Cloude, Optik, **75**, 26 (1986).
6. D.G. Collins, W.G. Blattner, M.B. Wells, and H.G. Horak, Appl. Optics, **11**, 2684, (1972).
7. J.F. Crifo, and A.V. Rodionov, Icarus, **127**, 319 (1997a).
8. J.F. Crifo, and A.V. Rodionov, Icarus, **129**, 72 (1997b).
9. Edoh, O. "Optical properties of carbon from the far infrared to the far ultraviolet". Ph.D. Dissertation, Univ. of Arizona (1983).
10. E. Hadamcik, A.C. Levasseur-Regourd, and J.B. Renard, Earth, Moon, and Planets, **78**, 359 (1997).
11. R. Hellmich, Astron. Astrophys. **93**, 341 (1981).
12. J.W. Hovenier, and C.V.M. Van der Mee, J. Quant. Spectrosc. Rad. Transfer, **55**, 649 (1996).
13. T.J. Jones, and R.D. Gehrz, Icarus, **143**, 338 (2000).
14. H.U. Keller, and 17 colleagues, Nature, **321**, 320 (1986).
15. N.N. Kiselev, Solar System Res., **33**, 158 (1999).
16. N.N. Kiselev, and F.P. Velichko, Icarus, **133**, 286 (1998).
17. Kissel, J., and 22 colleagues, Nature, **321**, 280 (1986).
18. A. Li, and J.M. Greenberg, Astrophys. J., **498**, L83 (1998).
19. N. Manset, and P. Bastien, Icarus, **145**, 203 (2000).
20. G.I. Marchuk, G.A. Mikhailov, M.A. Nazaraliev, R.A. Darbinjan, B.A. Kargin, and B.S. Elepov. "The Monte Carlo Methods in Atmospheric Optics". Springer-Verlag, Berlin (1980).
21. M.L. Marconi, and D.A. Mendis, Astrophys. J. **287**, 445 (1984).
22. C.G. Mason, R.D. Gehrz, T.J. Jones, C.E. Woodward, M.S. Hanner, and D.M. Williams, Astrophys. J., **549**, 635 (2001).
23. K. Metz, and R. Haefner, Astron. Astrophys., **187**, 539 (1987).
24. F. Moreno, O. Muñoz, J.J. López-Moreno, A. Molina, and J.L. Ortiz, Icarus, in press (2002).
25. O. Muñoz, H. Volten, J.F. de Haan, W. Vassen, and J.W. Hovenier, Astron. Astrophys., **360**, 777 (2000).
26. O. Muñoz, H. Volten, and J.W. Hovenier, this book.
27. G.N. Plass, and G.W. Kattawar, J. Atmos. Sci. 28, 1187 (1971).
28. V.K. Rosenbush, N.M. Shakhovskoj, and A.E. Rosenbush, Earth, Moon, and Planets, **78**, 381 (1997).
29. H. Salo, Icarus, **76**, 253 (1988).
30. P. Tanga, A. Cellino, and M. Di Martino, Earth, Moon, and Planets, **78**, 359 (1997).
31. H.C. Van de Hulst. "Light Scattering by Small Particles", Dover, New York (1957).
32. H. Volten, O. Muñoz, E. Rol, J.F. de Haan, W. Vassen, J.W. Hovenier, K. Muinonen, K., and T. Nousiainen, J. Geophys. Res., **106**, 17375 (2001).
33. P.A. Yanamandra-Fisher, and M.S. Hanner, Icarus **138**, 107 (1999).

ATMOSPHERIC EXTINCTION DERIVED FROM COMETARY OBSERVATIONS

J. SVOREŇ[1] , J. ŽIŽŇOVSKÝ[1] , Z. MIKULÁŠEK[2] , AND J. TREMKO[1]

[1]*Astronomical Institute, Slovak Academy of Sciences, Skalnaté Pleso Observatory, SK – 059 60 Tatranská Lomnica, The Slovak Republic*

[2] *Institute of Theoretical Physics and Astrophysics, Masaryk University, Kotlářská 2, CZ-611 37 Brno, Czech Republic*

Abstract. Special filters for photoelectric photometry of comets enable the study of atmospheric extinction not only in the broad band photometric systems but also in the intermediate band IHW/IAU system covering wavelengths from 365 to 514 nm. Extinction coefficients, which are characteristics of an atmospheric state, are by-products of observations of comets. Contrary to the observation of variable stars, comets have to be very frequently observed a few degrees above the horizon and so wide intervals of air masses are used in the calculation of characteristic values. Extinction measurements obtained during the observations of bright comets 1P/Halley in 1985/86, 23P/Brorsen-Metcalf in 1989 and C/Austin in 1990 are used in this study. 78 measurments in 21 observational nights cover all the parts of year with the exception of spring with cloudy weather in our mountain site and June with the shortest observational nights. Average extinction coefficients are derived in the paper and some conclusions of common validity (in central European conditions) are formulated. The results of IHW/IAU extinction study have been compared with the results of *UBV* extinction coefficients gained mostly as a by-product of the photoelectric photometry of variable stars (407 nights in the period from 1962 to 1995). Seasonal variations of atmospheric extinction are discussed.

G. Videen and M. Kocifaj (eds.), Optics of Cosmic Dust, 183–190.

1. Introduction

All the astronomical measurements done through the atmosphere from the Earth surface in the visible spectral window are considerably biased by the variable transparency of the Earth's atmosphere due to presence of molecules, and aerosol and dust particles. A stage of atmosphere is well characterized by the extinction coefficients, which are usually obtained as by-products of photoelectric measurements made in various photometric systems. Extinction coefficients in a particular colour *c*, *K*(*c*), represent the extinction of the light expressed in magnitudes per air mass *X*. A number of observatories have analyzed their photometric measurements for the atmospheric extinction and published the data describing the effect statistically concerning its time variations and nature. The classical broad band standard *UBV* photometric system was used in almost all the sets of observations.

In this paper we present extinction measurements made at the Skalnaté Pleso Observatory. This site has a very advantageous position for measuring extinction as the second factor acting on quality of astronomical observations i.e. light pollution of the night sky is very low due to high altitude of 1783 meters. It is sufficiently far from any industrial developments and some remote urban lights on the horizon do not interfere with its night observations. The location of the Skalnaté Pleso Observatory is as follows:

$\lambda = -1^{h}20^{m}58.70^{s}$,
$\varphi = +49°11'20.0''$,
$h = 1783$ m a.s.l.,
$\rho = 0.99836$ of the equatorial radius of the Earth.

The photometric measurements were performed using a 0.6-m f/12.5 mirror telescope equipped with a standard type photoelectric photometer. A full description of the instrument can be found in a previously published paper [2].

2. Measurements in the cometary filter set

Special filters for photoelectric photometry of comets enable the study of atmospheric extinction not only in the broad band standard *UBV* photometric system but also in the intermediate band IHW/IAU system covering wavelengths from 365 to 514 nm. The filters were made in 1983 for the International Halley Watch campaign and are still widely used to observe comets. Five basic IHW filters and CO^+ one are installed in the optical part of the photometer. The filters are centered at the main cometary emission bands *CN*, C_2, C_3 and CO^+ and two spectral regions for continuum study. The characteristics of the filters are listed in Tab. 1.

Additionally supplementary set of filters (mainly for infrared) has been produced. Contrary to the observation of variable stars, comets have to be very frequently observed a few degrees above the horizon and so wide intervals of air masses are used in the calculation of characteristic values. Extinction measurements obtained during the observations of bright comets 1P/Halley in 1985/86 [5,6,7], 23P/Brorsen-Metcalf in 1989 [8], and C/Austin in 1990 [9] are used in this study (the tables of results for individual nights and filters are accessible from the web *http://www.ta3.sk/~astrsven/*).

Table 1: Characteristics of the IHW/IAU filters

filter	central wave-length in nm	half-power width in nm	transmittance at peak in %
U continuum	365.0	8.0	34
CN	387.1	4.9	24
C_3	406.0	7.4	48
CO^+	426.0	6.5	45
B continuum	484.5	6.9	73
C_2	514.0	9.1	65

The average values of the coefficients are listed in the Tab. 2 and the information about the wavelength dependency in Fig. 2.

The observations used did not make it possible to determine the seasonal variation of the coefficients. However some conclusions of common validity can be formulated:

Individual values show large night to night changes of the extinction. These rapid variations of atmospheric conditions (accompanied by changes of temperature) precluded applying any mean values of the coefficients during a chosen observational interval. This conclusion is commonly valid, at least for the European climate with frequent changes of meteorological conditions.

The derived values of coefficients show not only large night to night changes of the atmosphere transparency, but also large changes of the extinction during the same night. On November 16, 1985 the extinction changed by a factor of two during a time interval as short as 1.5 hours.

It is not usual in European climatic conditions to determine the brightness of celestial objects through more than 3 air masses. The short duration of the astronomical night at the end of May meant that the Comet Austin was observed through larger air masses. The largest air mass at which the comet was measured was 2.47, but the largest air mass for the comparison star (HD 189 340) was 4.46. Air mass was calculated by applying Hardie's polynomial expression [1]:

$$X = \sec z - 0.0018167(\sec z - 1) - 0.002875(\sec z - 1)^2 - 0.008083\,(\sec z)^3 \quad (1)$$

It is clear that at such air masses only observations during very clear nights (and in conjunction with the high-altitude position of the observatory) can be used for further reductions. Therefore, only the observations obtained under excellent atmospheric conditions on the night of May 27/28, 1990, are presented here.

It was shown that the dispersion of data is low, extinction data (magnitude versus air mass) can be fitted linearly and coefficients of correlation are from 0.99 to 1. We can conclude that during extraordinary favourable circumstances it is also possible in central European conditions to obtain data low on the horizon with a usable precision.

3. UBV measurements

The extinction coefficients in *UBV* for 407 nights were measured at the Skalnaté Pleso Observatory in 1962-95. The whole set of the observations is expressed in the broad band standard *UBV* photometric system. The observation method can be divided into two groups according to the two periods in which the observations were obtained. The first part of the measurements, made up to the end of 1974, was performed within the framework of the observational programs directed at studying variable stars. These observations were generally not accompanied with special extinction measurements and this is why only mean extinction coefficients could be derived from them for each observing night.

In 1975, the methodology of obtaining the observations was changed and replaced by the numerical computations of the extinction coefficients from the colour system measurements. From time to time, the more reliable determinations were supplemented with special extinction measurements based on a list of selected standard stars. Thus, these extinction coefficients are based on four stars of different spectral types, two observed near zenith and two at large zenith distances. As a result the corresponding extinction coefficients are about four times more accurate than those obtained in the previous period.

The extensive set of the measurements is not included in this paper because they can be accessible from the web: *http://www.ta3.sk/~ziga/ftp/extin*.

We can examine the relationship between extinction coefficients simultaneously measured in different colours. All graphs displaying the relation between extinction coefficients obtained in *V*, *B* and *U* (as example see Fig. 1) prove:

1) A very strong correlation exists among simultaneously established extinction coefficients ($r \approx 0.9$).

2) The scatter of these relationships is apparently less than mutual variations, consequently observed variations of extinction are very probably real.
3) The observed relationships among extinction coefficients can be satisfactorily represented by simple straight lines, which do not intersect the origin.

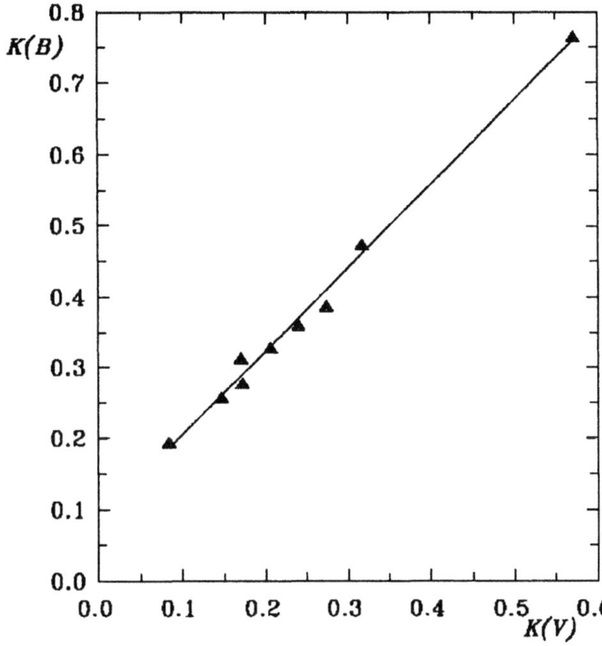

Figure 1: The correlation between the extinction coefficients K_V and K_B following from the measurements made at Skalnaté Pleso. Each of the points is an average of about 20 neighbouring doubles of measurements.

4. Results and conclusions

An extensive collection of *UBV* observations was analyzed in detail before [3,4]. That is why only a small part of main obtained results is reviewed here. The atmospheric extinction $E(\lambda)$ at the Skalnaté Pleso Observatory can be satisfactorily described by a standard two-component model consisting of a more or less constant component given by Rayleigh scattering $R(\lambda)$ and a variable component $D(\lambda)$:

$$E(\lambda) = R(\lambda) + D(\lambda), \tag{2}$$

where

$$\mathrm{R}(\lambda) = 0.327\left(\frac{\lambda}{416}\right)^{-4} \frac{P(h)}{P(0)} \; . \tag{3}$$

All extinction coefficients are expressed in magnitudes per air mass, wavelengths are expressed in nanometers. The mean ratio of the barometric pressure $P(h)$ of a station at an elevation h (in meters) above sea level to the pressure at sea level $P(0)$ can be approximated using a simple model of the so called standard Earth atmosphere with a characteristic height of 7996 m.

$$\mathrm{P}(h)/\mathrm{P}(0) = \exp(-h/7996) \tag{4}$$

The ratio is thus a function of the altitude h of the observing station, h in meters. The altitude of the Skalnaté Pleso Observatory is: h(SP) = 1783 m above sea level.

$$R(\lambda) = 0.262\left(\frac{\lambda}{416}\right)^{-4} \tag{5}$$

The spectral dependence of the variable term $D(\lambda)$ can be expressed by a power function:

$$D(\lambda) = D\left(\frac{\lambda}{416}\right)^{-\alpha} \tag{6}$$

where D corresponds to the extinction at 416 nm. It follows from equation [4] that $\alpha = 1.04 \pm 0.04$. Using a standard two-component model a prediction of the spectral dependence of the mean value $\overline{E(\lambda)}$ is as follows

$$\overline{E(\lambda)} = \overline{D}\left(\frac{\lambda}{416}\right)^{-1.04} + 0.262\left(\frac{\lambda}{416}\right)^{-4} \tag{7}$$

The value $\overline{D} = (0.197 \pm 0.010)$ mag/air mass was found using the least squares method.

- The observed dependence of mean values of atmospheric extinction on wavelength can be satisfactorily described by the two-component model formulated in [4]. The Skalnaté Pleso extinction in UC, CN, C_3 and CO^+ filters is influenced by the Rayleigh scattering, the extinction in BC and C_2 filters is slightly affected by the scattering of aerosol particles.
- Seasonal variations of the dust extinction at the Skalnaté Pleso Observatory seem to be controlled by the annual variations in the position of the layer of temperature inversion. The atmosphere is very dusty under this layer and relatively very clear above it. In summer time, the inversion frequently ascends above the observatory, which is the main cause of the observed more or less sinusoidal variations of extinction with the minimum around January (3 ± 10). At that time, the dust extinction decreases to (76 ± 6)% of its average value and a half-year earlier it climbs to (146 ± 12)%. Systematic measurements of the concentration of aerosols in the Earth atmosphere

conducted by The Solar and Ozone Laboratory of the Czech Hydrometeorological Institute in Hradec Králové also showed the seasonal course with a summer maximum (120% of the annual average) and a winter minimum (80%) [10].

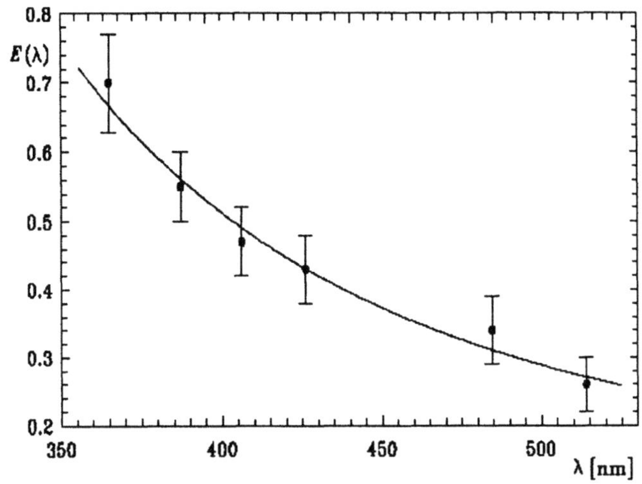

Figure 2: Dependence of average extinction coefficients $E\{\lambda\}$ on wavelength λ (full line - the model prediction). The errorbars correspond to the uncertainty of the particular average.

Table 2 : Average extinction coefficients August 1985 – May 1990

filter	wavelength in nm	number of nights	mean extinction coefficient	model prediction	share of dust
UC	365.0	14	0.70±0.07	0.67	37%
CN	387.1	18	0.55±0.05	0.56	36%
C_3	406.0	16	0.47±0.05	0.49	38%
CO^+	426.0	8	0.43±0.05	0.43	44%
BC	484.5	14	0.34±0.04	0.31	59%
C_2	514.0	8	0.26±0.04	0.27	58%

- It was found that observations namely in the *U* colour becomes more and more badly affected by the growing extinction. Nevertheless, it is not reasonable to quit the near ultraviolet measurements when studying hot stars, as these measurements inform us about the astrophysically crucial region behind the Balmer jump. Also studies of comets are very incomplete without the observations of the strong *CN* band in the near ultraviolet spectral region. Problems with the near future extinction at the Skalnaté Pleso Observatory thanks to its high altitude are not very topical.

5. Acknowledgments

The authors thank the staff of observatory who participated over three decades in different observational programmes from which the observed extinction data was extracted. This research was partly supported by VEGA grants Nos. 1023, 7107, 1157 and Grant Agency of Czech Republic, grant No. 205/02/0445.

References

1. R. Hardie, Astronomical Techniques, ed. W.A. Hiltner, Chicago, 178 (1962).
2. L. Klocok, J. Žižňovský, J. Zverko, Contrib. Astron. Obs. Skalnaté Pleso **16**, 43 (1987).
3. Z. Mikulášek, J. Papoušek, M. Vetešník, J. Tremko, J. Žižňovský, Contrib. Astron. Obs. Skalnaté Pleso **30**, 89 (2000).
4. Z. Mikulášek, J. Papoušek, J. Tremko, J. Žižňovský, Contrib. Astron. Obs. Skalnaté Pleso **31**, 90 (2001).
5. J. Svoreň, "In: Exploration of Halley's Comet III" (ESA SP-250, Noordwijk, 1986, pp. 39-43).
6. J. Svoreň, "In: Interplanetary Matter 2" (10th European Regional Astronomy Meeting, Astron. Inst., Ondřejov, 1987, pp. 75-78).
7. J. Svoreň, Bull. Astron. Inst. Czechosl. **42**, 16 (1991).
8. J. Svoreň, Planet. Space Sci. **42**, 269 (1994).
9. J. Svoreň, Planet. Space Sci. **44**, 1001 (1996).
10. K. Vaníček, private comm. (2000).

PHOTOMETRIC AND POLARIMETRIC OPPOSITION PHENOMENA EXHIBITED BY SOLAR SYSTEM BODIES

VERA ROSENBUSH,[1] NIKOLAI KISELEV,[2] VICTOR AVRAMCHUK,[1] AND MICHAEL MISHCHENKO[3]
[1]*Main Astronomical Observatory of the National Academy of Sciences, 27 Zabolotny Street, 03680 Kyiv, Ukraine*
[2]*Institute of Astronomy, Kharkiv National University, 35 Sumskaya Street, 61022 Kharkiv, Ukraine*
[3]*Goddard Institute for Space Studies, 2880 Broadway, New York, NY 10025, USA*

Abstract. Ground-based and spacecraft observations of atmosphereless solar system bodies (planets, satellites, planetary rings, and asteroids) and cometary and interplanetary dust particles have provided a wealth of new photometric and polarimetric data over a wide range of phase angles and wavelengths. This chapter reviews the progress in the study of the photometric and polarimetric phase effects observed near opposition. We also present the results of recent polarimetric observations of the Galilean satellites of Jupiter, Iapetus, and the asteroid 64 Angelina at very small phase angles. Analyses of the available data allow us to determine the parameters of the brightness and polarization opposition effects and investigate correlations between them. The results obtained may form the observational basis for detailed theoretical modeling and interpretation of the phase effects.

1. Introduction

Measurements of the intensity and polarization of scattered light as functions of the phase angle are a primary source of information about the physical properties of the surfaces of solar system bodies. Such observations are especially informative near the opposition, i.e., at small phase angles. Many atmosphereless solar system bodies (ASSBs) exhibit an opposition brightening, the so-called brightness opposition effect (BOE), as a nonlinear brightness increase toward smaller phase angles. Negative values of the degree of linear polarization at

G. Videen and M. Kocifaj (eds.), Optics of Cosmic Dust, 191–224.

backscattering angles are also a common property of rather different objects such as asteroids, satellites, planets, comets, interplanetary dust particles, and planetary rings. Both effects have also been observed for many laboratory samples [1].

A number of physical mechanisms have been proposed to explain the photometric opposition effect and negative polarization resulting from the interaction of light with porous, powder-like surface layers or rough surfaces of the ASSBs [2–5]. Presently, the so-called coherent backscattering mechanism (CBM) and the mutual shadowing mechanism (SM) are believed to be the primary candidates to explain the observed opposition phenomena. Both astronomical observations of ASSBs and laboratory measurements of structural analogs of planetary regoliths may provide important tests of theoretical models.

Until now, the origin of the BOE and negative polarization is not yet fully understood, the lack of high-accuracy photometric and polarimetric data for various objects at small phase angles being one of the main reasons. Observations at the smallest phase angles ($\leq 2°$) are very rare. Furthermore, the polarization at these angles is often weak (several tenths of a percent), thereby necessitating very high measurement accuracy and making most of the available data poorly suitable for studies of the opposition effects.

This chapter covers three closely related subjects: (i) a brief review of the existing photometric and polarimetric data for ASSBs at small phase angles; (ii) a more detailed discussion of most recent polarimetric observations at phase angles approaching zero; and (iii) a discussion of how to determine the parameters of the BOE and negative polarization and study their relationships.

2. Basic definitions

In this chapter, we use the standard photometric and polarimetric terminology adopted in planetary astrophysics. The opposition region is defined as the range of phase angles from about 25° down to $0°$. The phase angle, α, is the angular distance between the Sun and the observer as seen from the astronomical object.

Usually, the BOE means a nonlinear surge in brightness, measured either in magnitude scale (m) units or in intensity (I) units, at small phase angles. For several high-albedo objects, the BOE is observed in the form of a very sharp and narrow intensity peak often called the opposition spike. For phase angles $\alpha_{\mathrm{opp}} < \alpha \leq 25°$, the brightness, measured in magnitudes, increases linearly with decreasing phase angle, where α_{opp} is the phase angle at which the BOE begins. The curves showing the dependence of the distance-corrected brightness on the phase angle are called photometric phase curves. The slope of the linear part of the phase curve is called the phase coefficient: $\beta = d\mathrm{m}/d\alpha$.

For the quantitative description of the BOE, the following parameters are useful: the amplitude Δm (or the enhancement factor ζ) and the half-width at half-maximum $HWHM_B$ of the opposition surge; the background intensity I_b calculated by extrapolating the brightness linearly to zero phase angle; β; and α_{opp}. There are two definitions of the amplitude of the BOE. The first one is the ratio of two intensities, $I(\alpha_1)/I(\alpha_2)$, where $0 \le \alpha_1 \le \alpha_2 \le \alpha_{opp}$. This definition is often used for the quantitative description of the BOE when the lack of observations does not allow one to plot a detailed phase curve. The second one is the brightness increase relative to I_b. In theoretical studies of the BOE, the enhancement factor ζ is usually defined as the ratio of the backscattered intensity at exactly the backscattering direction to the background intensity.

Unpolarized light scattered by a particulate surface becomes partially linearly polarized. By definition, the degree of linear polarization is $P = (I_\perp - I_\parallel)/(I_\perp + I_\parallel)$, where $I_\perp$ and $I_\parallel$ are the intensity components with the electric vectors perpendicular and parallel to the scattering plane. The latter is defined by the Sun, the observer, and the object. The polarization degree changes with phase angle and produces a specific so-called polarization phase curve $P(\alpha)$. It often has a negative branch ($I_\perp < I_\parallel$) near opposition and a positive branch ($I_\perp > I_\parallel$) at phase angles $\alpha > \alpha_{inv}$. Angle α_{inv} is called the inversion angle and is the phase angle at which the polarization changes sign ($I_\perp \approx I_\parallel$). The parameters of the negative polarization branch are α_{inv}, P_{min} (the minimal polarization value), and α_{min} (the phase angle of minimal polarization). We also use the parameter $0 < HWHM_P < \alpha_{min}$, specifying the phase angle at which the polarization value is equal to $\frac{1}{2}P_{min}$. All these parameters depend on the physical and chemical characteristics of the scattering particles and on the macroscopic roughness profile of the particulate surface.

The behavior of the negative polarization branch at backscattering angles ($\alpha < 2°$) is very intriguing and poorly understood. A very asymmetric negative branch is observed for high-albedo Saturn's rings, while the dark Moon shows a regular, nearly parabolic negative branch [6]. Mishchenko [7] introduced the term "polarization opposition effect" (POE) to indicate situations when the polarization of the backscattered light, being zero at zero phase angle, becomes negative and reaches its maximal negative value at a very small phase angle (less than 1°). He further assumed that both the extremely narrow BOE and the POE for high-albedo objects are caused by coherent backscattering. This approach implies the possible presence of a separate broad negative minimum of polarization centered at a significantly larger phase angle and generated by an optical mechanism potentially different from coherent backscattering. Therefore, the resulting polarization curve at backscattering angles can have different shape depending on the relative contributions of different mechanisms and can range from a single POE spike to a single, nearly parabolic negative polarization branch. In contrast, Shkuratov [8] and Muinonen *et al.* [4] believe that the CBM can explain all types of negative polarization branches observed for ASSBs and interplanetary and cometary dust.

3. Observations

3.1 Limitations of photometric and polarimetric observations

The specific illumination/observation geometry often does not allow one to observe an object at both small and large phase angles. For example, the smallest accessible phase angles in ground-based observations of the Moon (outside of an eclipse) and Mars are about 0.5° and 1.2°, respectively. Furthermore, the parameters characterizing the BOE at extremely small phase angles may not be reliable because the brightness spike can be broadened due to the finite angular size of the solar disk [9], which should be taken into account for such objects as the Moon, Mars, and the satellites of Jupiter and Saturn. Ground-based observations of such distant objects as the satellites of Jupiter, Saturn, and, especially, Neptune and Uranus are quite difficult because of the narrow phase angle coverage: from 0° to 12° for Jupiter, from 0° to 6° for Saturn, from 0° to 3° for Uranus, and from 0° to 2° for Neptune. In the case of satellites rotating along low orbits such as Phobos, Deimos, Tethys, Dione, Rhea, Titania, Oberon, and others, the scattered light from the primary planet and, potentially, planetary rings can also be a major problem in obtaining precise photometric and polarimetric data.

Furthermore, the brightness and polarization of a satellite can change with orbital phase (the so-called longitude effect). This can be caused by the heterogeneity of the local albedo over the surface of the satellite. To investigate the phase dependence of the brightness and polarization, averaged measurements for the leading and the trailing hemisphere of a satellite are usually used. This averaging can lead to a significant scatter of data points. Therefore, the separation of variations caused by the longitude effect from those caused by changing solar phase angle should be done separately for each orbital longitude.

There are also brightness (and possibly polarization) variations caused by the rotation of irregularly shaped objects such as Hyperion, Nereid, and some asteroids. In such cases, the orientation of the pole, which may cause brightness variations in different apparitions, must be taken into account.

3.2 Photometric observations

The Moon. The study of the lunar disk is important, since the Moon is used as a standard for interpreting observations of other planets, satellites, and asteroids. Moreover, the disk of the Moon can be easily resolved, and the distribution of the surface brightness can be mapped as a function of the phase angle and the local angles of incidence and reflection. In fact, the majority of the photometric studies of the Moon have been done for specific small areas rather than for the whole lunar disk.

The lunar regolith is an example of a surface that exhibits prominent backscattering. The first rather accurate phase curve of the disk-integrated intensity in the visible spectral range was obtained by Russell [10]. The most accurate dependence of the disk-integrated brightness on the phase angle was derived by Rougier [11] on the basis of photoelectric observations. The existence of the BOE was demonstrated by Gehrels *et al.* [12]. Lane and Irvine [13] observed the whole Moon in ten narrow spectral bands as well as in the UBV filters within the phase angle range 6.6°–118.6°.

A sharp increase of brightness near opposition for 31 lunar regions was first discovered by Barabashev [14] in 1922. Many photometric observations of the lunar formations covering a wide range of wavelengths have been collected in [15–20]. These studies have established that the brightness phase curves for different surface features on the Moon behave differently depending on the location on the disk and, thus, on the structure and composition of the surface. However, all of them show a nonlinear surge in brightness near opposition. According to Gehrels *et al.* [12], there is a 10–15% increase in brightness between phase angles 1.4° and 0.7°. The total brightening within the phase angle range from 5° to 0° may reach 50%–100%. Colorimetric investigations of different lunar regions have demonstrated the spectral dependence of the BOE as well as variations of its value with surface type. An extensive review of lunar photometry performed by different authors, and its interpretation was given by Hapke [21].

Detailed investigations of the BOE were continued during the Apollo program [22–24]. Apollo 8 photographs provided the first opportunity to determine the lunar photometric function in the vicinity of zero phase without extrapolation. The opposition brightening was found to be 1.3–1.4 as the phase angle decreased from 8° to 0°. Differences in the BOE amplitude for maria (7%) and for highlands (19%) at the smallest phase angles (from 1.5° to 0°) were discovered. There was no obvious correlation between the degree of brightening and the albedo or other characteristics of surface.

Recently, observations performed during the Clementine mission have been analyzed, and the BOE was investigated in detail in the visible and near infrared spectral ranges [25–29]. According to Kreslavsky *et al.* [29], the brightness phase curves for regions in Sinus Medii and Mare Fecunditatis clearly flatten at phase angles < 0.25°, which is due to the significant angular size of the solar disk as viewed from the Moon. The average amplitude of the lunar opposition surge in the range of phase angles from 1° to 0° is approximately 10%. There is no obvious wavelength dependence of the opposition surge for the phase angle range 0°–1.4°.

Mercury. Danjon [30,31] performed the first reliable and detailed measurements of the phase dependence of brightness for Mercury at phase angles from 3° to 123°. The most important result of these observations is that there is a clear opposition brightness increase at phase angles less that about 12°. CCD disk-integrated photometric data were recently obtained with the SOHO spacecraft in the phase angle ranges $2° < \alpha < 22°$ and $160° < \alpha < 178°$ [32]. The combination of these data with ground-based observations revealed that the brightness surge for Mercury exceeds 40% between phase angles 10° and 2°.

Mars. This planet is not an atmosphereless body, but the optical thickness of the Martian atmosphere during the periods of high transparency is very small. Therefore, the atmospheric effects are usually negligible in the red light (λ >500 nm), and the light reflected by the surface is dominant. However, dust storms, which often occur near opposition, may cause a significant atmospheric scattering contribution leading to strong brightness variations. The latter can suppress or even totally obscure the BOE.

The BOE for Mars was first detected by Vaucouleurs [33,34] during the 1958 opposition and was subsequently confirmed by O'Leary [35] and Bugaenko *et al.* [36]. Extensive observations of Mars during the 1971–75 oppositions were reported by Aleksandrov *et al.* [37], who also found a brightness increase at small phase angles. The extrapolation of the observed brightness to zero phase angle yields the BOE amplitude of about 0.1–0.3^m at different wavelengths [34].

BOE measurements for different morphological formations of Mars at phase angles down to 0.1° were obtained by the Viking Orbiter 1 spacecraft [38]. A sharp brightness increase (10%) was observed at phase angles $\alpha < 3°$ for flat, bright and dark areas in the Chryse Acidalia, Arabia, and Elysium Planitia regions. The opposition peak was stronger and narrower for high-albedo areas than for dark features. A weak, if any, opposition effect was found for the Syrtis Major regions. Thorpe [39] concluded that the BOE for Mars does exist and depends on the surface albedo and porosity.

The brightness increase for Pavonis Mons was measured in eight spectral bands (λ = 315 – 550 nm) using Phobos-2 observations at phase angles 0° and 15.7° [40]. The amplitude, defined as the ratio $I(0°)/I(15.7°)$ and determined after correcting for the atmospheric effects, was found to be about 1.7. It depended weakly on the wavelength within the spectral range λ = 315–445 nm, but decreased substantially at λ = 445–550 nm.

Satellites of Mars. Photometric data for Phobos and Deimos based on Mariner-9 observations cover a restricted range of large phase angles (18°–83°) [41]. A more extensive coverage was obtained from the Viking observations: 1.3°–122° for Phobos and 0.5°–122° for Deimos [42,43]. These data allowed, for the first time, to study the BOEs for both satellites of Mars and the wavelength dependence of the photometric phase curves between 445 nm and 593 nm. The phase curves exhibited a linear part between 20° and 80° with phase coefficients about 0.024 mag/deg for Phobos and 0.020 mag/deg for Deimos, and a conspicuous brightness surge at phase angles less than 20°. The phase curves for the Martian satellites were also compiled for phase angles up to 125° from ground-based and spacecraft photometric observations [44]. These investigations have shown that the phase curves for Phobos and Deimos are very similar. This result contradicts previous indications that the phase curve for Phobos is steeper than that for Deimos [42].

Disk-resolved images of Phobos taken by the Phobos 2 spacecraft at phase angles less than 30° showed that rims of many grooves and craters are usually 20–30% brighter than the surrounding areas, whereas the brightness contrast is much less evident at larger phase angles [45–47]. The Phobos disk-integrated photometry with the Hubble Space Telescope (HST) at phase angles ranging from 10.6° to 40.5° allowed one to track the start of the opposition brightening [48].

Galilean satellites. All Galilean satellites of Jupiter show unequivocal BOEs [49–53]. According to [54], the amplitude of the opposition effect is 0.17^m for Io, 0.09^m for Europa, 0.07^m for Ganymede, and 0.25^m and 0.13^m for the leading and the trailing side of Callisto, respectively. Thompson and Lockwood [55] discovered an extremely narrow spike in brightness (about 0.2^m) at a phase angle $\alpha \approx 0.2°$ for both hemispheres of Europa, as shown in Fig. 1a. As one can see, the photometric phase curves are essentially the same over the phase angle range 0.5°–12° for both hemispheres of Europa.

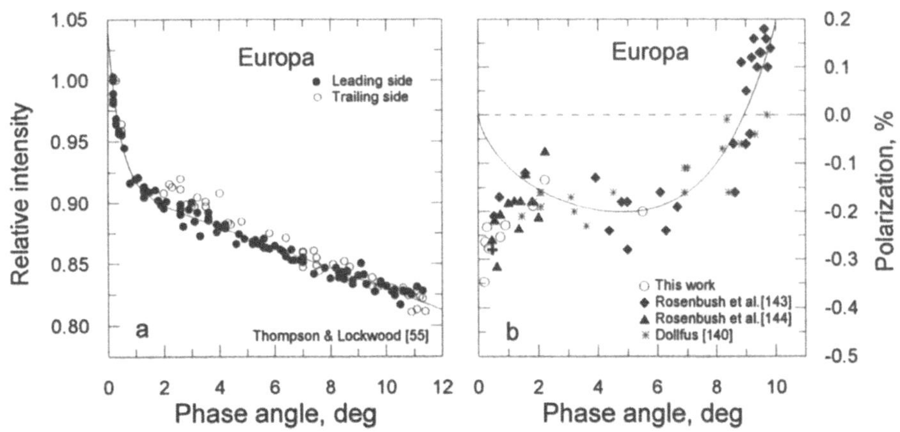

Figure 1: The brightness (a) and polarization (b) opposition effects for Europa in the V band. The brightness data are fitted with an exponential-linear function [56]; the polarization data are fitted with a trigonometric polynomial [57] (solid curves).

The leading/trailing hemisphere dichotomy for the Galilean satellites was already known from the observations of Stebbins [49] and Stebbins and Jacobsen [50]. The leading hemispheres of Io, Europa, and Ganymede are brighter than the trailing hemispheres in the entire phase angle range 0.5°–12°. For Callisto, however, the trend is reverse at large phase angles: the trailing side is brighter than the leading one (Fig. 2a). Near the opposition, there is almost no difference in the brightness of the two sides due to the fact that the opposition surge of the dark side is much stronger than that of the bright side. Note that there is also an albedo asymmetry of the Galilean satellites in the infrared [58] and in the ultraviolet [59].

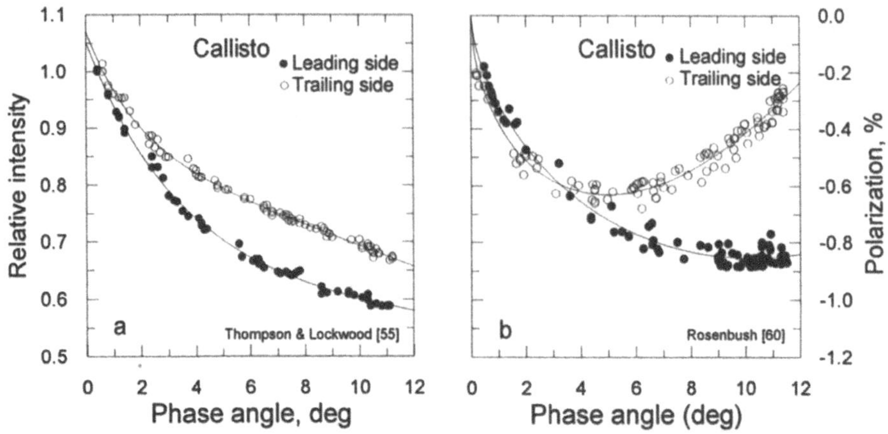

Figure 2: The BOE (a) and the negative polarization branch (b) for the leading and trailing hemispheres of Callisto in the V band. The variations of polarization with orbital longitude are taken into account [60]. The fitted functions (solid curves) are analogous to those in Fig. 1.

Systematic observations of Io showed brightness changes of up to a few percent per year not observed for the other Galilean satellites [53]. They are most likely caused by albedo changes associated with active volcanic processes. Until now, there are no carefully measured photometric phase curves for both the leading and the trailing side of Io with a good phase angle coverage near opposition and a small scatter of data points. Nevertheless, the available data allow one to conclude that despite the similar albedos for Io (0.62) and Europa (0.68), the phase coefficient for Io (0.022) is considerably greater than that for Europa (0.006) [54].

Spacecraft observations marked a new stage in the study of the Galilean satellites by extending the range of phase angles up to 160°. Domingue and Verbiscer [61] combined the results of previous telescopic observations [51,52, 62] with the Voyager data and constructed photometric phase curves for the leading and trailing hemispheres of the satellites to examine any possible hemisphere dichotomies. They have confirmed that Europa and Callisto have a strong hemisphere dichotomy in their rotational lightcurve. Ganymede has a hemisphere brightness difference in the same sense as Europa, but not as pronounced.

Studies of the BOE for several features on the surface of Europa by the Galileo spacecraft are especially interesting [63]. According to the measurements in three spectral bands (410, 560, and 990 nm), all terrains exhibit narrow ($HWHM_B$ < 0.2°) BOEs, but with different amplitudes. The amplitudes for most abundant materials on Europa are about 1.5 times greater than those predicted from ground-based observations. Low-albedo materials at dark locations exhibit unusually strong opposition effects, up to four times stronger than bright icy terrains.

Satellites of Saturn. In spite of the significant volume of ground-based [64–66] and spacecraft [67–71] data, the BOEs for the satellites of Saturn are very poorly studied. The data appear to be too sparse to construct detailed photometric phase curves. Noland *et al.* [64], Franklin and Cook [65], and Franz and Millis [66] found BOEs for all satellites, although their quantitative descriptions are absent. On the basis of the available data, only the geometric albedo, phase coefficients, and brightness differences between the leading and trailing hemispheres of the inner satellites Mimas, Enceladus, Tethys, Dione, and Rhea were determined [67]. The leading hemispheres of the inner satellites Tethys, Dione, and Rhea are brighter than their trailing sides, whereas for Enceladus and Mimas an opposite trend was found. These satellites have very high geometric albedos (0.77 for Mimas, 1.04 for Enceladus, 0.80 for Tethys, 0.55 for Dione, and 0.65 for Rhea [72]), thereby indicating icy surfaces with a very low fraction of dark opaque materials. The Voyager spacecraft images have established that the surface of Enceladus is strongly backscattering with normal reflectance >1.1, i.e., substantially greater than that of any typical natural surface such as freshly fallen snow.

Of all distant satellites of Saturn, extensive photometric data exist only for Hyperion and Iapetus. The sharp and irregular shape of Hyperion [71], probably produced by fragmentation of a larger parent body, causes its asynchronous rotation. In comparison with high-albedo inner satellites, Hyperion has low geometric albedo, which changes from 0.17 to 0.28 in line with visibility aspect. Earth-based photometric observations of this satellite in the phase angle range 0.23°–5.90° were reported in [65,73,74]. The Voyager data [71] extended this range up to 80°. The BOE amplitude determined by Degewij *et al.* [73] is 0.38^m in the V filter. The phase coefficient determined from the spacecraft data is 0.013 ± 0.007 mag/deg.

Iapetus is known to have a very unusual brightness asymmetry between the leading (geometric albedo p_V = 0.05) and the trailing (p_V = 0.60) hemisphere [69,75]. Therefore, the linear phase coefficient strongly varies with longitude from 0.028 mag/deg for the bright side to 0.068 mag/deg for the dark side. A nonlinear surge was found for both sides of the satellite [64,65]. Ground-based [64–66,74,76,77] and spacecraft [69] data allowed one to derive the phase curves for the bright and dark sides of Iapetus at phase angles from 0.23° to 90°.

Analyses of all available data have shown that detailed photometric phase curves can be constructed only for Tethys [67,78,79], Hyperion, and Iapetus. It should be noted that the phase curves for Iapetus can be obtained only for a very narrow range of longitudes: 270°–300° for the bright and 80°–92° for the dark side.

Saturn's rings. The unusual increase of brightness of the Saturn system near opposition was observed for the first time by G. Müller [80], who attributed the phenomenon to the rings and explained it by the effect of mutual shadowing among the ring particles. Using these observations, Seeliger [81] made the first quantitative determination of the opposition effect. Subsequently, Schönberg [82] and Hertzsprung [83] performed separate observations of rings A and B and studied their respective phase-angle brightness dependencies in the phase angle range from 0.12° to 6.45°. Schönberg [84] did not find any systematic features in the spectral brightness dependence of the B ring.

The work by Franklin and Cook [85] has become a classical study of the BOE. They obtained differentiated brightness curves for Saturn and for the A and B rings in the phase angle range 0.09°–5.68° in the B and V filters. The scatter of individual data points is so small that these results can be considered standard. No significant difference between the two rings was detected.

Numerous and extensive observations were performed in order to investigate the potential dependence of the photometric phase curve for Saturn's rings on declination of the Earth with respect to the ring plane (the so-called tilt effect) and on the orbital phase angle (the azimuthal effect) at different wavelengths [86–90]. The most important results of these papers can be summarized as follows: there was no significant difference between the shape of the phase curves for the A and B rings in different spectral bands [89]; no significant distinction was found between the western and the eastern cusp of the rings [86]; the phase curves for the cusps were remarkably similar at all ring tilts, wavelengths, and radial distances from Saturn [90].

The center of Saturn is often used as a spectrophotometric standard for studying the brightness of the rings. Therefore, it is very important to investigate the intrinsic opposition effect of Saturn. Avramchuk (unpublished data) has shown that the amplitude and $HWHM_B$ of the BOE for Saturn may reach 0.03^m and 0.49° in the V filter, respectively. The phase coefficient is found to be 0.010 mag/deg, which is smaller than that for the rings (0.037 mag/deg).

Satellites of Uranus. In 1983, Brown and Cruikshank [91] reported evidence of a strong opposition brightness surge of Ariel, Titania, Oberon, and possibly Umbriel at near-infrared wavelengths (1.43–2.57 μm). Subsequently, Goguen *et al.* [92] performed V filter photometry of Titania and Oberon in the phase angle range $0.06° < \alpha < 3.1°$. It was found that the amplitudes of orbital lightcurves of satellites were less than 0.1^m. The phase coefficients for Titania and Oberon ($\beta =$ 0.10 ± 0.02 mag/deg) were comparable with those observed for asteroids at similar phase angles. The increase of brightness was found to be 0.4^m in the phase angle range from 3.04° down to 0.06°. The measurements from Voyager and Voyager 2 extended the phase angle range up to 139° [93]. Buratti *et al.* [94] concluded that the five largest satellites of Uranus form a unique class of low-albedo, spectrally flat objects weakly different from each other. According to [95], the amplitudes of the opposition surges were very large (0.40^m–0.75^m), with the exception of Umbriel (0.14^m).

Extensive photometric data for four Uranian rings and 16 satellites were obtained with the HST in 1997 [96,97]. The phase angle range was 0.03°–2.8° and the 25 different filters covered the wavelength range from 0.27 to 2 μm. The data by Karkoschka [97] near $\alpha = 0°$ showed phase curves much steeper than those observed previously for each satellites.

Satellites of Neptune. In spite of the narrow phase-angle coverage (< 2°), ground-based observations of these satellites are suitable for studies of the BOE. The only satellite, for which the phase dependence of brightness has been examined in detail, is Nereid [95]. This is a small, chaotically rotating satellite with an irregular shape, which displays brightness variations at time scales from hours to days or even months [98]. The lightcurve obtained by Schaefer and Tourtellotte [95] in 1998 shows variations of up to 0.52^m with a single peak coinciding with opposition. Therefore, it is possible that this peak is mainly caused by the opposition effect. The phase curve displays an opposition brightening with an amplitude of 0.44^m, which puts Nereid in the class of objects with very strong opposition surges.

The V filter photometry of Triton [92] showed a weak brightness increase ($\beta = 0.03 \pm 0.03$ mag/deg) with an amplitude of 0.05^m.

Asteroids. Gehrels [99] was the first to observe the opposition effect for asteroids, namely, for an S-type asteroid 20 Massalia (Fig. 3a). Subsequently, Harris *et al.* [100] found an opposition spike for E-type asteroids 44 Nysa and 64 Angelina. The BOE is also typical of low-albedo asteroids [101]. At present, a significant number of photometric observations of asteroids have been published, but only a few of them have a good phase-angle coverage and are suitable for a study of the BOE [100–112]. The BOE parameters vary with taxonomic class of asteroids [113]. The average phase coefficients are 0.042, 0.033, 0.029, and 0.019 mag/deg for C-, M-, S-, and E-type asteroids, respectively [114]. The BOE typically starts at phase angles 3°–6° [115]. There are indications that the BOE parameters can vary slightly within one type of asteroid [116].

The photometric properties of near-Earth asteroids (NEAs) are very similar to those of the main belt asteroids [117]. Specifically, S-type asteroids of both families have the same average phase coefficient $\beta = 0.030 \pm 0.007$ mag/deg. The BOE is only known for two NEAs [118,119].

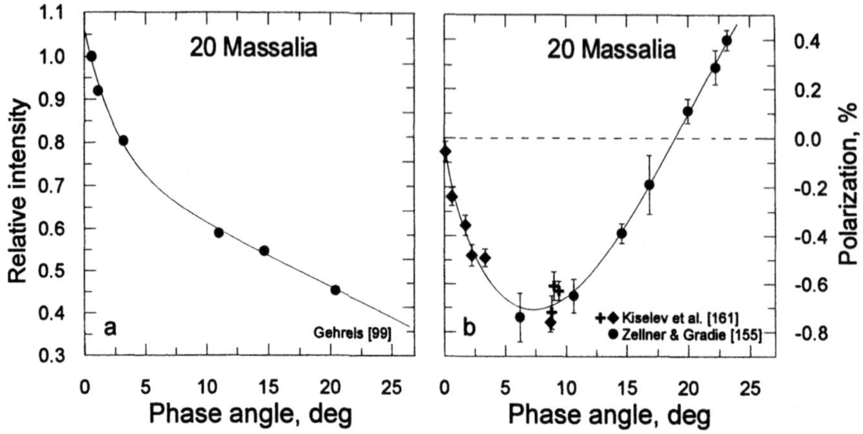

Figure 3: (a) The BOE and (b) the negative polarization branch for the S-type asteroid 20 Massalia. The fitted functions (solid curves) are analogous to those in Fig. 1.

Comets. Measurements of the brightness of cometary dust near opposition are still very rare. There is only one comet, 28P/Neujmin 1, for which photometric observations have been performed at phase angles less than 1° (0.8° < α < 8°) [120]. Available observational data and their interpretation can be found in [121,122]. Schleicher *et al.* [122] indicated that the phase dependence of the comet Halley dust has a pronounced curvature over the observed range 1.5°–66° (Fig. 4a). The phase coefficient is 0.045 mag/deg near opposition but becomes nearly flat at angles greater than about 30°. The brightness at 0° is 2.9 times that at the minimum near 50°. Fernandez *et al.* [123] determined the brightness phase dependence of the Encke nucleus. The phase coefficient is 0.06 mag/deg, thereby implying one of the steepest slopes known for any small solar system body.

Interplanetary dust. The photometric opposition effect is also observed for interplanetary dust particles (IDPs) [124]. Renard *et al.* [125] obtained the phase-angle dependence of the IDP brightness in the ecliptic plane within the phase angle range 0° ≤ α ≤ 90° and at the solar distance 1.5 AU. According to [126,127], the decrease in brightness of the zodiacal light with increasing phase angle is of the order of 0.015 mag/deg. The enhancement factor is about 2 at phase angles below 7°.

3.3 Polarimetric observations

The Moon. Starting with Arago, polarimetric observations of the Moon have been carried out by many investigators. However, a hundred years after Arago, B. Lyot [6] discovered that the disk-integrated polarization of the Moon as well as that of its local regions becomes negative at phase angles about 23°. Detailed reviews of the polarimetric properties of the Moon can be found in [20,128–131]. Summarizing these studies, one can conclude that the main properties of the negative polarization branch for the Moon are the following: i) the negative polarization branch for the whole Moon is almost symmetric with parameters $P_{min} = -1.2\%$, $\alpha_{min} = 11°$, and $\alpha_{inv} = 23.6°$; ii) the lunar objects of different morphology exhibit different values of P_{min} within the range from –0.6% to –1.4%. The low negative polarization values are characteristic of bright highlands, whereas dark maria have greater $|P_{min}|$; iii) the dependence of P_{min} on the surface albedo for various lunar formations has a horse-shoe shape: at low albedos, $|P_{min}|$ increases with albedo, but then decreases when the albedo exceeds a certain threshold; iv) the absolute value of P_{min} for bright craters is independent of the wavelength, but slightly increases with wavelength for dark material; iv) the inversion angle slightly increases with wavelength.

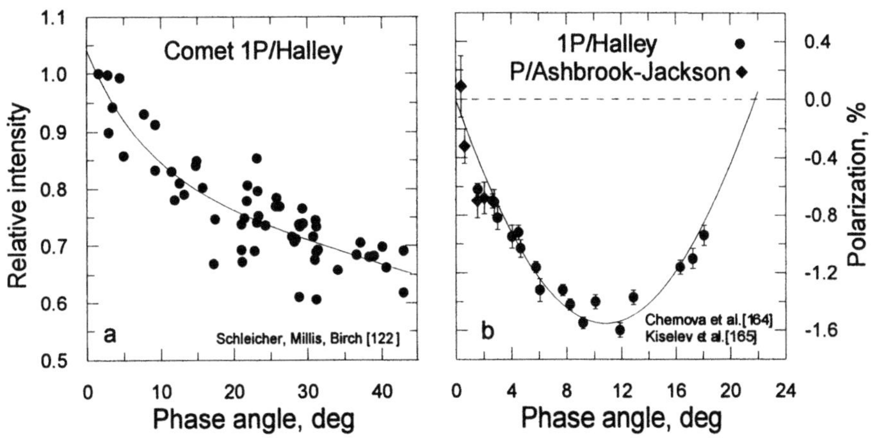

Figure 4: The BOE (a) and negative polarization branch (b) for comet 1P/Halley. The fitted functions (solid curves) are analogous to those in Fig. 1.

Mercury. The first observations of Mercury near opposition (5° < α < 22°) were made by Lyot [132] in 1930. The disk-integrated polarization was very similar to that for the waxing and waning Moon. Dollfus and Auriete [133] extended the polarization curve for Mercury down to the phase angle 4° in several spectral bands. They confirmed that the polarization parameters for Mercury, P_{min} = –1.4%, α_{min} = 11°, and α_{inv} = 25°, were close to those of the Moon.

Mars. Initially Lyot [6] and later Dollfus [134], Dollfus and Focas [135], Morozhenko [136], and Bugaenko *et al.* [137] measured polarization of the whole Mars as well as of different regions near opposition. The results can be summarized as follows: i) the polarization curves are different for different oppositions; ii) P_{min} is about –1% and α_{min} is about 12°; iii) the inversion angle increases with wavelength and is in the range 24°–29°. Thus, the inversion angle for Mars is among the largest values ever measured for ASSBs; iv) the absolute value of the degree of polarization is greater for dark features (maria) then for bright features (continents); v) the degree of polarization of the dark features strongly depends on the latitude and season.

Deimos. Zellner [138] discovered a well-developed negative polarization branch for Deimos, with P_{min} close to –1.5%, α_{min} close to 10°, and an inversion angle of about 19.5°.

Galilean satellites. Several polarimetric data sets for the Galilean satellites have been obtained with UBVR filters at phase angles ranging from 11.8° to nearly 0° [139–144]. The phase-angle, orbital, and spectral dependencies of polarization were studied. There are some systematic differences in the depth of the negative polarization branch for Io, Europa, Ganymede, and Callisto. Large orbital variations in the degree of polarization were found for Callisto (Fig. 2b) and Io, whereas smaller longitudinal effects have been detected for Ganymede and Europa. The minimal polarization values for the trailing hemispheres of Io, Europa, and Ganymede are systematically higher than those for the respective leading hemispheres. However, the leading hemisphere of Callisto is characterized by considerably higher values of polarization and larger α_{min} in all spectral bands as compared to those for the trailing hemisphere. The separation of the phase and longitudinal effects in polarization was performed for Ganymede [140] and Callisto [140, 60].

For the first time, the presence of the polarization opposition effect in the form of a sharp spike of negative polarization centered at a very small phase angle of $\alpha \approx 0.5°-0.7°$ and superimposed on the regular negative polarization branch was detected for Io, Europa, and Ganymede [143–145]. This phase angle is comparable to the width of the brightness peak observed for Europa (Fig. 1b), thereby indicating that both opposition phenomena are likely to be produced by the coherent backscattering mechanism [143]. For Callisto, a sharp negative polarization spike at $\alpha < 1°$ was not detected (Fig. 2b).

Satellites of Saturn. There are sparse polarization measurements for three satellites of Saturn: Dione, Rhea, and Iapetus [146,147]. The results are sufficient to estimate the depth of the negative polarization branch, but the phase-angle dependence of polarization cannot be determined. Dione and Rhea have similar values of P_{min} close to –0.4%. For Iapetus, the negative branch was found to be much deeper for the dark leading hemisphere (P_{min} = –1.3%) than for the bright trailing side (P_{min} = –0.2%).

Saturn's rings. Lyot [4] measured polarization of Saturn's rings A and B at very small phase angles. The polarization curve for the B ring was found to be highly asymmetric and differed dramatically from the nearly parabolic negative polarization branch typical of most ASSBs. Extensive investigations of the ring polarization were also performed in [148–151]. Johnson *et al.* [150] measured a detailed profile of negative polarization for the A and B rings before and after opposition in the phase angle range 0.5°–6°. The $P(\alpha)$ curve was asymmetric with $P_{\min} = -0.5\%$ at a phase angle of about 3°. They called the extrapolated sharp change of polarization at very small phase angles ($0° < \alpha < 0.5°$) "the hypothesized polarimetric opposition effect." Johnson *et al.* concluded that both the BOE and the POE are caused by light-scattering processes that occur within optically active regolith layers of individual ring particles and are not due to multiple reflections of light and mutual shadowing by different particles.

Dollfus [151] indicated that the observed polarization appears to be the result of several effects. He corrected the polarimetric measurements of the rings for the effect of illumination from the Saturn's globe and for the multiply scattered component and derived the polarization curve of the B ring caused by direct reflection of sunlight by the ring particles. In his original work Dollfus [151] approximated the polarization curve $P(\alpha)$ by an asymmetric curve with $P_{\min} \approx -0.35\%$ at $\alpha_{\min} \approx 2°$. Later, Dollfus re-analyzed those data and found a narrow spike of negative polarization centered at a phase angle $\alpha \approx 0.5°$ [143]. Thus, the POE, which was first unequivocally detected for the Galilean satellites, was also observed for the Saturn's B ring.

Asteroids. Lyot [152] was the first to discover polarization of light scattered by asteroids. He found that asteroids 1 Ceres and 4 Vesta exhibit the same negative polarization at small phase angles as the Moon, Mars, and Mercury. An extensive program of polarimetric observations has been carried out by Zellner, Gradie, and Gehrels [153–155]. As a result, the negative polarization branch for many asteroids was studied in detail. The parameters $P_{\min}$, $\alpha_{\min}$, and α_{inv}, were found to be different between the main composition types of asteroids S, C, M, and E.

Polarimetric studies of asteroids were actively pursued in the former Soviet Union [156] and, in recent years, were continued in the Kharkiv National University [157]. The main purpose of this research was to classify asteroids according to their $P_{\min}$ values and to study the spectral dependence of $P_{\min}$. It has been found that the absolute value of $P_{\min}$ increases with wavelength for moderate-albedo asteroids (types S, M, and V), whereas $P_{\min}$ of low-albedo asteroids (types G, B, CP, and F) decreases with wavelength. Several asteroids have been observed near opposition ($0.1° < \alpha < 3.3°$) [158–161]. However, only the observations of 20 Massalia allow one to investigate the behavior of polarization near opposition in detail. All available data for Massalia are shown in Fig. 3b. Obviously, the negative polarization branch is slightly asymmetric and has no peculiarities at very small phase angles.

Comets. Negative polarization of sunlight scattered by cometary dust was discovered by Kiselev and Chernova [162,163] in 1976 and is a typical feature of comets. 1P/Halley is a unique comet for which the detailed negative branch of polarization was observed with high accuracy at phase angles down to 1.57° [164]. It has a symmetric shape with $P_{\min} = -1.5\%$ and $\alpha_{\min} = 11°$ in the green continuum filter. Only one comet, 47P/Ashbrook-Jackson, was observed at smaller phase angles down to 0.36° [165]. The polarization phase curves for the comets Halley and Ashbrook-Jackson are plotted in Fig. 4b. There is possibly a small depression of polarization at a phase angle about 2°.

Interplanetary dust. Polarization observations of the zodiacal light have been performed by various authors. Wolstencroft and Rose [166] detected negative polarization near opposition with $P_{\min}$ close to –2%, which is similar to the values observed for comets. However, the inversion angle was only of the order of 12°, i.e., significantly smaller than that for comets. Thus, one should not exclude the possibility that physical properties of interplanetary dust particles are not identical to those of cometary dust [167,168].

4. Recent polarimetry of satellites and an asteroid

As we could see from our review of polarimetric observations of ASSBs, the measurements at very small phase angles ($\alpha < 1°$) are rare, and the detailed behavior of polarization is still poorly known. Below we present and discuss the results of the most recent polarization observations of the Galilean satellites, Iapetus, and the asteroid 64 Angelina near opposition. These high-albedo objects were specifically selected for a study of the POE. For some of them (Europa and 64 Angelina), a spike-like BOE is also found. This is in agreement with the theoretical prediction by Mishchenko [7], according to which the POE and the sharp BOE can be two spectacular manifestations of the same optical phenomenon, viz., coherent backscattering.

Galilean satellites. CCD imaging polarimetry of the Galilean satellites was carried out with the 70-cm telescope of the Astronomical Observatory of the Kharkiv National University during the period 9–25 September 1998 over the phase angle range $0.4° - 3°$ in the V filter [144,169]. During the 2000 opposition, electrophotopolarimetric observations of the Galilean satellites were performed with the same telescope in the UBVR filters [145]. The minimal phase angle reached was as small as 0.20°. For the first time, the polarization phase dependence for the Galilean satellites near opposition was studied in detail and with high accuracy (about ±0.02%). Figures 5a and 5b illustrate the behavior of polarization for Europa (before and after the opposition) and Ganymede in different spectral bands. As one can see, there is a sharp spike of negative polarization for Europa (with an amplitude of 0.35% at a phase angle of 0.20°). The angular profile of the observed polarization feature resembles the theoretical angular profile caused by the CBM for nonabsorbing Rayleigh particles [7,170] and the profile measured for a particulate surface composed of microscopic MgO grains [6]. Similar results were obtained in the laboratory for other high-albedo fine powder samples [1]. The behavior of polarization for Ganymede near opposition is significantly different from that for Europa. The POE may be in the form of either a well-separated peak or a highly asymmetric curve. However, previous observations indicate the likely superposition of a separate narrow peak (P_{min} = –0.38% and α_{min} = 0.6°) and a regular negative branch.

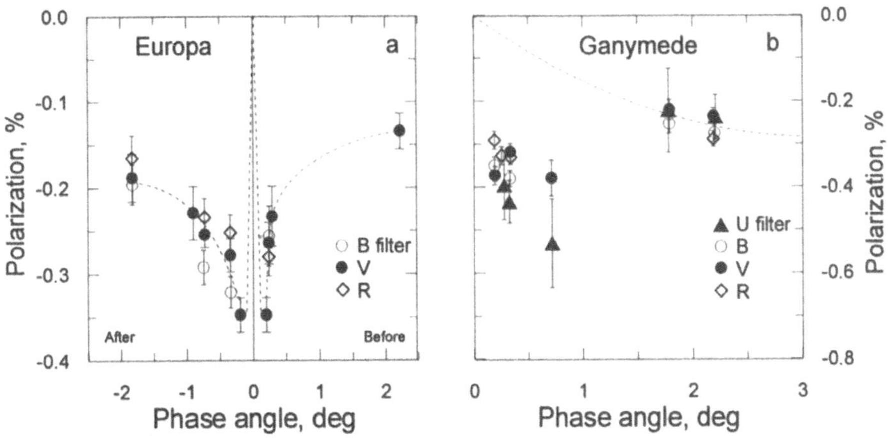

Figure 5: The polarization opposition effect for Europa (a) and Ganymede (b) in the UBVR filters. The dashed curve for Ganymede is a trigonometric fit of all previous observations [140–144].

The results of all polarization measurements for Europa obtained by different observers [140–144] are summarized in Fig. 1b. The analysis of the available data shows that Io, Europa, and Ganymede have a sharp spike of negative polarization centered at a very small phase angle (< 1°) and superimposed on the regular negative polarization branch. It is interesting that the satellites demonstrate spikes of different shape. Callisto (Fig. 2b) exhibits only the regular negative polarization branch for the leading hemisphere and a slightly asymmetric phase curve for the trailing hemisphere without any features near opposition [60].

Iapetus. Observations of Iapetus were carried out with the five-channel UBVRI photopolarimeter installed on the 1.25-m telescope of the Crimean Astrophysical Observatory [137]. Iapetus was observed from 11 October 1998 to 18 January 1999 at phase angles up to 6.1°. The minimal phase angle reached was as small as 0.30°. The phase-angle dependence of polarization for the leading (dark) and trailing (bright) hemispheres of Iapetus is shown in Fig. 6. The phase curve for the dark side (Fig. 6a) is asymmetric with P_{min} = –1.1% at α_{min} =3.5° and does not show the POE. The bright side (Fig. 6b) reveals a high degree of negative polarization (up to –0.8% at α = 1°); whereas, polarization is approximately

–0.2% at phase angles 5°–6°. One may expect that the negative polarization branch for the bright side of Iapetus either contains a separate POE peak or is a highly asymmetric curve.

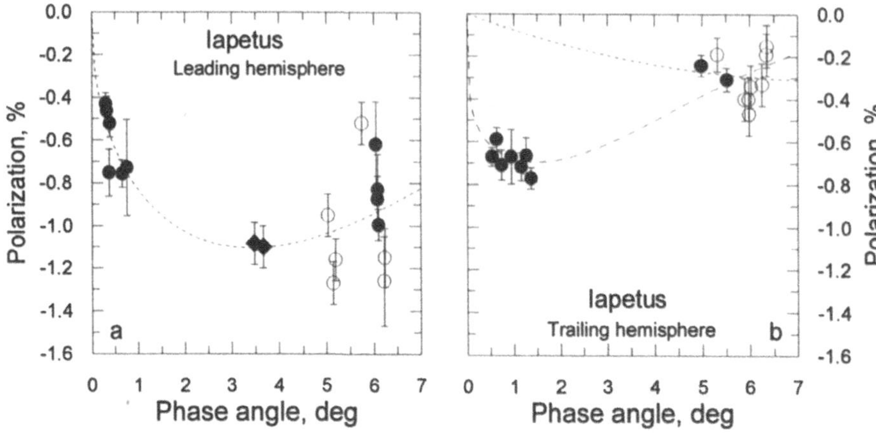

Figure 6: The polarization of the leading (a) and trailing (b) hemispheres of Iapetus in the R filter. Filled circles show data by Rosenbush *et al.* [144]; filled diamonds – those by Jockers [144]; open circles – those by Zellner [146,171].

64 Angelina. During the 1995, 1999, and 2000–2001 oppositions, the UBVRI polarimetry of the E-type asteroid 64 Angelina was performed with the same equipment as the observations of Iapetus [144,145]. The phase-angle dependence of polarization in different filters clearly shows the presence of the POE in the form of a narrow peak of negative polarization centered at $\alpha \sim 1.5°$ and superimposed on the regular negative polarization branch (Fig. 7). The amplitude of the POE is apparition-dependent. Similar apparition dependence of the brightness for Angelina had already been noticed by Poutanen [172]. In addition, there is a distinct difference between the POE amplitudes before and after the 1999 opposition. As one can see in Fig. 7, the $P_{\min}$ value of the polarization opposition spike also depends on wavelength. The angular half-width of the polarization spike is comparable to that of the brightness opposition peak [100]. Thus, it appears likely that the polarization spike is a manifestation of the coherent backscattering of sunlight by the Angelina surface.

The totality of the observations is consistent with the theory of coherent backscattering. According to this theory, high-albedo particulate surfaces should exhibit a strong and very sharp intensity peak centered at zero phase angle [173]. Furthermore, the narrow BOE should be accompanied by the POE provided that the surface grains are comparable to or smaller than the wavelength [7,170]. All objects for which we have discovered a sharp spike of negative polarization have a high albedo (64 Angelina – 0.55; Io – 0.61; Europa – 0.64; Ganymede – 0.42) and show a strong and narrow brightness peak [52,55,100]. In agreement with the theory, the angular semi-widths and the angular positions of the polarization spikes relative to the exact backscattering direction are comparable to the angular semi-widths of the respective photometric opposition peaks.

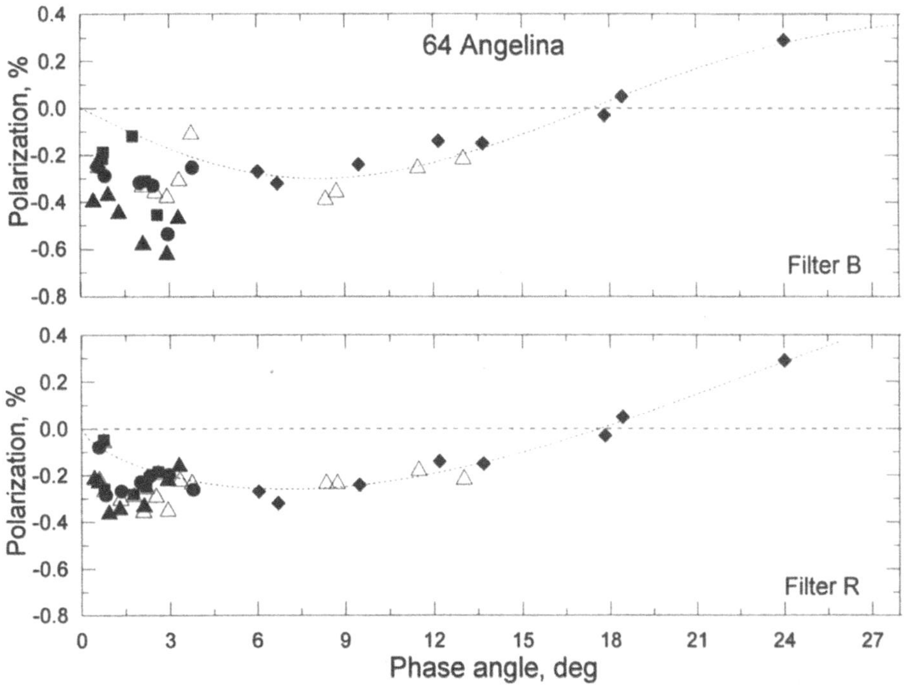

Figure 7: The POE for 64 Angelina in the B and R filters. The dashed curve is a trigonometric polynomial fit to the regular negative polarization branch [57]. Filled circles show the data obtained during the 2000 opposition; filled and open triangles show the data obtained before and after the 1999 opposition, respectively; filled squares show the results collected during the 1995 opposition. Diamonds are the data by Zellner and Gradie [155].

However, the position and amplitude of the observed polarization spikes and the width and amplitude of the brightness peaks can be different for different objects, e.g., for Europa and 64 Angelina. Therefore, it is reasonable to assume that even within the framework of the same optical mechanism (coherent backscattering), the natural variability in particle parameters (such as size, shape, refractive index, porosity) can change the parameters of the BOE and POE.

5. Relationships between photometric and polarimetric opposition effects

Photometric and polarimetric characteristics of light scattered by ASSB surfaces should be interrelated because they are possibly the result of the same optical processes. In particular, it is interesting to study such interrelations in the range of small phase angles, where both the brightness surge and the negative polarization branch are observed. As follows from the existing theory, two primary physical mechanisms, coherent backscattering and shadow hiding, may explain some of the observed phenomena near opposition. Relative contributions of both mechanisms may depend on the physical properties (composition, porosity, structure, particle sizes, etc.) of the surface layers as well as on the geometry of light scattering. Comparisons of the parameters of the observed photometric and polarization effects may provide some evidence as to what mechanism dominates in specific circumstances.

There are some indications that the BOE and the negative polarization are interrelated. Specifically, Kolokolova [174] found correlations between some parameters of both effects for some ASSBs, but she also pointed out that the number of objects analyzed was not large enough to derive definitive conclusions. A correlation between $P_{\min}$ and the slope of the brightness phase curve at small phase angles for some lunar surface features was detected by Shkuratov [20], who also found indications of a common nature of the negative polarization branch and the BOE. The currently available volume of data sets, including the new polarization data, allows us to revisit this issue.

5.1 Observational data and approximating functions

We have analyzed the entire volume of the available observational data for different objects in order to select objects with a sufficient phase angle coverage and a small scatter of data points in brightness and polarization phase curves. The extrapolation of brightness phase curves to zero phase angle depends on the choice of approximating function and may lead to inaccurate estimates of the BOE parameters. We used a modified four-parameter exponential-linear function by Keränen *et al.* [56]:

$$I_{\text{fit}}(\alpha) = I_s \exp\left(-\frac{\alpha}{1.45 \times \text{HWHM}_B}\right) + I_b + \beta\alpha$$

where I_s is the amplitude of the BOE, I_b is the background intensity, β is the slope of the linear part, and HWHM_B is the semi-width at half-maximum of the BOE. For well-defined phase curves, this function yields the parameters with good accuracy, but it rather strongly depends on the smallest phase angle for which observational data are available. For some objects observed at phase angles down to a few hundredths of a degree (e. g., Europa, Oberon, Titania, Nereid), we have compared the parameters obtained through the exponential-linear fit and directly from observations without any approximation. The agreement is excellent. The main advantage of the exponential-linear fit is that it yields simultaneously all parameters describing the opposition effect and the linear part of the photometric phase curve.

The intensity extrapolated to zero phase angle may be in error because the detailed shape of the opposition surge in the nearest vicinity of opposition ($\alpha < 1°$) is still unknown for the majority of the objects. The finite angular size of the Sun should be taken into account for such objects as the Moon, Mars, and the satellites of Jupiter and Saturn. The observed photometric phase curves do not form a single family and show significant diversity. For some bodies (e.g., Callisto, Phobos, Deimos), the curvature of the brightness phase dependence changes rather slowly over the entire phase angle range, which does not allow one to identify reliably the linear part. For several dark asteroids (e.g., C- and P-type asteroids), the parameters of the BOE are poorly defined due to the weakness of the opposition surge, and one can see only the linear part of the phase curve.

Table 1 lists the objects analyzed and the disk-integrated or hemisphere-integrated parameters of the BOE: ζ, HWHM$_B$, β, and α_{opp}. These parameters were also derived for some features of the Europa's, Martian, and lunar surface and are listed in Table 2. The tables also list the corresponding geometric albedo values p_v and provide references to the relevant literature sources. The estimated retrieval errors are as follows: 0.01–0.08 for ζ; 0.1°–0.7° for HWHM$_B$; 0.0001–0.004 for β; and 0.1°–1° for α_{opp}.

In order to determine the parameters of the negative polarization branch P_{min}, α_{min}, and α_{inv}, the observations were fitted with a trigonometric polynomial [57]

$$P_{fit}(\alpha) = b \sin^{c_1}(\alpha) \cos^{c_2}(\alpha/2) \sin(\alpha - \alpha_{inv})$$

where the parameters b, c_1, c_2, and α_{inv} are determined using a non-linear least-squares fitting technique. The parameter c_2 describes the asymmetry of the polarization phase curve at large phase angles and is usually small, thereby making the term $\cos^{c_2}(\alpha/2)$ close to unity. The disk-integrated or hemisphere-integrated parameters P_{min}, α_{min}, and α_{inv} are listed in Table 3 along with the corresponding values of p_v and HWHM$_P$ and references to the relevant literature sources.

5.2 Correlations between BOE and POE parameters

Figure 8 illustrates the relationships between the BOE parameters ζ, HWHM$_B$, β, and α_{opp} and the geometric albedo p_v for different ASSBs (the notation is explained in Fig. 9; the letters denote the respective asteroid classes). Two branches on the $\zeta - p_v$ diagram (Fig. 8a) are clearly distinguishable for bright and dark objects. As a consequence, two objects with a very low and a very high albedo may exhibit the same enhancement factor. This is an indication that the BOE for dark and bright objects is likely to be caused by two different optical mechanisms, presumably by the SM and the CBM, respectively. The CBM is a multiple-scattering phenomenon and is more relevant to bright surfaces. On the

Table 1: Parameters of the brightness opposition effect for various ASSBs

Object	p_v	ζ	$HWHM_B$ deg	β, mag/deg	α_{opp}, deg	Reference
Moon	0.12	1.63	1.11	0.0219	6.7	[11,12]
Mars	0.12	1.47	3.21	0.0133	16.9	[37]
Phobos	0.05	2.12	5.03	0.0234	14.8	[42]
Deimos	0.06	1.95	6.14	0.0206	17.3	[42]
Io	0.62	1.20	0.83	0.0235	4.7	[52,175]
Europa, leading	0.68	1.14	0.30	0.0105	1.1	[55]
Europa, trailing	0.68	1.27	0.16	0.0129	0.78	[55]
Ganymede, leading	0.44	1.26	0.25	0.0107	1.3	[52]
Callisto, leading	0.19	1.65	2.35	0.0099	12.8	[55]
Callisto, trailing	0.19	1.23	1.24	0.0255	5.4	[55]
Saturn's Ring A	0.75	1.32	0.26	0.0403	1.07	[89]
Saturn's Ring B	0.82	1.27	0.31	0.0332	1.37	[89]
Ring A+B	0.74	1.28	0.32	0.0368	1.22	[85]
Tethys	0.79	1.39	0.35	0.0157	1.6	[64,66,72,78,79]
Iapetus, leading	0.05	1.68	2.79	0.0115	15.2	[64,65,76,77]
Iapetus, trailng	0.60	1.41	0.14	0.0236	0.7	[64,65,76,77]
Hyperion	0.25	1.51	0.34	0.0096	1.8	[65,70,73]
Titania	0.27	1.40	0.57	0.0210	2.2	[92,93,97]
Oberon	0.23	1.58	0.47	0.0171	2.8	[92,93,97]
Nereid	0.16	1.43	0.38	0.0800	1.9	[95,98]
44 Nysa, type E	0.55	1.28	0.80	0.0205	3.6	[100]
64 Angelina, E	0.48	1.31	0.85	0.0198	3.9	[100]
214 Aschera, E	0.52	1.15	1.07	0.0289	4.2	[107]
317 Roxane, E	0.49	1.22	0.72	0.0238	3.2	[108]
5 Astraea, S	0.23	1.33	2.58	0.0257	12.0	[106]
20 Massalia, S	0.21	1.41	1.77	0.0306	8.9	[99]
29 Amphitrite, S	0.18	1.51	2.19	0.0242	11.3	[176]
30 Urania, S	0.17	1.72	1.85	0.0221	10.2	[177]
79 Eurynome, S	0.26	1.50	1.37	0.0219	7.0	[111]
126 Velleda, S	0.17	1.42	0.80	0.0267	4.0	[116]
695 Bella, S	0.16	1.22	2.20	0.0358	10.5	[108]
133 Cyrene, SR	0.26	1.42	1.09	0.0305	5.5	[103]
16 Psyche, M	0.10	1.58	1.61	0.0222	8.9	[178]
22 Kalliope, M	0.14	1.43	1.68	0.0266	8.4	[179]
55 Pandora, M	0.13	1.36	1.31	0.0297	6.3	[109]
69 Hesperia, M	0.14	1.46	2.04	0.0247	10.2	[104]
110 Lydia, M	0.18	1.32	2.10	0.0272	10.6	[180]
201 Penelope, M	0.16	1.49	1.99	0.0243	10.3	[181]
83 Beatrix, M	0.09	1.40	2.26	0.0313	11.3	[182]
1 Ceres, G	0.11	1.30	1.90	0.0395	9.4	[183]
19 Fortuna, G	0.04	1.30	3.11	0.0271	15.3	[176]
130 Elektra, G	0.08	1.35	1.76	0.0270	9.0	[111]
10 Hygiea, C	0.07	1.36	–	0.0366	10	[112]
47 Aglaja, C	0.08	1.21	1.72	0.0355	7.1	[158]
165 Loreley, C	0.08	1.22	2.52	0.0402	10.6	[108]
344 Desiderata, C	0.04	1.19	2.50	0.0409	10.4	[110]
379 Huena, C?	0.08	1.41	0.64	0.0404	3.4	[108]
59 Elpis, CP	0.04	1.26	3.24	0.0398	14.3	[111]
50 Virginia, P	0.04	1.39	3.22	0.0414	16.0	[184]
102 Miriam, P	0.05	1.26	2.46	0.0434	10.9	[184]
276 Adelheid, PC	0.04	1.05	1.24	0.0504	4.2	[110]

Table 2: Parameters of the brightness opposition effect for some features of the Europa's, Martian, and lunar surfaces

Object		p_V	ζ	$HWHN_B$ deg	β, mag/deg	α_{opp}, deg
Features of the lunar surface [12,15–17,27]						
Wood's Region		0.07	5.4	0.44	0.0142	3.5
Platon		0.07	1.7	3.44	0.0176	18.6
Mare Imbrium		0.08	2.36	0.79	0.0176	5.1
Sinus Media		0.08	1.41	0.51	0.0157	3.1
Features of the Martian surface [38,39]						
Bright areas,	λ539 nm	0.21	1.18	1.81	0.0133	7.7
	λ592 nm	0.30	1.16	0.86	0.0120	3.5
Flat areas 1,	λ539 nm	0.20	1.16	1.19	0.0144	4.8
	λ592 nm	0.29	1.15	0.77	0.0109	3.1
Flat areas 2,	λ592 nm	0.29	1.10	0.30	0.0117	1.0
Dark areas,	λ539 nm	0.18	1.14	1.82	0.0117	7.0
	λ592 nm	0.21	1.08	1.09	0.0064	3.5
Crater Tails,	λ592 nm	0.29	1.14	0.85	0.0139	9.2
Arabia BR,	λ592 nm	0.34	1.14	1.29	0.0163	5.0
Arabia FL,	λ592 nm	0.34	1.09	0.44	0.0266	1.4
Syrtis Major, BR,	λ592 nm	0.22	1.20	1.35	0.0009	5.8
Syrtis Major FL,	λ592 nm	0.18	1.05	1.81	0.0119	4.5
Chryse BR,	λ592 nm	0.30	1.15	1.03	0.0094	3.9
Chryse BFL,	λ592 nm	0.29	1.14	0.49	0.0073	1.9
Chryse AFL,	λ592 nm	0.29	1.12	1.09	0.0112	3.9
Chryse DA,	λ592 nm	0.21	1.11	1.71	0.0039	5.8
Mare Erythraeum,	λ592 nm	0.22	1.15	2.16	0.0005	8.5
Features of the Europa's surface [63]						
IR-bright icy		0.96	1.56	0.08	0.0340	0.50
IR-dark icy		0.93	1.57	0.08	0.0332	0.48
Dark lineaments		0.74	1.82	0.08	0.0217	0.49
Dark spot		0.48	1.68	0.09	0.0524	0.44
Dark ridge		0.64	1.65	0.09	0.0242	0.51
IR-dark icy ridge		0.77	1.48	0.12	0.0220	0.63

other hand, the SM is a single-scattering phenomenon, and as such is more relevant to dark surfaces. It is, thus, expected that the CBM is stronger for bright surfaces, whereas the SM is stronger for dark surfaces. For all dark objects, with the exception of Phobos, Deimos, and some features on the Moon (see Tables 1 and 2), the values of ζ do not exceed 1.8, whereas for all bright objects $\zeta < 1.6$. These values are in a good quantitative agreement with theoretical predictions [186,173,170].

Low-albedo objects display a great scatter in ζ, which may be attributed to varying porosity of regoliths [186]. For bright objects, the multiple-scattering contribution increases with increasing albedo and raises the enhancement factor, which is in agreement with the theory of the CBM. For moderate-albedo objects ($p_V \approx 0.3$), both the SM and CBM can contribute to the opposition effect. A non-monotonic dependence of the BOE amplitude on albedo for asteroids, which was found in [115], is only one part of the two-branch dependence for a wide class of objects.

Table 3: Parameters of the negative polarization branch for various ASSBs

Object	P_{min}, %	α_{min}, deg	α_{inv}, deg	$HWHM_B$, deg	References
Whole Moon	−1.3	10	22.7	2.1	[130]
Moon, Mare Imbrium	−1.0	11.5	23.5	3.2	[130]
Mercury	−1.4	10	25		[133]
Mars	−1.0	12	24–29		[135]
Deimos	−1.5	10.3	19.5	4.0	[138]
Io	−0.21	3.5	9.7	1.0	[185]
Europa	−0.22	5.2	8.6	1.5	[185]
Ganymede	−0.29	3.3	8.9	0.97	[185]
Callisto, leading side	−0.85	10.1	26	1.72	[60]
Callisto, trailing side	−0.63	4.9	14	0.65	[60]
Dione	−0.4				[141]
Rhea	−0.4				[141]
Iapetus, leading side	−1.1	3.0		0.36	This work
Iapetus, trailing side	−0.2				This work
Comets	−1.6	11.2	21.5	2.6	[164]
Interplanetary Dust	≈−2		≈12		[167]
Saturn's Ring B	−0.38	2	7	0.3	[151]
	Asteroids: [160, this work]				
1 Ceres	−1.71	7.2	17.8	1.68	
5 Astraea	−0.70	8.2	19.1	2.13	
16 Psyche	−1.43	9.8	25.3	2.84	
19 Fortuna	−1.81	10.6	21.7	3.17	
20 Massalia	−0.64	6.7	20.1	0.73	
24 Themis	−1.76	8.7	21.1	2.93	
29 Amphitrite	−0.90	10.4	21.8	4.1	
30 Urania	−0.80	8.7	19.7	2.57	
44 Nysa	−0.31	5.1	17.3	0.68	
47 Aglaja	−1.46	8.4	17.8	2.78	
55 Pandora	−0.95	8.5	17.3	1.28	
64 Angelina	−0.32	6.3	17.6	1.13	

One of the main characteristics of the BOE is $HWHM_B$. As shown by Mishchenko in the framework of the CBM theory [187], $HWHM_B$ can depend strongly on the optical properties of the scattering medium. Figure 8b shows that $HWHM_B$ is inversely proportional to the geometric albedo. The CBM is most effective at small phase angles. Therefore, high-albedo objects should have small $HWHM_B$ values, as is indeed observed. For low-albedo bodies, the large scatter of $HWHM_B$ data points occurs due to variations in physical properties of the surface, mainly the porosity and texture. A similar relationship is observed between α_{opp} and p_V (Fig. 8d). It is a consequence of a strong linear correlation between the $HWHM_B$ and the phase angle at which the BOE starts.

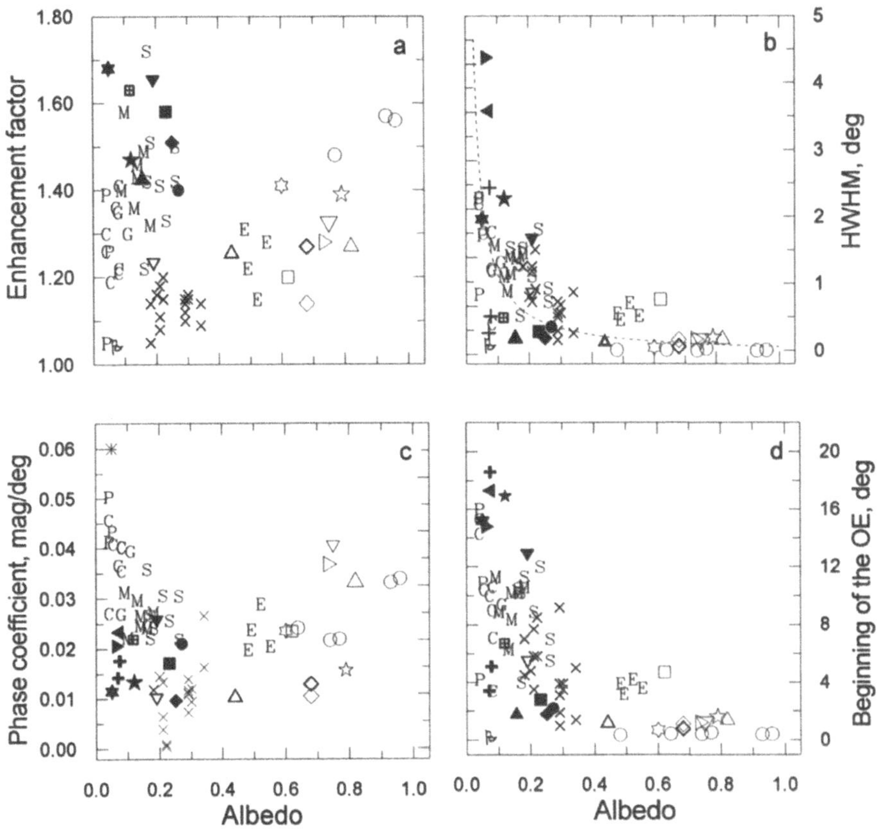

Figure 8: ζ, $HWHM_B$, β, and α_{opp} versus p_v for a representative selection of ASSBs.

A two-branch dependence of the phase coefficient on the geometric albedo is clearly seen in Fig. 8c. Again, bright and dark objects show different behavior. For low-albedo bodies, the phase coefficient decreases with increasing albedo. Obviously, the growth of the multiple-scattering component results in the attenuation of shadows. For the same reason, high-albedo objects show, in general, lower values of the phase coefficient than low-albedo objects. However, there is an increase of β with increasing albedo for moderate- and high-albedo objects.

If the BOE and the negative polarization have a common origin, then there must be a relationship between the photometric and polarimetric parameters. Indeed, there is a direct correlation between the angular widths of the BOE and POE, as Fig. 10a demonstrates. The bright and dark objects are well separated. Note that the region with HWHM $\leq 0.5°$ corresponds to objects which exhibit a pronounced opposition spike in brightness and a well-separated negative polarization spike. The region with $0.5° \leq$ HWHM $\leq 1.5°$ is occupied by bright objects with an opposition spike in brightness and a regular asymmetric negative polarization branch. Low-albedo objects have large angular half-widths, between

1.5° and 3.5°. It is possible that the observed angular half-widths for these objects are the result of simultaneous contributions from the SM and the CBM. Mishchenko [187] showed theoretically that the angular half-width of the coherent opposition effects for silicate surfaces can exceed a few degrees. And indeed, 64 Angelina has the largest values of both $HWHM_B$ and $HWHM_P$ among the high-albedo objects.

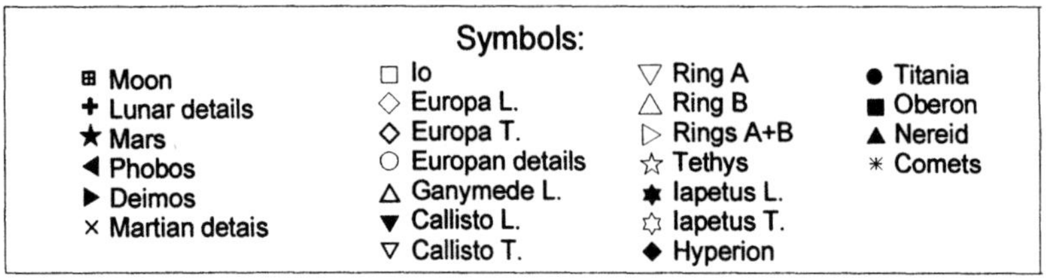

Figure 9: Symbols used to denote different objects in Figs. 8 and 10.

The parameters ζ and $P_{\min}$ are essentially the amplitudes of the BOE and the POE, respectively. Unfortunately, the existing theory of coherent backscattering does not predict a well-defined correlation between them. Figure 10b suggests that this relationship can, in fact, be non-monotonic. Additional observations are needed before more definitive conclusions can be reached.

Figure 10c is a scatter plot of $\alpha_{\min}$ versus α_{opp}. A direct correlation between these parameters is rather evident. Furthermore, the objects are obviously segregated into two groups with high and low albedos. Objects exhibiting the POE form a separate group. Figure 10d shows $|P_{\min}|$ versus the geometric albedo. It is evident that $|P_{\min}| \propto 1/p_v$. A very similar dependence was found by Zellner [146].

The dependence of $HWHM_B$ on the wavelength may be a good test of whether CBM can explain the opposition spikes exhibited by high-albedo objects [187–189]. The good agreement between the theoretical and experimental results in Fig. 11 suggests that at least for Saturn's rings, the opposition brightness spike is undoubtedly caused by coherent backscattering.

6. Summary

Our analyses of the available photometric and polarimetric data collected near opposition lead to the following conclusions.

- Many different classes of ASSBs (planets, satellites, rings, and asteroids as well as cometary and interplanetary dust particles) exhibit a brightness

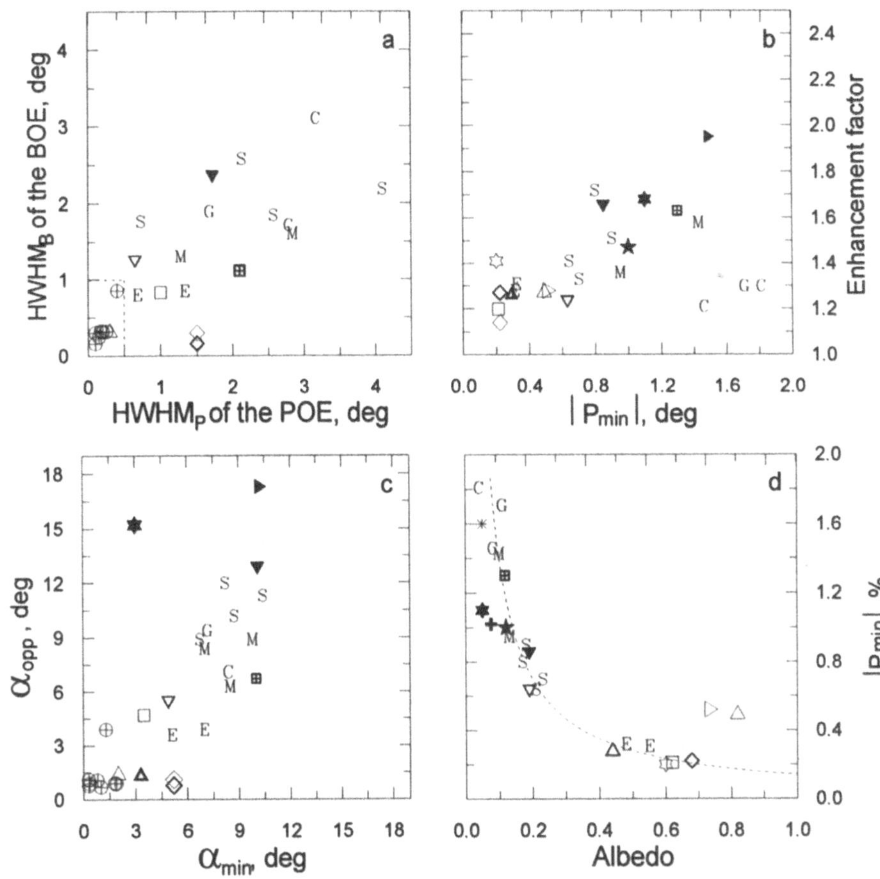

Figure 10: Photometric parameters $HWHM_B$, ζ, α_{opp}, and p_v versus negative polarization branch parameters $HWHM_P$, P_{min}, and α_{min} for various ASSBs.

opposition effect and the negative polarization branch at small phase angles. There are significant longitudinal, latitudinal and regional variations in brightness and polarization, which cause peculiarities in the photometric and polarimetric phase curves observed both for entire objects (asteroids, satellites) and for disk-resolved areas (the Moon, Mars, some satellites). The observed photometric phase curves show a great diversity. For some objects (e.g., icy satellites, Saturn's rings, asteroids 44 Nysa and 64 Angelina), a very narrow opposition brightness peak is observed. For some objects (e. g., Callisto, Phobos, Deimos), the curvature of the photometric phase dependence changes rather slowly over a broad range of phase angles, which does not allow one to distinguish a linear part. For some dark objects (e. g., C- and P-type asteroids), the opposition intensity surge is very weak, and one can see only the linear part of the phase curve. These features are related to the physical and chemical properties of the scattering particles.

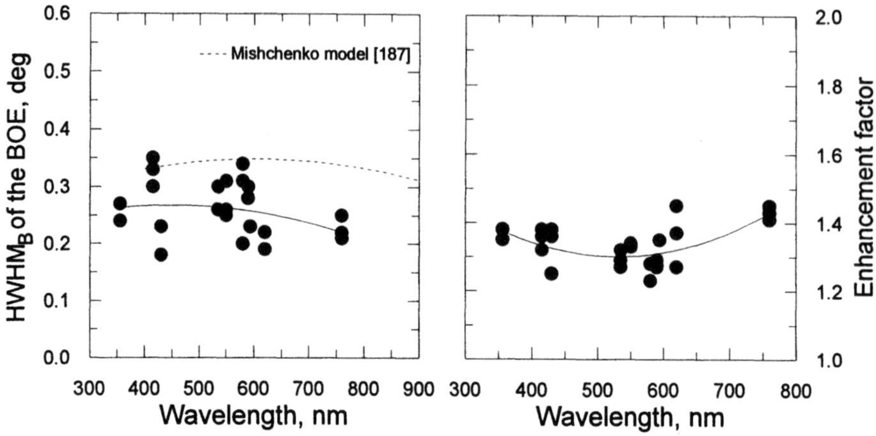

Figure 11: Wavelength dependence of the observed $\mathrm{HWHM_B}$ (a) and enhancement factor (b) for Saturn's rings. The dashed curve shows the theoretical dependence of the $\mathrm{HWHM_B}$ calculated by Mishchenko [187].

- The shape of the polarization phase curves close to opposition ($\alpha \leq 2°$) was not studied well until quite recently. Our latest observations indicate unambiguously the presence of the POE in the form of a separate peak of negative polarization superimposed on the regular (perhaps slightly asymmetric) negative polarization branch for the Galilean satellites of Jupiter and asteroid 64 Angelina. The lack of observations for the trailing hemisphere of Iapetus does not allow us to determine the shape of the negative polarization branch: it is either highly asymmetric or features a separate peak at very small phase angles. One can expect that the differences in the shape of the polarization minima and their angular locations and amplitudes are caused by varying physical and chemical properties of the scattering particles such as size, refractive index, shape, and porosity. A range of surface types can contribute to the disk-integrated polarization. Therefore, the regular negative polarization branch produced by various kinds of larger grains and/or by surface irregularities can co-exist with the POE peak caused by small grains. Depending on the relative contributions of different mechanisms to the total polarization, the resulting polarization phase curve at small phase angles can vary from object to object.

- The photometric and polarimetric behavior of different ASSBs near opposition is highly variable. The amplitude of the BOE lies in the range 1.1–1.8 in disk-integrated measurements (1.1–5.4 in disk-resolved observations), the half-width at half-maximum varies from 0.7° to 6.1° (from 0.08° to 3.4° in disk-resolved data), the phase angle where the BOE starts varies from 0.7° to 17° (from 0.4° to 19° for disk-resolved features), and the slope of the linear part varies from 0.01 to 0.06 mag/deg (from 0.0005 to 0.05 mag/deg for disk-resolved features). It should be noted that Phobos and Deimos differ significantly from all other ASSBs studied. The parameters of the negative polarization branch $|P_{\min}|$, $\alpha_{\min}$, α_{inv}, and HWHM_P are usually in the ranges $0.2-2.1\%$, $2°-12°$, $7°-29°$, and $0.3°-4°$, respectively.

- We have found strong relationships between different parameters of the BOE as well as between the characteristics of the BOE and the negative polarization branch based on a large sample of data for objects with different physical and chemical properties. This result is important because it may suggest the common origin of the photometric and polarization opposition phenomena. Furthermore, these relationships call for an obvious segregation of all ASSBs studied into two groups of high-albedo and low-albedo objects. This result indicates unequivocally the different roles played by the CBM and the SM in the formation of the photometric and polarimetric opposition phenomena for high- and low-albedo objects, respectively.

7. Acknowledgments

We thank Yu. Shkuratov for useful suggestions which improved the manuscript. N. Kiselev and M. Mishchenko appreciate funding from INTAS Grant No. 1999–00652 and the NASA's Radiation Sciences Program, respectively.

References

1. Yu. Shkuratov, A. Ovcharenko, E. Zubko, *et al.*, Icarus, in press (2002).
2. B. Hapke, "Theory of Reflectance and Emittance Spectroscopy" (Cambridge University Press, Cambridge, 1993).
3. Yu. G. Shkuratov, K. Muinonen, E. Bowell, et al., Earth Moon Planets **65**, 201 (1994).
4. K. Muinonen, J. Piironen, Yu. G. Shkuratov, et al., in R. Binzel (Ed.), "Asteroids III," in press (University of Arizona Press, Tucson, AZ, 2002).
5. see chapter by M. Mishchenko, V. Tishkovetz, and P. Litvinov, "Exact Results of the Vector Theory of Coherent Backscattering from Discrete Random Media: An Overview."
6. B. Lyot, NASA Tech. Transl. F-187 (NASA, Washington, D.C., 1964).
7. M. I. Mishchenko, Astrophys. J. **411**, 351 (1993).
8. Yu. G. Shkuratov, Solar Syst. Res. **28**, 23 (1994).
9. Yu. G. Shkuratov, Astron. Vestnik **25**, 71 (1991).
10. H. N. Russel, Astrophys. J. **43**, 103 (1916)
11. G. Rougier, Ann. L'Obs. Strasbourg **2**, 205 (1933).
12. T. Gehrels, T. Coffeen, and D. Owings, Astrophys. J. **69**, 826 (1964).
13. A. Lane and W. Irvine, Astron. J. **78**, 267 (1973).
14. N. P. Barabashev, Astron Nachr. **217**, 445 (1922).
15. V. A. Fedorets, Uchenye Zapiski Kharkov Univ. **42**, 49 (1952).

16. R. L. Wildey and H. Pohn, Astron. J. **69**, 1964 (1964).
17. J. Van Diggellen, Planet. Space Sci. **13**, 271 (1965).
18. K. Peacock, Icarus **9**, 16 (1968).
19. W. M. Irvine and A. P. Lane, Astron. J. **78**, 267 (1973).
20. Yu. G. Shkuratov, N. V. Opanasenko, and M. A. Kreslavsky, Icarus **95**, 283 (1992).
21. B. Hapke, in Z. Kopal (Ed.) "Physics and Astronomy of the Moon," (Academic Press, New York, London, 1971).
22. H. A. Pohn, H. W. Radin, and R. L. Wildey, Astron. J. **157**, L193 (1969).
23. H. A. Pohn, R. L. Wildey, and T. W. Offield, NASA SP-272 (1971).
24. R. L. Wildey, NASA SP-315 (1972).
25. S. Nozette et al., Science **266**, 1835 (1994).
26. B. J. Buratti, J. K. Hiller, and M. Wang, Icarus **124**, 490 (1996) .
27. Yu. Shkuratov, M. Kreslavsky, A. Ovcharenko, et al., Icarus **141**, 132 (1999).
28. J. Hiller, B. Buratti, and K. Hill , Icarus **141**, 205 (1999) .
29. M. A. Kreslavsky, Yu. G. Shkuratov, Yu. I. Velikodsky, et al., J. Geophys. Res. **105**, 20281 (2000).
30. A. Danjon, Bul. Astron. (Paris) **14**, 315 (1949).
31. A. Danjon, Bul. Astron. (Paris) **17**, 363 (1954).
32. A. Mallama, D. Wang, and R. A. Howard, Icarus **155**, 253 (2002).
33. G. Vaucouleours, Planet. Space Sci. **2**, 26 (1959).
34. G. Vaucouleours, in A. Dollfus (Ed.), "Surfaces and Interiors of Planets and Satellites", p. 267 (Academic Press, London, 1970).
35. B. T. O'Leary, Astrophys. J. **149**, L147 (1967).
36. L. A. Bugaenko, I. K. Koval, and A. V. Morozenko, Astron. Tsirk. **434** (1976).
37. Yu. V. Aleksandrov, D. F. Lupishko, and T. A. Lupishko, "Absolute Photometry of Mars in 1971, 1973, 1975" (Vishcha Shkola, Kharkiv, 1977).
38. T. E. Thorpe, Icarus **37**, 389 (1979).
39. T. E. Thorpe, Icarus **47**, 398 (1982).
40. Yu. G. Shkuratov, D. G. Stankevich, A. A. Ovcharenko, et al., Solar Syst. Res. **32**, 90 (1998).
41. M. Noland and J. Veverka, Icarus **28**, 405 (1976).
42. K. P. Klaasen, T. C. Duxbury, and J. Veverka, J. Geophys. Res. **84**, 8478 (1979).
43. J. Veverka and J. A. Burns, Ann. Rev. Earth Planet. Sci. **8**, 527 (1980).
44. K. D. Pang, G. W. Rhoads, G. A. Hanover, et al., J. Geophys. Res. **88**, 2475 (1983).
45. G. A. Avanesov, B. I. Bonev, F. Kempe, et al., Nature **341**, 585 (1989).
46. G. Avanesov, B. Zhukov, Ya. Ziman, et al., Planet. Space Sci. **39**, 281 (1991).
47. Yu. G. Shkuratov, N. V. Opanasenko, A. T. Basilevsky, et al., Planet. Space Sci. **39**, 341 (1991).
48. B. A. Cantor, M. J. Wolff, P. C. Thomas, et al., Icarus **142**, 414 (1999).
49. J. Stebbins, Lick Obs. Bull. **13**, 1 (1927).
50. J. Stebbins and T. S. Jacobsen, Lick Obs. Bull. **13**, 180 (1928).
51. D. Morrison, N. D. Morrison, and A. R. Lazarewicz, Icarus **23**, 399 (1974).
52. R. L. Millis and D. T. Thompson, Icarus **26**, 408 (1975).
53. G. W. Lockwood, D. T. Thompson, and K. Lumme, Astron. J. **86**, 961 (1980).

54. J. Veverka, in J. A. Burns (Ed.). "Planetary Satellites," p. 171 (University of Arizona Press, Tucson, AZ, 1977).
55. D. T. Thompson and G. W. Lockwood, J. Geophys. Res. **97**, 14761 (1992).
56. S. Keränen, K. Muinonen, and J. Piironen, in G. Videen, Q. Fu, and P. Chýlek (Eds.) "Light Scattering by Nonspherical Particles: Halifax Contributions," p. 279 (Army Research Laboratory, Adelphi, MD, 2000).
57. K. Lumme and K. Muinonen, in A. Milani, M. Di Martino, and A. Cellino (Eds.), "Asteroids, Comets, Meteors 1993," p. 194 (Kluwer, Dordrecht, 1993).
58. J. B. Pollack, F. C. Witteborn, E. F. Erickson, D. W. Strecker, B. J. Baldwin, and T. E. Bunch, Icarus **36**, 271 (1978).
59. A. L. Line, R. M. Nelson, and D. L. Matson, Nature **292**, 38 (1981).
60. V. K. Rosenbush, Icarus, in press (2002).
61. D. L. Domingue and A. Verbiscer, Icarus **128**, 49 (1997).
62. D. L. Domingue, B. W. Hapke, G.W. Lockwood, and D. T. Thompson, Icarus **90**, 30 (1991).
63. P. Helfenstein, N. Currier, B. E. Clark, J. Veverka, et al., Icarus **135**, 41 (1998).
64. M. Noland, J. Veverka, D. Morrison, D. P. Cruikshang, et. al., Icarus **23**, 334 (1974).
65. F. A. Franklin, and A. F. Cook, Icarus **23**, 355 (1974).
66. O. Franz and R. L. Millis, Icarus **24**, 433 (1975).
67. B. J. Buratti and J. Veverka, Icarus **58**, 254 (1984).
68. B. J. Buratti, Icarus **59**, 395 (1984).
69. S. W. Squyres, B. Buratti, J. Veverka, and C. Sagan, Icarus **59**, 426 (1984).
70. D. J. Tholen and B. Zellner, Icarus **58**, 246 (1984).
71. R. Thomas and J. Veverka, Icarus **64**, 414 (1985).
72. B. J. Buratti, J. A. Mosher, and T.V. Johnson, Icarus **87**, 339 (1990).
73. J. Degewij, L. E. Andersson, and B. Zellner, Icarus **44**, 520 (1980).
74. D. J. Tholen and B. Zellner, Icarus **53**, 341 (1983).
75. J. Goguen, H. Hammel, D. P. Cruikshank, and W. K. Hartmann, Bull. Am. Astron. Soc. **15**, 854 (1983).
76. R. L. Millis, Icarus **31**, 81 (1977).
77. R. W. Payne, J. British Astron. Ass. **81**, 123 (1971).
78. G. N. Blair and N. F. Owen, Icarus **22**, 224 (1974).
79. C. Blanco and S. Catalano, Astron. Astrophys. **33**, 105 (1974).
80. G. Müller, Publ. Obs. Potsdam **8**, 193 (1893).
81. H. Seeliger, Abh. Bayer. Akad. Wiss. Math. Naturwiss. Kl. II, **18**, 1 (1887).
82. E. Schönberg, Ann. Acad. Sci. Fennicae Ser. A, **16**, 5 (1922).
83. E. Hertzsprung, Astron. Nachr. **208**, 81 (1919).
84. E. Schönberg, Viertelyaresschrift der Astronomischen Gesellschaft **68**, 387 (1933).
85. F. A. Franklin and A. F. Cook, Astron. J. **70**, 704 (1965).
86. K. Lumme and W. M. Irvin, Astron. J. **81**, 865 (1976).
87. L. W. Esposito and K. Lumme, Icarus **31**, 157 (1977).
88. K. Lumme, L. W. Esposito, W. D. Benton, and W. A. Baum, Astron. J. **84**, 1402 (1979).
89. L. W. Esposito, K. Lumme, W. D. Benton, and L. J. Martin, Astron. J. **84**, 1408 (1979).

90. W. T. Thompson, K. Lumme, W. M. Irvin, et al., Icarus **46**, 187 (1981).
91. R. H. Brown and D. R. Cruikshank, Icarus **55**, 83 (1983).
92. J. D. Goguen and H. B. Hammel, Icarus **77**, 239 (1989).
93. J. Veverka, P. Thomas, P. Helfenstein, et al., J. Geophys. Res. **92**, 14895 (1987).
94. B. Buratti, F. Wong, and J. Mosher, Icarus **84**, 203 (1990).
95. B. E. Schaefer and S. W. Tourtellotte, Icarus **151**, 112 (2001).
96. E. Karkoschka, Icarus **125**, 348 (1997).
97. E. Karkoschka, Icarus **151**, 51 (2001).
98. B. E. Schaefer and M. W. Schaefer, Icarus **146**, 541 (2000).
99. T. Gehrels, Astrophys. J. **123**, 331 (1956).
100. A. W. Harris, J. W. Young, L. Contreiras, et al. Icarus **81**, 365 (1989).
101. A. W. Harris and J. W. Young, Lunar Planet Sci. Conf. XIX **2**, 447 (1989).
102. A. W. Harris and J. W. Young, Icarus **54**, 59 (1983).
103. A. W. Harris, M. Carlsson, J. W. Young, and C.-I. Lagerkvist, Icarus **58**, 377 (1984).
104. M. Poutanen, E. Bowell, L. J. Martin, and D. T. Thompson, Astron. Astrophys. Suppl. **61**, 291 (1985).
105. A. W. Harris, J. W. Young, E. Bowell, et al. Icarus **77**, 171 (1989).
106. C.-I. Lagerkvist, P. Magnusson, I. P. Williams, et al. Astron. Astrophys. Suppl. **78**, 519 (1989).
107. C.-I. Lagerkvist and P. Magnusson, Astron. Astrophys. Suppl. **86**, 119 (1990).
108. A. W. Harris, J. W. Young, T. Dockweiler, et. al., Icarus **95**, 115 (1992).
109. V. G. Shevchenko, Yu. N. Krugly, D. F. Lupishko, et. al., Astron. Vestnik **27**, 75 (1993).
110. J. Piironen, E. Bowell, A. Erikson, and P. Magnusson, Astron. Astrophys. Suppl. **106**, 587 (1994).
111. V. G. Shevchenko, V. G. Chorny, and A. V. Kalashnikov, Astron. Astrophys. Suppl. **115**, 1 (1996).
112. V. G. Shevchenko, Solar Syst. Res. **31**, 219 (1997).
113. L. M. French, Bull. Am. Astron. Soc. **19**, 850 (1987).
114. V. G. Shevchenko and D. F. Lupishko, Solar Syst. Res. **32**, 250 (1998).
115. I. N. Belskaya and V. G. Shevchenko, Icarus **147**, 94 (2000).
116. A. N. Dovgopol, Yu. N. Krugly, and V. G. Shevchenko, Acta. Astron. **42**, 67 (1992).
117. D. F. Lupishko and M. Di Martino, Planet. Space Sci. **46**, 47 (1998).
118. Yu. G. Krugly and V. G. Shevchenko, Lunar Planet Sci. Conf. **XXX**, 1595 (1999).
119. W. Z. Wisniewski, M. A. Barucci, M. Fulchignoni, et al. Icarus **101**, 213 (1993).
120. C. E. Delahodde, K. J. Meech, O. R. Hainaut, and E. Dotto, Astron. Astrophys. **376**, 672 (2001).
121. K. J. Meech and D. C. Jewitt, Astron. Astrophys. **187**, 585 (1987).
122. D. G. Schleicher, R. L. Millis, and P. V. Birch, Icarus **132**, 397 (1998).
123. Y. R. Fernandez , C. M. Lisse, H. U.Kaufl, et al., Icarus **147**, 145 (2000).
124. R. Dumont and F. Sanchez, Astron. Astrophys. **51**, 393 (1976).
125. J. B. Renard, R. Dumond, and A. C. Levasseur-Regourd, Astron. Astrophys. **304**, 602 (1995).
126. A. C. Levasseur-Regourd, J. B. Renard, and R. Dumond, in A. C. Levasseur-Regourd and H. Hasegawa (Eds.), "Origin and Evolution of Interplanetary Dust," p. 191 (Kluwer, Tokyo, 1991).
127. P. L. Lamy and J.-M. Perrin, Astron. Astrophys. **163**, 269 (1986).

128. G. V. Coyne and S. F. Pellicori, Astron. J. **75**, 54 (1970).
129. A. Dollfus and E. Bowell, Astron. Astrophys. **10**, 29 (1971).
130. O. I. Kvartskhelia, Bull. Abastumani Astrophys. Obs. **64**, 3 (1988).
131. N. V. Opanasenko and Yu. G. Shkuratov, Solar Syst. Res. **28**, 233 (1994).
132. B. Lyot, C. R. Acad. Sci. **20**, (1930).
133. A. Dollfus and M. Auriete, Icarus **23**, 465 (1974).
134. A. Dollfus, in G. P. Kuiper and B. M. Middlehurst (Eds.), "Planets and Satellites," p. 306 (University of Chicsgo Press, Chicago, IL, 1961).
135. A. Dollfus and J. Focas, Astron. Astrophys. **2**, 63 (1969).
136. A. V. Morozhenko, Astron. Zh. **50**, 1057 (1973).
137. O. I. Bugaenko, A. V. Morozhenko, and E. G. Yanovitskij, in T. Gehrels (Ed.), "Planets, Stars, and Nebulae Studied with Photopolarimetry," p. 599 (University of Arizona Press, Tucson, AZ, 1974).
138. B . Zellner, Astron. J. **77**, 183 (1972).
139. J. Veverka, Icarus **14**, 355 (1971).
140. A. Dollfus, Icarus **25**, 416 (1975).
141. J. Veverka, in J. A. Burns (Ed.). "Planetary satellites," p. 210 (University of Arizona Press, Tucson, AZ, 1977).
142. R. A. Chigladze, PhD thesis, Abastumany Astrophys. Obs. (1989).
143. V. K. Rosenbush, V. V. Avramchuk, A. E. Rosenbush, and M. I. Mishchenko, Astrophys. J. **487**, 402 (1997).
144. V. K. Rosenbush, N. N. Kiselev, K. Jockers, et al., Kinem. Phys. Celest. Bodies. Suppl. Ser. No. 3, 227 (2000).
145. V. K. Rosenbush, N. N. Kiselev, V. V. Avramchuk, et al., in B. Gustafson, L. Kolokolova, and G. Videen (Eds.), "Electromagnetics and Light Scattering by Nonspherical Particles," p. 279 (Army Research Laboratory, Adelphi, MD, 2002).
146. B. Zellner, Astrophys. J. **174**, L107 (1972).
147. E. Bowell and B. Zellner, in T. Gehrels (Ed.), "Planets, Stars, and Nebulae Studied with Photopolarimetry," p. 381 (University of Arizona Press, Tucson, AZ, 1974).
148. J. C. Kemp and R. E. Murphy, Astrophys. J. **189**, 679 (1973).
149. A. Dollfus, Icarus **37**, 404 (1979).
150. P. E. Johnson, J. C. Kemp, R. King, et al., Nature **283**, 146 (1980).
151. A. Dollfus, Icarus **124**, 96 (1996).
152. B. Lyot, C. R. Acad. Sci. **199**, 774 (1934).
153. B. Zellner and J. Gradie, Icarus **28**, 117 (1974).
154. B. Zellner, T. Gehrels, and J. Gradie, Astron. J. **79**, 1100 (1974).
155. B. Zellner and J. Gradie, Astron. J. **81**, 262 (1976).
156. A. Dollfus, M. Wolf, J. E. Geake, et al., in M. P. Binzel, T. Gehrels, and M. S. Matthews (Eds.), "Asteroids-II," p. 594 (University of Arizona Press, Tucson, AZ, 1989).
157. D. F. Lupishko, Doctoral thesis, Kharkiv Univ. (1999).
158. G. P. Chernova, D. F. Lupishko, V. G. Shevchenko, et al., Kinem. Phys. Celest. Bodies **7**, 20 (1991).

159. D. F. Lupishko, N. N. Kiselev, and G. P. Chernova, Kinem. Phys. Celest. Bodies **10**, 40 (1994).

160. D. F. Lupishko and C. B. Vasilyev, Kinem. Phys. Celest. Bodies **13**, 17 (1997) (http://pdssbn.astro.umd.edu/sbnhtml/).

161. I. N. Belskaya, N. N. Kiselev, N. M. Shakhovskoj, et al., in "Abstracts: Asteroids, Comets, Meteors," p. 59 (Cornell University, Ithaca, NY, 1999).

162. N. N. Kiselev and G. P. Chernova, Sov. Astron. **22**, 607 (1978).

163. N. N. Kiselev and G. P. Chernova, Icarus **48**, 473 (1981).

164. G. P. Chernova, N. N. Kiselev, and K. Jockers, Icarus **103**, 144 (1993).

165. N. N. Kiselev, N. M. Shakhovskoy, and Yu. S. Efimov, Icarus **120**, 408 (1996).

166. R. D. Wolstencroft and L. J. Rose, Astrophys. J. **147**, 271 (1967).

167. A. C. Levasseur-Regourd, R Dumond, and J. B. Renard, Icarus **86**, 264 (1990).

168. A. C. Levasseur-Regourd, J. B. Renard, and R Dumond, in A. C. Levasseur-Regourd and H. Hasegawa (Eds.), "Origin and Evolution of Interplanetary Dust," p. 191 (Kluwer, Tokyo, 1991).

169. V. V. Korokhin, S. A. Beletsky, Yu. I. Velikodsky et al., Kinem. Phys. Celest. Bodies **16**, 80 (2000).

170. M. I. Mishchenko, J.-M. Luck, and T. M., Nieuwenhuizen, J. Opt. Soc. Am. A **17**, 888.

171. J. Veverka, in J. A. Burns (Ed.), "Planetary Satellites," p. 244 (University of Arizona Press, Tucson, AZ, 1977).

172. M. Poutanen, in C.-I. Lagerkvist and H Rickman (Eds.), "Asteroids, Comets, Meteors," p. 45 (Uppsala University, Uppsala, 1983).

173. M. I. Mishchenko and J. M. Dlugach, Mon. Not. Royal Astron. Soc. 254, 15P (1992); M. I. Mishchenko and J. M. Dlugach, Planet. Space Sci. **41**, 173 (1993).

174. O. Kolokolova, in A. V. Morozhenko (Ed.), "Photometric and Polarimetric Investigations of Celestial Bodies," p. 38 (Naukova Dumka, Kyiv, 1985).

175. D. Morrison and N. D. Morrison, in J. A. Burns (Ed.), "Planetary Satellites," p. 363 (University of Arizona Press, Tucson, AZ, 1977).

176. D. F. Lupishko, F. A. Tupieva, F. P. Velichko, et al., Astron. Vestnik **15**, 25 (1981).

177. W. Z. Wisniewski, T. M. Michowski, A. W. Harris, and R. S. McMillan, Icarus **126**, 395 (1997).

178. E. F. Tedesko, R. C. Taylor, J. Drummond, et al., Icarus **54**, 30 (1983).

179. F. Scaltriti, V. Zappala, and R. Stanzel, Icarus **34**, 93 (1978).

180. R. C. Taylor, T. Gehrels, and A. B. Silvester, Astron. J. **76**, 141 (1971).

181. V. G. Shevchenko, PhD thesis, Kharkiv University (1997).

182. C.-I. Lagerkvist and P. Magnusson, Astron. Astrophys. Suppl. **108**, 143 (1994).

183. E. F. Tedesko, R. C. Taylor, J. Drummond, et al., Icarus **54**, 23 (1983).

184. V. G. Shevchenko, I. N. Belskaya, V. G. Chiorny, et al., Planet. Space Sci. **45**, 1615 (1997).

185. V. K. Rosenbush, V. V. Avramchuk, Astron. Vestnik **33**, 307 (1999).

186. B. Hapke, Icarus **67**, 264 (1986).

187. M. I. Mishchenko, Astrophys. Space Sci. **194**, 327 (1992).

188. B. Hapke, R. Nelson, and W. Smithe, Icarus **133**, 89 (1998).

EXPERIMENTAL MODELING OF OPPOSITION EFFECT AND NEGATIVE POLARIZATION OF REGOLITH-LIKE SURFACES

Yu. G. SHKURATOV and A. V. OVCHARENKO
Astronomical Observatory of Kharkov National University. 35 Sumskaya St. Kharkov. 61022. Ukraine.

Abstract. We present results of laboratory photometric and polarimetric measurements of samples simulating the structure of planetary regoliths at small phase angles. We study a suite of samples that are characterized by a variety of structures and albedos. Metallic powders usually exhibit a wide branch of negative polarization independent of particle size. Unlike dielectric powders, metallic particulate surfaces show negative polarization even when they are placed in water. We demonstrate a strong particle-size dependence of the negative branch of polarization for powdered dielectric surfaces. For dielectric powders, both the opposition effect and negative polarization become more prominent when the samples are compressed. We show the opposition effect and negative polarization for rough and polished solid surfaces with microcrystalline structure. Our measurements show that the degree of polarization for dielectric surfaces depends on the surface tilt; even at small phase angles the polarization degree for tilted surfaces can be non-zero. The negative polarization depends strongly on micro- and macro-structured optical inhomogeneities of scattering surfaces. Double-minimum branches of negative polarization are found in this series of measurements.

1. Introduction

Many natural and artificial surfaces with complicated micro-structure reveal strong backscatter and accompanying negative polarization. In particular, these two phenomena are observed for the regoliths of the Moon, asteroids, Mars, Mercury, planetary satellites, and many simulates of regoliths at phase angles

G. Videen and M. Kocifaj (eds.), Optics of Cosmic Dust, 225–238.

$\alpha < 20°$ [1-10]. There is a greater diversity of photometric and polarimetric phase curves for laboratory samples than for planetary surfaces.

The highly reflective surfaces can often exhibit additional opposition phenomena at $\alpha < 5°$: a conspicuous narrow brightness opposition spike and a corresponding narrow negative polarization branch at phase angles of several degrees. This is observed for bright asteroids and satellites [11-13] and laboratory simulates [9, 10, 14-17].

Physical understanding of all these phenomena is still insufficient [see also Muinonen *et al.*, Rosenbush *et al.*, Mishchenko *et al.*, this volume]. Three mechanisms are generally exploited to explain them: the classical shadow-hiding mechanism, the coherent backscatter (or interference) mechanism, and the single-particle scatter when individual particles have a peak of backscatter and an accompanying negative polarization.

The shadow-hiding mechanism has been studied extensively, e. g., [18]. This can provide an explanation of the backscatter effect but is not able to describe the negative polarization effect [8, 10, 18]. The coherent backscatter is recognized as a mechanism explaining both opposition effect and negative polarization [10, 19-29]. Unfortunately, there is yet no generally accepted theory of these effects allowing routine calculations and fitting to experimental data. Therefore computer modeling of light scatter with the coherent backscatter mechanism seems to be the most promising approach [10, 30]. This does not use complicated mathematics and imitates physical experiments. The third mechanism that can explain the phenomena is single-particle scattering by particles having the backscatter features at small phase angles. This was shown experimentally [31-33] and theoretically [34, 35]. This mechanism is able even to explain sharp brightness opposition spikes, but only in special cases when surfaces consist, for instance, of spherical or cubic particles of suitable size.

There are few experimental data that show which mechanism and in what cases can give rise, or additionally contribute, to the brightness opposition effect and negative polarization. The main goal of this chapter is to systematize important experimental data for verification of theoretical models of the phenomena. For this purpose we use new photometric and polarimetric measurements as well as results from our previous works, e. g., [8-10].

2. Equipment and samples

For this work we use two laboratory reflectance photometer/polarimeters. They allow measurements of powders. We describe these instruments below as well as the characteristics of the samples used in our investigations.

2.1 Instruments

The laboratory photometer/polarimeters are exploited to study phase dependences of reflectance and degree of linear polarization of powdered surfaces. One of the instruments provides measurements in the phase angle range 3°- 160°; the other one provides measurements of samples at smaller phase angles from 0.2° to 17°. A detailed description of these instruments can be found elsewhere [7-10], so we only provide a brief description here.

We use halogen lamps in both instruments as the light source, i.e. we work with initially unpolarized light. For this study, we select a wide spectral band centered at $\lambda = 0.65$ μm. In some cases we use another spectral band, $\lambda = 0.43$ μm in addition. Most of our measurements are made at standard illumination/observation geometry, when samples are viewed with the detector along the sample normal. The rotated light source arm changes the incidence angle. Sample albedo is measured and determined at $\alpha = 2°$ relative to a compressed Halon sample that we used for our reflectance standard [36]. The small-phase-angle-range photometer/polarimeter is calibrated against a comparable JPL NASA instrument [37]. The size of all samples in our measurements with the wide-phase-angle polarimeter is approximately 10 mm. For the small-phase-angle polarimeter these are approximately 60 mm.

The laboratory polarimeters require thorough calibration of instrumental polarization. The polarization of the receiver part of the polarimeter is determined by a diffusing light source in place of the sample; this light source is a lamp covered by two frosted glass plates. Residual polarization from the source is about 0.05%. Measurements obtained by rotating the diffusing light source allow the measurement of instrumental polarization of the receiver. The degree of the polarization is about 0.03%. We incorporate these values in our data processing routines. Determining and correcting the instrumental polarization associated with the light source of the polarimeter is the most difficult problem. The polarization of the halogen lamp can be several percent. We compensate much of the polarization by inserting a tilted glass plate, decreasing the instrumental polarization to several tenths of percent. The rest of parasitic polarization is compensated with a Lyot depolarizer.

The average accuracy of our measurements lies within 0.05%, including random and systematic errors. The accuracy is somewhat less when extremely dark surfaces are measured. We must especially note the problem of reproducing measurement results for the same material samples prepared at different times. Typically the difference is within 0.05%, however, in some cases it is noticeably more. Thus, the accuracy near 0.05% is an acceptable tolerance for laboratory polarimetric measurements.

2.2 Samples

To better understand the nature of the opposition phenomena in a wider physical context we have measured many natural and artificial samples with controlled parameters. For instance, we examined smoke deposits of MgO and carbon soot (samples with very different albedo), a SiO_2 ultra-disperse powder, and size-separates of Al_2O_3 powder.

The magnesium oxide sample was prepared by incineration of metallic Mg. Smoke was precipitated on a glass plate. We obtain an optically thick layer of MgO coating. Albedo of this sample is near 100%. Photos taken with an electron microscope show that the particles of smoked MgO are noticeably smaller than 1 μm. Comparatively large particles with cubic form can also be seen. Soot deposits are produced by incineration of black rubber. Albedo of this sample is very small, 2.6%. Electron-microphotographs of a soot sample show that all structural units of the soot surface are significantly smaller than 1 μm. We also study a few aluminum oxide powders that have been used in experiments [10, 15, 16]. Albedo of these samples is very high, up to 95%. The fine alumina has isometric-in-average micron-size particles of random shapes. As an example of structure formed with superfine scatterers we use a SiO_2 powder consisting of particles with sizes of about 10 nm that are condensed from the gas phase. Albedo of this sample is not more than 80%. We measure solid surfaces; for example, a glass-ceramic plate consisting of small micro-crystals. One surface of this plate is rough with a characteristic roughness scale of about 50 μm; the other side is polished. The sizes of microcrystals on the plate are different, but predominantly smaller than the wavelength. The glass-ceramic plate has a high albedo, near 100%. We measure mixtures of chalk and soot in different proportions. Chalk is the micro-structured calcite remains of plankton algae that have fine micro-structure. To prepare the mixtures we grind dry chalk and soot powders in a mortar.

3. Results and discussion

We present results of our measurements, investigating factors that influence the photometric and polarimetric effects. Among these factors are choice of material (dielectric or metal), volume density of powdered samples, surface relief of solid samples, particle size, optical micro and macro inhomogeneities of powdered surfaces, illumination and observation geometry.

3.1 Behavior of dielectrics and metals

Dielectric and metal powders exhibit completely different behavior of the brightness opposition effect and negative polarization. Metallic powders usually reveal negative polarization even when particle size is much larger than the wavelength. Dielectric particulate surfaces do not show the negative polarization under this condition if the standard geometry of illumination and observation is used [8, 10, 38, 39].

To illustrate the difference between the behavior of metallic and dielectric materials we present in Figure 1 data for an iron powder (particle size about 30 μm) and glass powder (particle size < 0.5 μm) in dry state and in water. The immersed glass-powder sample does not reveal any negative polarization. On the other hand, the iron sample shows a prominent branch of negative polarization, which has a complicated form: at large phase angles where the polarization typically becomes positive, the polarization decreases again, becoming more negative. This behavior is related to outgoing light refraction on the immersion boundary that gives negative polarization at large phase angles.

The different behavior of metallic and dielectric powders is quite natural: dielectrics in water have too small a relative refractive index to show noticeable multiple scatter, metallic powders give rise to significant multiple scatter in water due to their high imaginary part of refractive index.

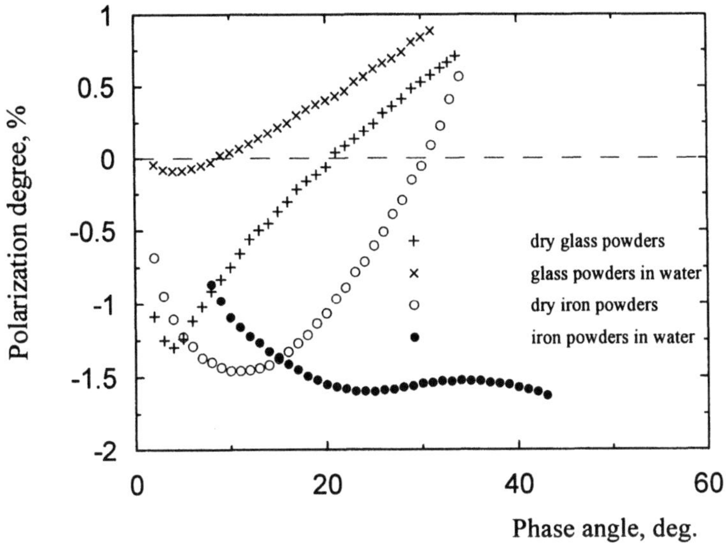

Figure 1. Polarization phase curves for fine-grained glass and coarse-grained iron powders in water and in dry state at $\lambda = 0.65\ \mu m$.

3.2 Compression and albedo effects

The volume density of a medium can strongly influence the opposition effect and negative polarization. We obtain results for smoked MgO coatings before and after compression by drying-in-alcohol. Large changes are observed after compression, e.g., the negative polarization minimum shifts to large phase angles. Only the volume density is varied in this experiment, nevertheless the polarimetric opposition effect of MgO is transformed to a more symmetric negative polarization branch covering a wider phase-angle range (see Figure 2). This can perhaps be treated as evidence of the fact that there are no separate polarimetric phenomena at small (up to 3°) and relatively large (up to 20°) phase angles. We note also that at large phase angles the brightness phase curve of the compressed sample is noticeably steeper than in case of the uncompressed surface. This is an unexpected result, since the compressed sample has a more even surface, than the initial one, and, hence, the shadow-hiding effect should be revealed to a lesser degree. We may conclude that the reason for the difference can be the coherent backscatter enhancement, which can, as it turns out, be extended up to phase angles much more than several degrees. This somewhat contradicts the common opinion. We consider the described results as important tests for theoretical models.

Experiments with alcohol-drying compression are carried out with a superfine powder of SiO_2 having particulates with sizes on the order of 10 nm (see short curves in Figure 2). This initially very fluffy sample produces an unusual beak-shaped negative polarization branch with a minimum occurring at phase angles smaller than 0.2° and an inversion angle at 0.5°. This is the shortest inversion angle found in our measurements. After compressing the SiO_2 powder, its volume density decreases by several orders of magnitude and the negative polarization branch increases in depth and angular width.

The sample containing extremely small particles is, hence, a good model to test theories that use Rayleigh indicatrix of single-particle scatter and deal with extremely fluffy surfaces (e. g., such a theory is presented in [29]).

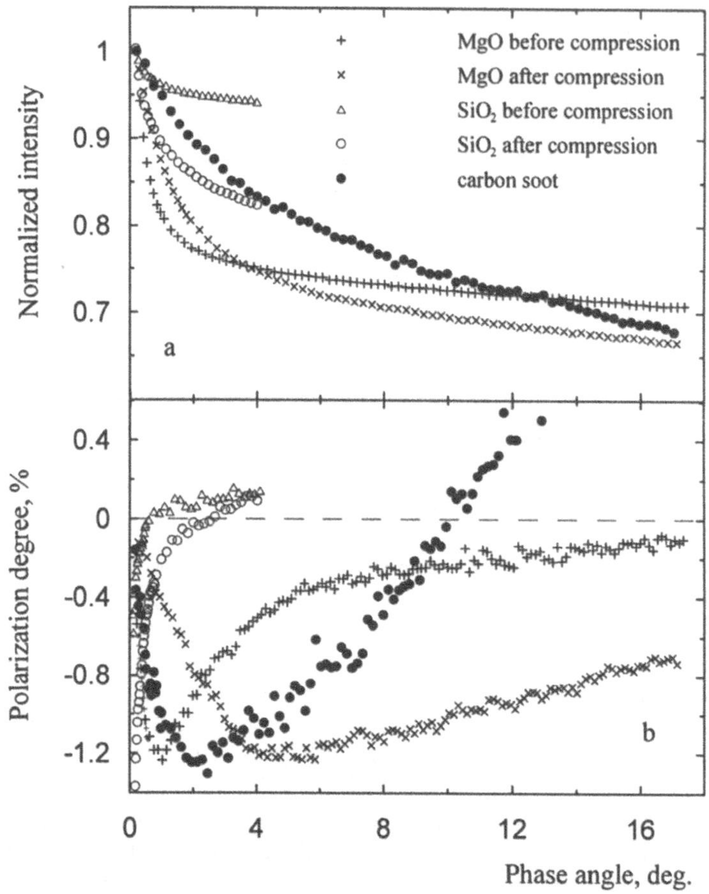

Figure 2. Phase functions of brightness and degree of polarization for a superfine powder of SiO_2 and smoked MgO coatings before and after compressing through drying alcohol, measured at λ = 0.65 μm. Data for carbon soot are also presented.

We emphasize that our experiments with the SiO_2 and MgO powders clearly show existence of polarization inversion at very small phase angles, contradicting some model calculations [29].

It is interesting to compare data for fine powders with different albedo. Figure 2 presents photometric and polarimetric data for carbon soot. One can compare brightness phase curves for MgO (a high reflective surface) and soot (a very dark surface). The MgO sample shows significantly stronger opposition effect than the dark one does. On the other hand, the polarization minima of both samples are rather close to each other. This shows that the negative polarization is due to a number of low orders of scatter; whereas, all orders of scatter contribute to the brightness opposition peak.

3.3 Surface relief and particle-size effects

The brightness opposition effect and negative polarization are revealed not only for powdered but also for solid surfaces. We studied a glass-ceramic plate with microcrystalline structure that exhibits negative polarization branches.

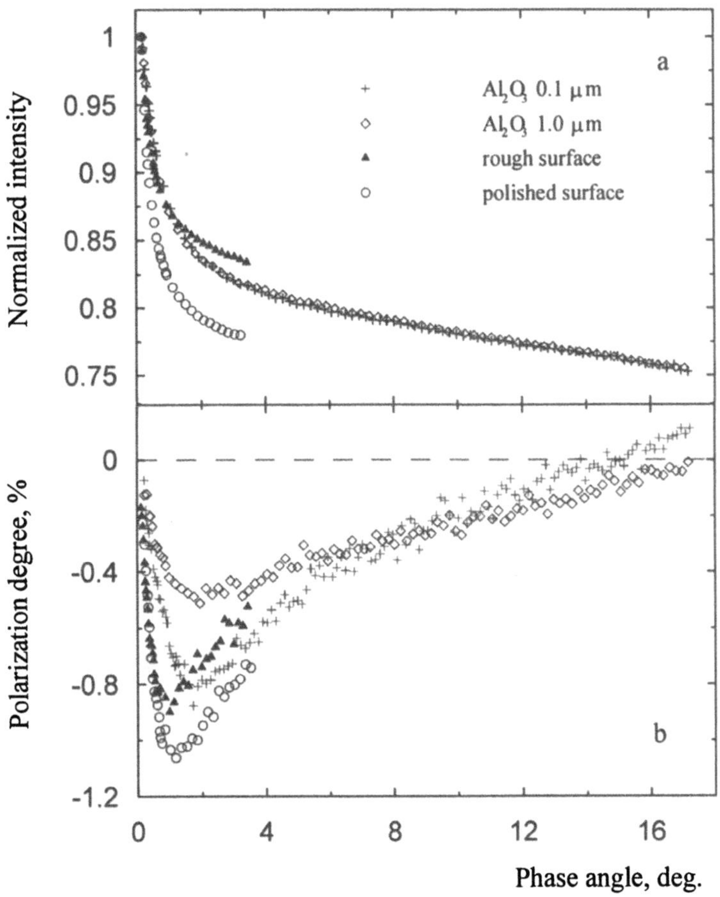

Figure 3. Brightness and polarization phase curves for particle-size separations of aluminum oxide powder (long branches) and for the polished and dull surfaces of a glass-ceramic plate (short branches) measured at $\lambda = 0.65$ μm.

We measure both sides of the plate, one of which is polished and the other of which is rough. We tilt the polished surface slightly to avoid the specular peak. The negative polarization branches are similar to the ones we measure from the MgO sample (see Figure 3). The data for both sides of the glass-ceramic plate illustrate the dominance of surface internal micro-structure on the opposition effect and negative polarization. However, surface relief is also important. The introduction of surface relief reduces both effects. Similar results are obtained for solid rocks [40, 41].

We also study the influence of particle size on the opposition effect and negative polarization (see Figure 3). We examine two particle-size populations of Al_2O_3 with average sizes of 1.0 μm and 0.1 μm. The coarser powders reveal subdued negative polarization branches. On the other hand, both samples show amazingly coincident brightness phase curves. This result implies a mechanism other than coherent backscatter. It is likely that in this case a single-particle-scatter contribution to negative polarization must be taken into consideration. It is necessary to note that the particle-size populations have somewhat different volume densities.

3.4 Optical micro-structure inhomogeneities

Optical inhomogeneities of all scales affect the opposition phenomena significantly [42]. For instance, mixtures of MgO and soot powders give, in some case, negative polarization down to -3.5% and the inversion angle up to 45° [8, 42]. Figure 4 illustrates this effect for a micro-structured MgO and Fe_2O_3 mixture obtained by drying an alcohol suspension. The data are obtained with our wide-phase-angle polarimeter. As one can see, the mixture in blue light demonstrates the deepest negative polarization branch. This corresponds to maximal optical contrast of MgO and Fe_2O_3 components, as the Fe_2O_3 powder is very red. In red light this contrast is minimal, but still exists. Therefore the "red" branch is weaker than "blue" one, but more prominent than in the case of individual components, MgO and Fe_2O_3 powders.

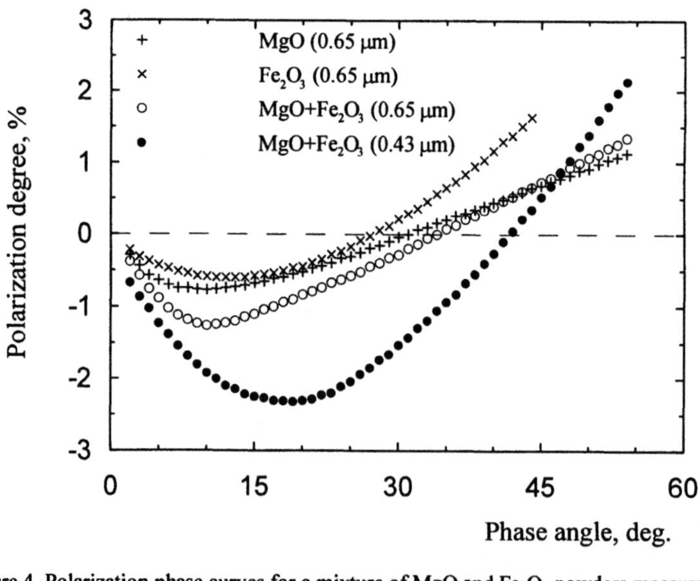

Figure 4. Polarization phase curves for a mixture of MgO and Fe_2O_3 powders measured at different wavelengths. The prominent branch corresponds to the mixture in blue light.

The effect of inhomogeneity on the negative polarization has no evident interpretation. This can be due to single-particle scatter, if it gives rise to the negative polarization. Actually the role of single-particle scatter of the bright component increases when the MgO particles are surrounded with dark (in blue light) Fe_2O_3 particles, as this surrounding suppresses the multiple scatter. It can be also due to scatter in small aggregates of MgO particles. Indeed, low orders of multiple scatter that just dominate in aggregates can give rise to a considerable negative polarization. To better understand the effect, combined single-particle [31-33] and powder measurements are needed.

3.5 Optical macro-structure inhomogeneities

We study the effect of macro-structure inhomogeneities on photometric and polarimetric phase dependences using composite samples.

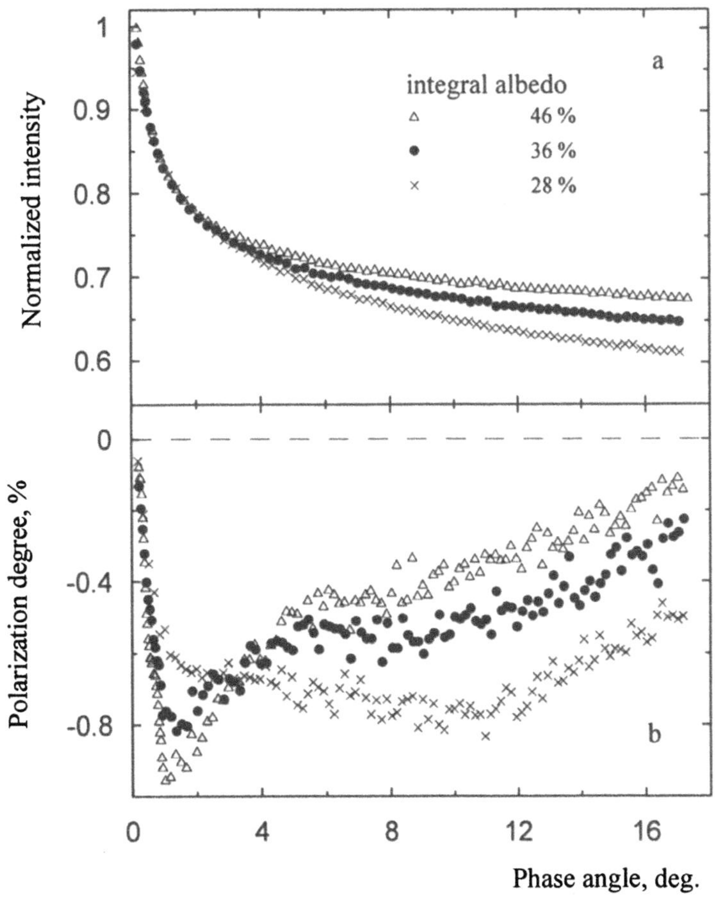

Figure 5. Photometric and polarimetric phase curves for composite samples. Resulting albedos of the samples are measured at $\lambda = 0.65$ μm. These samples produce a superposition of wide (mixture of chalk and soot) and narrow (MgO) negative polarization branches.

Each sample consists of two sub-samples. One of them (the host sample) is a mixture of soot and chalk with resulting albedo of about 20%. This sub-sample has a wide and deep negative polarization branch. The other one is an exchangeable small piece of cardboard with a thick slab of MgO that has a narrow branch of negative polarization. We prepare a few exchangeable sub-samples of different size that we place on the central part of the host sample.

Figure 5 shows phase measurements of the composite samples together with albedos of the composite samples. Three different superpositions of the narrow and wide branches are produced in which two local minima or an asymmetry in polarization phase dependence can be seen. The double-minimum feature was recently observed for the Galilean satellites [13]. We also tried to simulate this feature using the host sample covered with thin homogeneous layers of smoked MgO deposits, but we could not produce a clear double-minimum structure, only branches with different asymmetries.

Thus the presented measurements and observational data [13] suggest that the surfaces of the Galilean satellites contain different regions having different characteristic length scales. These regions may represent two types of surfaces: relatively dark materials having wide (regular) branches of negative polarization and bright materials with a MgO-like behavior of polarization. The first type would be, e.g., silicate components or dirty ices; the second one could be fresh hoar-frost-like deposits.

3.6 Illumination/observation geometry

Studies of the illumination/observation geometry on polarimetric phase curves may be useful in interpreting observations of asteroids and satellites, since brightness and polarization degree depend not only on the phase angle but also on illumination and observation parameters, such as the photometric longitude and latitude.

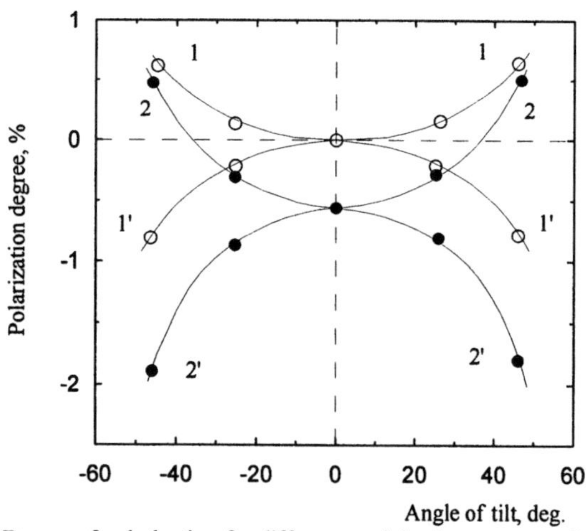

Figure 6. Degree of polarization for different particle-size separates: fraction < 0.5 μm (curves 1 and 1') and fraction ~ 100 μm (curves 2 and 2') of a transparent glass at the phase angle 5° and λ = 0.65 μm as a function of tilt angle. The branches going upward and downward correspond respectively to the latitude and longitude tilts.

For this study the two particle populations (the sizes < 0.5 μm and ~ 100 μm) of a transparent glass powder are analyzed using the wide-phase-angle instrument. Figure 6 shows brightness curves for the samples obtained at a constant phase angle of about 5° for the cases when the sample normal deviates from the scattering plane (latitude effect) and in the scattering plane (longitude effect). As one can see the tilts of the sample influence the polarization. Positive polarization shifts are produced when the slope of the sample normal deflects from the scattering plane. Tilts within the scattering plane produce negative shifts, which are almost the same for the inclinations toward the light source and in opposite direction.

These effects may have important consequences. A spherical, airless celestial body with a homogeneous surface can be represented as a spherical grid of infinitesimal plane areas each having different effective tilts. The set of tilted planes compensate each other in the pure backscatter direction, giving zero polarization of the integral scattered flux. If the body is irregularly shaped, or has oriented grooves, or has a crater large compared with its size, and/or the surface of the body is physically inhomogeneous (e. g., global albedo variations), the distribution and contribution of tilted planes may not perfectly compensate each other and the integral polarimetric observations of such a body can show non-zero polarization even at exactly zero phase angle.

4. Conclusions

We state here the following, more or less reliable, results that could be considered as a basis to test theoretical models of the opposition phenomena.

1. The polarization state of coarsely powdered dielectric and metallic surfaces differs significantly from each other at the standard illumination/observation geometry. Dielectric surfaces composed of particles having a characteristic structure scale on the order of a wavelength display a negative polarization branch over the range 0.2 – 17°. This effect disappears for larger structure scales of the same materials. This points out that models, using shadow-hiding to describe the negative polarization for dielectric powders, e.g., [43, 44], probably overestimate the effect. Metallic powders generally show well-developed negative polarization branches even when they are composed of relatively large particles. This implies that for metals there is a mechanism contributing to the negative polarization that differs from the mechanism relevant for dielectric powders.

2. Dielectric micro-structured powders show a dramatic dependence of backscatter and especially negative polarization on volume density. Increasing this value can transform the narrow-phase-angle effect to a wide-phase-angle one. In particular, increasing the volume density of MgO coatings shifts the minimum of negative polarization from 1° to 6° making the negative polarization branch more symmetric than the initially asymmetric one. The result supports arguments that the negative branches of polarization in the narrow ($\alpha < 3°$) and wide ($\alpha \sim 3 - 20°$) phase angle range have the same origin.

3. In many cases parameters of the photometric backscatter effect and negative polarization are closely correlated. However, we have found many examples of the typically observed relationships between the brightness backscatter effect and negative polarization that are not valid. In particular, the minima of the polarization branches of thick MgO coatings and smoked carbon soot are located very close to each other, but exhibit very different photometric behaviors near opposition. One can observe a prominent non-linear opposition spike for MgO and a comparatively linear surge for soot.

4. Optically contrasting micro-structured media reveal very pronounced negative polarization branches, shown by our experiments with the mixtures of MgO and Fe_2O_3 powders measured at different wavelengths: the negative polarization strengthens in blue light when optical contrast is maximal. These results allow us to continue a discussion on the physical mechanisms that contribute to the negative polarization supposing that in some cases single-particle scattering can noticeably contribute to this phenomenon in combination with the coherent backscatter enhancement mechanism.

5. Macro-structured surfaces can reveal complicated branches of negative polarization. If components of the surfaces have different negative polarization branches, e. g., narrow MgO-like and wide symmetric branches, their superposition can give rise to a branch with two minima. The double-minimum feature is observed for Galilean satellites and some bright asteroids [13]. Note that the surface variegations have been invoked to explain the photometric opposition effects of several classes of asteroids using computer modeling [45]. Thus, laboratory measurements and computer simulations of spotted surfaces seem to be important in a planetological context.

6. Negative polarization branches are very narrow and asymmetric for extremely fluffy dielectric powders consisting of nanometer-size particles. In this case the brightness opposition spike becomes too narrow to be measured in our experiments. A sample of nanophase SiO_2 exhibits unique properties having an inversion angle at 0.5°. This sample is a good analog of a rarified medium consisting of Rayleigh scatterers. Our measurements of the sample can be directly used to compare with theoretical models of the coherent backscatter phenomena working with the Rayleigh scatterers.

7. The degree of polarization for dielectric surfaces depends not only on the phase angle but also on the surface tilt. This is observed both for coarse and fine-grained surfaces. At exactly zero phase angle, the degree of polarization of tilted surfaces can be non-zero. A tilt of the surface normal in the scattering plane gives a parallel shift of the negative polarization branch to more negative values. A tilt in the perpendicular plane gives the same shift in the direction of positive polarization. An irregularly shaped asteroid or planetary satellite that is covered by an optically homogeneous surface can exhibit non-zero polarization in its integral polarimetric behavior at exactly zero phase angle.

5. Acknowledgments

These studies were supported by INTAS grant #2000-0792. The authors thank K. Muinonen for valuable remarks.

References

1. B. Lyot, Ann. Obs. Meudon. **8,** 1 (1929).
2. A. Dollfus and J. Focas, Ann. Astrophys. **2,** 63 (1969).
3. A. Dollfus and E. Bowell. Astron. Astrophys. **10,** 29 (1971).
4. A. Dollfus and M. Auriere, Icarus **23,** 465 (1974).
5. A. Dollfus and J. Geake, Proc. Lunar Sci. Conf. 6-th, 2749 (1975).
6. Yu. G. Shkuratov, L. A. Akimov, and V. P. Tishkovets, Solar System Res. **18,** 105 (1984).
7. Yu. G. Shkuratov, N. V. Opanasenko, and M. A. Kreslavsky, Icarus **95,** 283 (1992).
8. Yu. Shkuratov, K. Muinonen, E. Bowell, K. Lumme, J. Peltoniemi, M. Kreslavsky, D. Stankevich, V. Tishkovetz, N. Opanasenko, and L. Melkumova, The Earth, Moon, and Planets **65,** 201 (1994).
9. Yu. Shkuratov, M. Kreslavsky, A. Ovcharenko, D. Stankevich, E. Zubko, C. Pieters, and G. Arnold, Icarus **141**, 132 (1999).
10. Yu. Shkuratov, A. Ovcharenko, E. Zubko, O. Miloslavskaya, K. Muinonen, J. Piironen, R. Nelson, W. Smythe, V. Rosenbush, P. Helfenstein The opposition effect and negative polarization of structural analogs of planetary regoliths. Icarus. (2002). In press
11. A. Harris and 12 colleagues, Icarus **81,** 365 (1989).
12. P. Helfenstein and 21 colleagues, Icarus. **135**, 41 (1998).
13. V. Rosenbush, V. Avramchuk, and A. Rosenbush, Astrophys. Journal. **487,** 402 (1997).
14. P. Oetking, J. Geophys. Res. **71,** 2505 (1966).
15. R. Nelson, B. Hapke, W. Smythe, and L. Horn, Icarus **131,** 223 (1998).
16. R. Nelson, B. Hapke, W. Smythe, and L. Spilker, Icarus **147,** 545 (2000).
17. Ovcharenko and Yu.G. Shkuratov, Optics and Spectroscopy. **88,** 291 (2000).
18. B. Hapke, "Theory of reflectance and emittance spectroscopy" (Cambridge Univ. Press. 1993)
19. Yu. Shkuratov, Astronomicheskii Tsircular **1400**, 3 (Shternberg State Astron. Institute Moscow 1985).
20. Yu. G. Shkuratov, Kinematika i Fisika Nebesnykh Tel. **4,** 33 (1988).
21. Yu. G. Shkuratov, Astronomicheskii Vestnik **23**, 176 (1989).
22. Yu. G. Shkuratov, Solar System Res. **25**, 134 (1991).
23. K. Muinonen, In: "Proc. 1989 URSI Electromagnetic Theory Sympos". pp.428-430. (Stockholm, Sweden 1989).
24. K. Muinonen, "Light scattering by inhomogeneous media: backward enhancement and reversal of polarization". PhD-thesis. (University of Helsinki 1990).
26. K. Muinonen, In: "Asteroids, Comets and Meteors 1993" / Eds. A. Milani, M. Di Martino, and A. Cellino. pp. 271-296. (Kluwer Academic Publishers, Netherlands 1994).
27. M. I. Mishchenko, Astrophys. Space Sci. **194**, 327 (1992).
28. M. I. Mishchenko and J.M. Dlugach, Planet. Space Sci. **41,** 173 (1993).

29. M. I. Mishchenko, J. Opt. Soc. Am. **A17**. 888 (2000).

30. Muinonen, In: "The 6th International Conference on Electromagnetic and Light Scattering by Non-spherical Particles" (4-8 March 2002, Univ. of Florida, Gainesville, Florida, USA 2002).

31. H. Volten, O. Munoz, E. Rol, J. de Haan, W. Vassen, and J. Hovenier, K. Muinonen, T. Nousiainen, J. Geophys. Res. D, **106**, 17,375 (2001).

32. O. Munoz, H. Volten, J. de Haan, W. Vassen, and J. Hovenier, Astron. Astrophys. **360,** 777 (2000).

33. O. Munoz, H. Volten, J. de Haan, W. Vassen, and J. Hovenier, J. Geophys. Res. D, **106,** 22,833 (2001).

34. K. Lumme and J. Rahola. Astrophys. J. **425**, 653 (1994).

35. E. Zubko, M. Kreslavsky, and Yu. Shkuratov, Solar System Res., **33,** 296 (1999).

36. V. Weidner and J. Hsia, J. Opt. Soc. Am. **71,** 856 (1981).

37. R. Nelson, B. Hapke, W. Smyth, Yu. Shkuratov, A. Ovcharenko, and D. Stankevich, Lunar and Planet. Sci. 30-th. Abstract #2068. (LPI Houston 1999).

38. J. Geake and M. Geake Mon. Not. Roy. Ast. Soc. **245**, 46 (1990).

39. Yu. G. Shkuratov, L. A. Akimov, N. P. Stankevich, L. Ya. Melkumova, I. I. Latinina, and T. B. Bogdanova, Kinematika i Fisika Nebesnykh Tel. **3,** 32 (1987).

40. Yu. G. Shkuratov, L. Ya. Melkumova, and D. D. Badukov, Kinematika i Fisika Nebesnykh Tel. **4,** 11 (1988).

41. M. Shepard, and R. Arvidson, Icarus **141,** 172 (1999).

42. Yu. G. Shkuratov, Sov. Astron. Lett. **13,** No. 3, 182 (1987).

43. L. O. Kolokolova, Icarus. **84,** 305 (1990).

44. M. Wolff, Appl. Opt. **14,** 1395 (1975).

45. K. Muinonen, J. Piironen, Yu. Shkuratov, A. Ovcharenko, and B. Clark, in: “Asteroids III,” ed. W. Bottke, R. Binzel, A. Cellino, and P. Paolicchi (University of Arizona Press, Tucson, Arizona, 2002).

EXACT RESULTS OF THE VECTOR THEORY OF COHERENT BACKSCATTERING FROM DISCRETE RANDOM MEDIA: AN OVERVIEW

MICHAEL MISHCHENKO,[1] VICTOR TISHKOVETS,[2] AND PAVEL LITVINOV[3]
[1]*NASA Goddard Institute for Space Studies, 2880 Broadway, New York, NY 10025, USA*
[2]*Kharkiv Astronomical Observatory, 35 Sumskaya St., 61022 Kharkiv, Ukraine*
[3]*Institute of Radioastronomy of NASU, 4 Chervonopraporna St., Kharkiv, Ukraine*

Abstract. Coherent backscattering of light by discrete random media, otherwise known as weak photon localization, is a remarkable optical phenomenon caused by constructive interference of waves propagating along the same light-scattering paths but in opposite directions. A well-known manifestation of coherent backscattering is an intensity peak centered at exactly the backscattering direction. It also has been established that when the incident beam is unpolarized then the coherent backscattering intensity peak can be accompanied by a sharp asymmetric peak of negative polarization with a minimum centered at a very small phase angle. It has been suggested that coherent backscattering could be a contributor to some effects observed for solar system bodies in visible light and at radiowave frequencies [1–4]. However, accurate theoretical computations of weak photon localization based on first physical principles are difficult and have been used in analyses of planetary observations in only a handful of publications. This chapter briefly discusses manifestations of coherent backscattering and reviews the exact theory of this phenomenon and its applications to analyses of laboratory data and remote-sensing observations.

1. Introduction

It has been realized recently that diffuse, incoherent multiple scattering of electromagnetic waves by media composed of randomly positioned, discrete scattering particles is always accompanied by a phenomenon called coherent

G. Videen and M. Kocifaj (eds.), Optics of Cosmic Dust, 239–260.

backscattering (or weak photon localization). This phenomenon has been actively studied over the last two decades both experimentally and theoretically [5–8]. Furthermore, by comparing observational data with theoretical computations, it has been shown that coherent backscattering of sunlight by regolithic grains can be responsible for two spectacular natural phenomena exhibited by Saturn's rings and some bright atmosphereless solar system bodies, viz., the spike-like photometric and sharp, asymmetric polarization opposition effects [9–11]. It also has been suggested that diffuse and coherent backscattering of electromagnetic waves might explain unusual radar returns for icy outer planet satellites, polar caps on Mars and Mercury, and some terrestrial lava flows [3,12–16]. In this case, multiple scattering can be caused by discrete scatterers in a relatively transparent matrix, e.g., by particulate rock and/or ice surfaces in air or rocks and voids imbedded in ice.

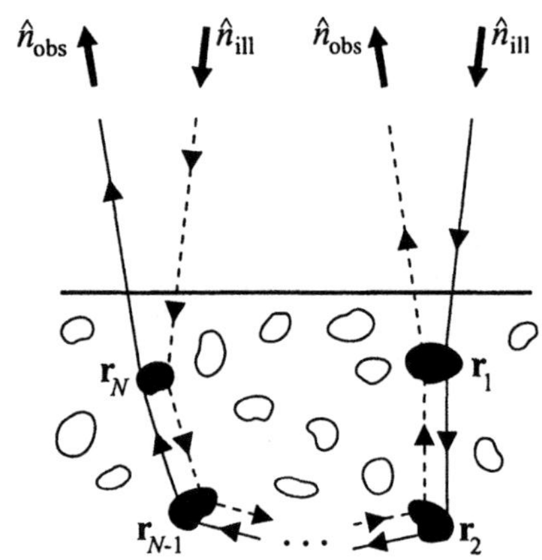

Figure 1: Schematic explanation of coherent backscattering.

To understand the physical origin of weak photon localization, let us consider a layer composed of discrete, randomly positioned scattering particles and illuminated by a parallel beam of light incident in the direction $\hat{n}_{ill}$ (Fig. 1). A distant observer measures the intensity of light reflected by the layer in the direction $\hat{n}_{obs}$. The reflected signal is a result of the contributions from waves scattered along various paths inside the layer involving different combinations of particles. Let us consider two conjugate scattering paths shown in Fig. 1 by solid and dashed lines. These paths go through the same group of N particles, denoted by their positions $\mathbf{r}_1$, $\mathbf{r}_2, \ldots,$ $\mathbf{r}_N$, but in opposite directions. The waves scattered along the two conjugate paths interfere, the interference being constructive or destructive depending on the phase difference $k(\mathbf{r}_N - \mathbf{r}_1)(\hat{n}_{ill} + \hat{n}_{obs})$, where $k = 2\pi/\lambda$ is the wave number and λ is the wavelength in the surrounding medium. If the observation direction is far from the exact backscattering direction given by $-\hat{n}_{ill}$, then the waves scattered along conjugate paths involving different groups of particles interfere in different ways, and the average effect of the interference is zero due to the randomness of particle positions. Consequently, the distant observer measures some average, incoherent intensity well described by the classical radiative transfer theory. However, at exactly the backscattering direction ($\hat{n}_{\rm obs} = -\hat{n}_{\rm ill}$), the phase difference between conjugate paths involving any group of particles is identically equal to zero and the interference is always constructive, thereby causing a coherent backscattering intensity peak.

Owing to its fundamental interference nature, coherent backscattering is a universal physical phenomenon, accompanies any multiple-scattering process, and occurs for particles of any size, shape, and refractive index. The angular width of the backscattering intensity peak can be very small for particles much smaller or much larger than the wavelength of the incident radiation [17], thereby often making the peak unobservable in passive remote sensing measurements. However, coherent backscattering almost always affects active observations with monostatic lidars and radars, e.g., radar observations of terrestrial and planetary surfaces.

Exact computations of the coherent backscattering effect based on solving the Maxwell equations are feasible only for few-particle clusters and is complicated by several factors. First, the scattering pattern for a monodisperse cluster of spheres in a fixed orientation is always heavily burdened by multiple interference maxima and minima and by an intricate resonance structure typical of single monodisperse spheres [18]. Second, the scattering pattern is further influenced by near-field effects, which result from component spheres in a closely packed cluster not being in the far-field zones of each other and can significantly suppress the coherent backscattering effect [19–21]. Third, simple trigonometry shows that the angular width of the coherent backscattering intensity peak is of the order of $1/k\langle d\rangle$, where $\langle d\rangle$ is the average distance between the cluster components. Therefore, this peak can be too broad to be reliably identified unless the cluster components are separated widely enough. However, increasing the distance between the cluster components diminishes the contribution of multiple scattering and, thus, the amplitude of the coherent backscattering peak, thereby making it difficult to detect. To smooth out the effect of the first factor and make the backscattering peak detectable, the scattering pattern must be averaged over particle sizes, cluster orientations, and distances between the components. Furthermore, the average distance between the cluster components must be much larger than the size of the components and the wavelength, but yet small enough so that the multiple-scattering contribution to the total signal remains significant.

The simplest case of two widely separated spheres with diameters much smaller than the wavelength was studied analytically in [22], but it was found that the amplitude of the coherent backscattering intensity peak was extremely small because of the weakness of the multiple-scattering contribution to the total scattered signal. Mishchenko [23] applied the exact superposition *T*-matrix method [18] to polydisperse, randomly oriented clusters composed of two equal wavelength-sized spheres. He computed the ratio of intensities scattered by the clusters and by two independent spheres of the same size assuming that the incident light was unpolarized. Figure 2 shows this ratio versus scattering angle (i.e., the angle between the incident and scattered beams) calculated for $k\langle d\rangle = 25$. The average component sphere size parameter is $k\langle a\rangle = 5$, where a is the sphere radius, and the relative refractive index is $m = 1.2$. The curve clearly shows an enhancement of intensity with an angular width comparable to $1/k\langle d\rangle$ and an amplitude of about 1.03. This feature persisted with varying $k\langle d\rangle$, $k\langle a\rangle$, and refractive index, thereby indicating that it was indeed caused by coherent backscattering.

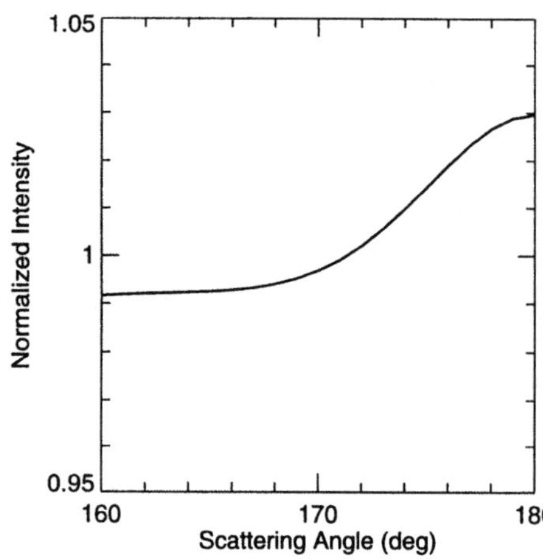

Figure 2: Coherent backscattering by polydisperse, randomly oriented two-sphere clusters.

The amplitude of the coherent backscattering peak can be significantly greater for scattering media composed of very large collections of particles (see Fig. 3 adapted from [24]). However, because of complicated multiple-scattering processes involved, exact physically-based computations of weak localization for such media are difficult and must often rely on one or more simplifying assumptions or must be replaced by approximate computations. The conventional theoretical tool for computing coherent backscattering has been the diffusion approximation [25,26]. However, although this approximation rather accurately predicts the angular profile of the backscattering intensity peak, it ignores the first-order-scattering contribution to the reflected light and, thus, cannot be used to compute the amplitude of the peak, i.e., the ratio of the intensity at the center of the peak to the incoherent background intensity. Furthermore, in some cases the main physical characteristics of a particulate medium (e.g., particle size, shape, and refractive index) cease to be explicit model parameters [27], thereby making difficult, if even possible, comparisons of model computations with results of controlled laboratory experiments. An additional complexifying factor is that accurate computations of the amplitude and polarization state of the backscattering peak must explicitly take into account the vector nature of light since polarization effects have been shown to be extremely important in coherent backscattering [7,11,28–32]. Another theoretical approach is based on Monte-Carlo computations of multiple scattering that keep track of the phase relations between the scattered waves [33–35; see also the chapter by Muinonen *et al.*]. However, such computations are time-consuming and must be extensively tested versus exact theoretical results.

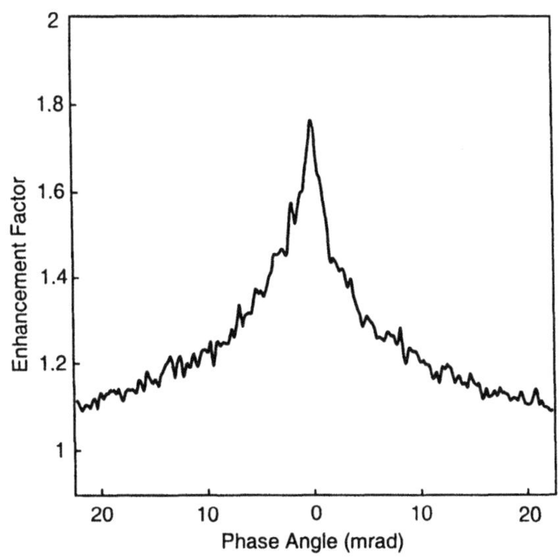

Figure 3: Angular profile of the coherent backscattering peak produced by a 1500-μm-thick slab of 9.6 vol% of 0.215-μm-diameter polystyrene spheres suspended in water. The slab was illuminated by a linearly polarized laser beam ($\lambda = 633\,\text{nm}$) incident perpendicularly to the slab surface. The scattering plane (i.e., the plane through the vectors $\hat{n}_{\text{ill}}$ and $\hat{n}_{\text{obs}}$) was fixed such that the electric vector of the incident beam vibrated in this plane. The detector measured the component of the backscattered intensity polarized parallel to the scattering plane. The curve shows the profile of the backscattered intensity normalized by the intensity of the incoherent background as a function of the phase angle α . The latter is defined as the angle between the vectors $\hat{n}_{\text{obs}}$ and $-\hat{n}_{\text{ill}}$.

In this chapter we review the exact theory of coherent backscattering for sparse discrete random media in which every scatterer is in the far-field zone of all other scatterers and the near-field effects can be neglected. The critical advantages of this theory are that polarization effects are fully taken into account and that particle size distribution, shape, refractive index, and number density are explicit model parameters, thereby facilitating comparisons of theoretical and experimental results. In the following section we will discuss the characteristics of weak photon localization at the exact backscattering direction, in which case it is possible to derive a simple relationship between the diffuse and coherent contributions to the total Stokes reflection matrix applicable to particles of any size, shape, and refractive index. In sections 3 and 4 we will consider the angular dependence of coherent backscattering in the case of normal incidence of light on a plane-parallel slab composed of discrete scattering particles. We will also mention, where appropriate, the relevance of theoretical calculations to specific manifestations of coherent backscattering observed in laboratory and remote-sensing measurements.

2. Exact backscattering direction

Consider a plane-parallel layer composed of discrete, randomly and sparsely positioned particles. Let the medium be illuminated by a parallel beam of quasi-monochromatic radiation incident in the direction $\hat{n}_{\rm ill}$ and characterized by the four-component Stokes column vector $\mathbf{I}_0$ so that I_0 is the incident energy flux per unit area perpendicular to the beam (we use the definition of the Stokes parameters I, Q, U, and V adopted in [18,36]). The Stokes vector $\mathbf{I}$ of radiation reflected by the medium in the direction $\hat{n}_{\rm obs}$ is given by

$$\mathbf{I} = \frac{1}{\pi}\mu_0 \mathbf{S}(\hat{n}_{\rm obs}, \hat{n}_{\rm ill})\mathbf{I}_0 \tag{1}$$

where μ_0 is the cosine of the zenith angle of the incidence direction and $\mathbf{S}(\hat{n}_{\rm obs}, \hat{n}_{\rm ill})$ is the 4×4 reflection matrix transforming the Stokes parameters of the incident light into those of the reflected light. Note that $\mathbf{I}_0$ and $\mathbf{I}$ are defined with respect to the meridional planes of the incidence and the scattering direction, respectively.

The reflection matrix can be decomposed as [7]

$$\mathbf{S}(\hat{n}_{\rm obs}, \hat{n}_{\rm ill}) = \mathbf{S}^1(\hat{n}_{\rm obs}, \hat{n}_{\rm ill}) + \mathbf{S}^{\rm L}(\hat{n}_{\rm obs}, \hat{n}_{\rm ill}) + \mathbf{S}^{\rm C}(\hat{n}_{\rm obs}, \hat{n}_{\rm ill}) \tag{2}$$

where 1 denotes the first-order-scattering component, L denotes the ladder component, and C denotes the cyclical component. The first-order-scattering and ladder components of the reflection matrix are rather slowly varying functions of the reflection direction, and their sum describes the diffuse, incoherent background of the reflected radiation. The cyclical component describes the effect of coherent backscattering, which is superposed on the diffuse background. Unlike $\mathbf{S}^1$ and $\mathbf{S}^{\rm L}$, the cyclical component deviates from zero only in the nearest vicinity of the backscattering direction given by $\hat{n}_{\rm obs} = -\hat{n}_{\rm ill}$ and vanishes outside the coherent backscattering peak.

Both $\mathbf{S}^1$ and $\mathbf{S}^{\mathrm{L}}$ can be found by solving numerically the vector radiative transfer equation [16,37–39]. The computation of the cyclical component $\mathbf{S}^{\mathrm{C}}(\hat{n}_{\mathrm{obs}}, \hat{n}_{\mathrm{ill}})$ for any $\hat{n}_{\mathrm{ill}}$ and $\hat{n}_{\mathrm{obs}}$ is a much more complicated problem which has been solved rigorously only for the case of nonabsorbing Rayleigh scattering (see the following section) and still awaits a general solution for particles of arbitrary size, shape, and refractive index. However, the rigorous relationship derived in [31,40] and based on Saxon's reciprocity relation for the scattering dyad [41] enables one to use the known ladder component of the reflection matrix for calculating the cyclical component at exactly the backscattering direction, i.e., the matrix $\mathbf{S}^{\mathrm{C}}(-\hat{n}_{\mathrm{ill}}, \hat{n}_{\mathrm{ill}})$. Specifically, assuming for simplicity that the scattering medium is macroscopically isotropic and mirror-symmetric [18,36], we have

$$\mathbf{S}^{\mathrm{C}}(-\hat{n}_{\mathrm{ill}}, \hat{n}_{\mathrm{ill}}) = \begin{bmatrix} S_{11}^{\mathrm{C}}(-\hat{n}_{\mathrm{ill}}, \hat{n}_{\mathrm{ill}}) & S_{12}^{\mathrm{L}}(-\hat{n}_{\mathrm{ill}}, \hat{n}_{\mathrm{ill}}) & 0 & 0 \\ S_{12}^{\mathrm{L}}(-\hat{n}_{\mathrm{ill}}, \hat{n}_{\mathrm{ill}}) & S_{22}^{\mathrm{C}}(-\hat{n}_{\mathrm{ill}}, \hat{n}_{\mathrm{ill}}) & 0 & 0 \\ 0 & 0 & S_{33}^{\mathrm{C}}(-\hat{n}_{\mathrm{ill}}, \hat{n}_{\mathrm{ill}}) & S_{34}^{\mathrm{L}}(-\hat{n}_{\mathrm{ill}}, \hat{n}_{\mathrm{ill}}) \\ 0 & 0 & -S_{34}^{\mathrm{L}}(-\hat{n}_{\mathrm{ill}}, \hat{n}_{\mathrm{ill}}) & S_{44}^{\mathrm{C}}(-\hat{n}_{\mathrm{ill}}, \hat{n}_{\mathrm{ill}}) \end{bmatrix} \tag{3}$$

where

$$S_{11}^{\mathrm{C}} = \tfrac{1}{2}(S_{11}^{\mathrm{L}} + S_{22}^{\mathrm{L}} - S_{33}^{\mathrm{L}} + S_{44}^{\mathrm{L}}) \tag{4}$$

$$S_{22}^{\mathrm{C}} = \tfrac{1}{2}(S_{11}^{\mathrm{L}} + S_{22}^{\mathrm{L}} + S_{33}^{\mathrm{L}} - S_{44}^{\mathrm{L}}) \tag{5}$$

$$S_{33}^{\mathrm{C}} = \tfrac{1}{2}(-S_{11}^{\mathrm{L}} + S_{22}^{\mathrm{L}} + S_{33}^{\mathrm{L}} + S_{44}^{\mathrm{L}}) \tag{6}$$

$$S_{44}^{\mathrm{C}} = \tfrac{1}{2}(S_{11}^{\mathrm{L}} - S_{22}^{\mathrm{L}} + S_{33}^{\mathrm{L}} + S_{44}^{\mathrm{L}}) \tag{7}$$

[the arguments $(-\hat{n}_{\mathrm{ill}}, \hat{n}_{\mathrm{ill}})$ are omitted for brevity]. Thus, the combined use of the vector radiative transfer theory and Eqs. (2)–(7) provides the full specification of the reflected radiation outside the peak [Eq. (2) with $\mathbf{S}^{\mathrm{C}}(\hat{n}_{\mathrm{obs}}, \hat{n}_{\mathrm{ill}}) = 0$] and at exactly the center of the backscattering peak.

Equations (3)–(7) represent one of a very few rigorous results of the vector theory of coherent backscattering by sparsely distributed, independently scattering particles and are an important source of benchmark theoretical results for arbitrary particles. Mishchenko [15,16,31] used this formulation to perform a detailed study of the analytical properties of various photometric and polarimetric characteristics of coherent backscattering and their qualitative and quantitative dependence on the particle size, shape, and refractive index. Extensive applications of this approach to model computations and analyses of laboratory and remote-sensing data are described in [9–11,16]. In particular, numerical calculations for specific scattering systems [40] show excellent quantitative agreement with the results of controlled laboratory experiments reported in [42,43].

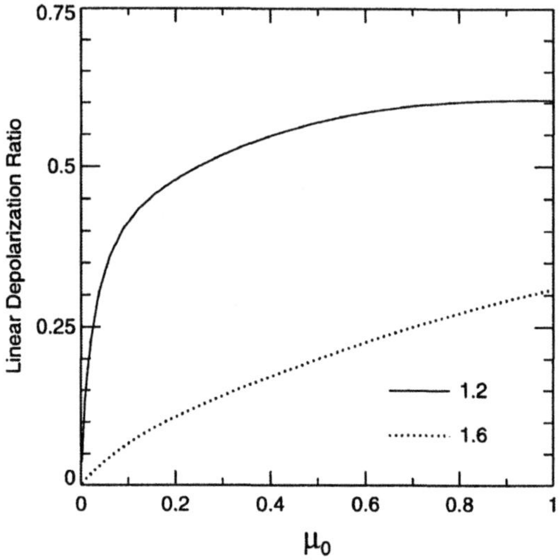

Figure 4: Linear depolarization ratio versus cosine of the incident zenith angle for backscattering by a semi-infinite homogeneous layer composed of polydisperse spherical particles with an effective size parameter of 10 and refractive indices 1.2 and 1.6.

As an example, Fig. 4 shows the linear depolarization ratio δ_L (also known in the radar literature as the linear polarization ratio μ_L) versus μ_0 for a semi-infinite layer composed of spherical particles [16]. δ_L is defined assuming that the incident light is linearly polarized parallel to the meridional plane of the incidence direction and is the ratio of the two orthogonal components of the total backscattered intensity: one polarized in the direction perpendicular to the meridional plane and the other polarized parallel to the plane. A deviation of δ_L from zero means that the backscattered light is partially depolarized. It is seen that δ_L is zero in the limit $\mu_0 \to 0$, i.e., when only the light scattered once contributes to the total reflected signal. This first-order-scattering contribution is not depolarized because the particles forming the medium are spherical [18]. However, for normal incidence ($\mu_0 = 1$) the multiple-scattering contribution to the total backscattered signal is significant and results in strong depolarization. Figure 5 shows analogous results for the circular depolarization ratio δ_C (also known as the circular polarization ratio μ_C). The latter is introduced assuming that the incident light is circularly polarized and is the ratio of two circular components of the total intensity of the backscattered light: one polarized in the same direction as the incident light and the other polarized in the opposite direction. Since for single scattering by a spherical particle $\delta_C \equiv 1$ [18], the large δ_C values shown in Fig. 5 for $\mu_0 \neq 0$ demonstrate a strong depolarizing effect of multiple scattering. The numerical results reported in [15,16] suggest that $\delta_L \leq 1$ and $\delta_C \geq \delta_L$. These properties of the depolarization ratios produced by multiple scattering are similar to those of the depolarization ratios caused by single scattering on nonspherical particles [18,36]. This factor can make it difficult in practice to distinguish between the two depolarization mechanisms.

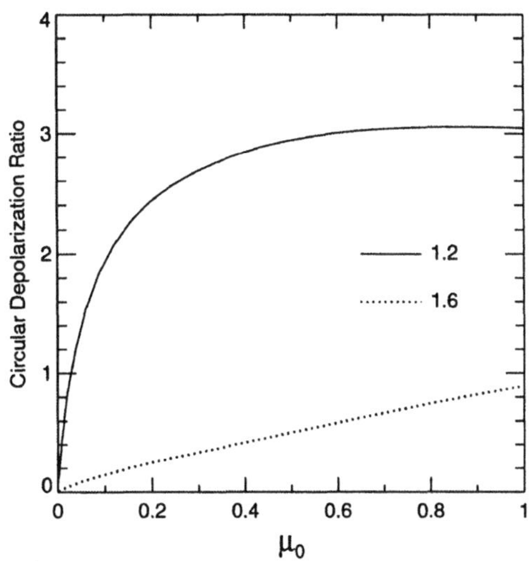

Figure 5: As in Fig. 4, but for the circular depolarization ratio.

Radar observations of icy outer planet satellites at centimeter wavelengths show that when illuminated and observed in linearly polarized radiation, $\delta_L \approx 0.5$, whereas when illuminated and observed in circularly polarized radiation, $\delta_C > 1$ [44,45]. The results of [15,16] show that the observed values of the depolarization ratios can be adequately explained by multiple scattering and localization of light in weakly absorbing discrete random media.

3. Conservative Rayleigh scattering

Another important exact result of the vector theory of coherent backscattering is the complete analytical solution for the case of normal incidence of light on a semi-infinite slab composed of nonabsorbing Rayleigh scatterers. This solution was first obtained by Ozrin [32] and was then re-derived in a more convenient form by Amik *et al.* [46].

Restricting the analysis to small phase angles α, where the first-order-scattering and ladder components of the total reflection matrix are approximately constant, and introducing the so-called angular parameter $q = kl\alpha$, where l is the photon transport mean free path in the scattering medium, one can rewrite the main formulas of [46] using the Stokes-vector representation of polarization as follows [47]:

$$\mathbf{S}^{1} = \tfrac{3}{16}\mathrm{diag}[1, 1, -1, -1] \tag{8}$$

$$\mathbf{S}^{\mathrm{L}} = \tfrac{1}{4}\mathrm{diag}[\gamma_{11}(0)+\gamma_{12}(0), \gamma_{11}(0)-\gamma_{12}(0), \gamma_{12}(0)-\gamma_{11}(0), \gamma_{44}(0)] - \mathbf{S}^{1} \tag{9}$$

$$\mathbf{S}^{\mathrm{C}}(q) = \begin{bmatrix} \widetilde{S}_{11}(q) & \widetilde{S}_{12}(q) & 0 & 0 \\ \widetilde{S}_{12}(q) & \widetilde{S}_{22}(q) & 0 & 0 \\ 0 & 0 & \widetilde{S}_{33}(q) & 0 \\ 0 & 0 & 0 & \widetilde{S}_{44}(q) \end{bmatrix} - \mathbf{S}^{1} \tag{10}$$

where

$$\widetilde{S}_{11}(q) = \tfrac{1}{8}[\gamma_{11}(q)+\gamma_{22}(q)-\gamma_{33}(q)+\gamma_{44}(q)] \tag{11}$$

$$\widetilde{S}_{22}(q) = \tfrac{1}{8}[\gamma_{11}(q)+\gamma_{22}(q)+\gamma_{33}(q)-\gamma_{44}(q)] \tag{12}$$

$$\widetilde{S}_{33}(q) = \tfrac{1}{8}[\gamma_{33}(q)+\gamma_{44}(q)] - \tfrac{1}{4}\gamma_{12}(q) \tag{13}$$

$$\widetilde{S}_{44}(q) = \tfrac{1}{8}[\gamma_{33}(q)+\gamma_{44}(q)] + \tfrac{1}{4}\gamma_{12}(q) \tag{14}$$

$$\widetilde{S}_{12}(q) = \tfrac{1}{8}[\gamma_{11}(q)-\gamma_{22}(q)] \tag{15}$$

Since the angular functions $\gamma_{ij}(q)$ can be computed numerically as described in detail in [46], Eqs. (2) and (8)–(15) can be used to calculate the full Stokes reflection matrix for any q. If, furthermore, the transport mean free path l is known, the reflection matrix can be expressed as a function of the phase angle α rather than the dimensionless angular parameter q. Note that the matrices $\mathbf{S}^{\mathrm{L}}$ and $\mathbf{S}^{\mathrm{C}}(0)$ given by Eqs. (9)–(15) fully comply with the general relationship derived for arbitrary macroscopically isotropic and mirror-symmetric scattering media and discussed in the previous section.

Assuming unpolarized incident light ($Q_0 = U_0 = V_0 = 0$), we define the intensity enhancement factor as the ratio of the total reflected intensity to the incoherent background value:

$$\zeta(q) = \frac{S_{11}^{1} + S_{11}^{\mathrm{L}} + S_{11}^{\mathrm{C}}(q)}{S_{11}^{1} + S_{11}^{\mathrm{L}}} \tag{16}$$

Equations (8)–(10) show that only the first two Stokes parameters of the reflected light can be nonzero. Therefore the degree of linear polarization can be defined as minus the ratio of the total reflected second Stokes parameter to the total reflected intensity:

$$P(q) = -\frac{Q(q)}{I(q)} = -\frac{S_{12}^{\mathrm{C}}(q)}{S_{11}^{1} + S_{11}^{\mathrm{L}} + S_{11}^{\mathrm{C}}(q)} \tag{17}$$

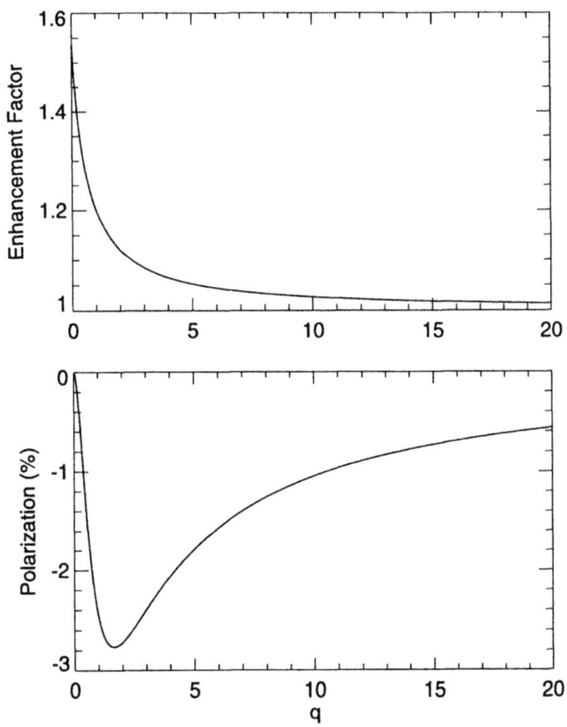

Figure 6: Angular profiles of the photometric enhancement factor and the degree of linear polarization of the reflected light for a half-space of nonabsorbing Rayleigh particles.

Figure 6 shows both the intensity enhancement factor and the degree of linear polarization as functions of the dimensionless angular parameter q [47]. The upper panel demonstrates the renowned intensity peak centered at exactly the opposition. The amplitude of the peak is $\zeta(0) \approx 1.5368$ and its half-width at half-maximum is $q_\zeta \approx 0.597$. Thus the relationship between the half-width at half-maximum of the backscattering intensity peak and the photon transport mean free path is given by $\alpha_\zeta \approx 0.597/kl$. The lower panel of Fig. 6 shows that the reflected polarization is zero at opposition, which is a consequence of azimuthal symmetry in the case of normal illumination and unpolarized incident light. However, with increasing q, polarization becomes negative, rapidly grows in absolute value, and reaches its minimal value

$P_{\min} \approx -2.765\%$ at a reflection direction very close to opposition ($q_P \approx 1.68$). The peak of negative polarization is highly asymmetric so that the half-minimal polarization value -1.383% is first reached at $q_{P,1} \approx 0.498$, which is even smaller than the value $q_\zeta \approx 0.597$ corresponding to the half-width at half-maximum of the backscattering intensity peak, and then at a much larger $q_{P,2} \approx 7.10$. This unusual behavior of polarization at near-backscattering angles was called in [11] the (asymmetric) polarization opposition effect (POE).

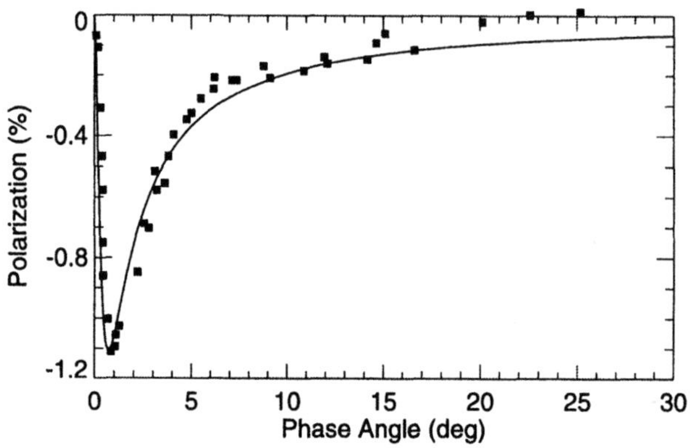

Figure 7: Polarization measurements for a particulate surface composed of microscopic magnesia particles (squares) and best-fit theoretical computations (solid curve).

Because lasers are the most frequently used source of illumination and usually generate linearly or circularly polarized light, explicit laboratory demonstrations of the sharp and asymmetric POE have been extremely rare. Apparently the first laboratory observation of the asymmetric POE was made by Lyot as long ago as in the 1920s [48], even though the physical explanation of this phenomenon was, of course, unknown at that time. Figure 7 shows Lyot's polarization measurements for a particulate surface obtained by burning a tape of magnesium under a glass plate until the deposit on the plate was completely opaque. Lyot described the observed polarization phase curve as "puzzling" and attributed it to the very small size of magnesia grains. Unfortunately, he did not measure the actual size of the grains and their packing density and thus did not provide the information necessary to compute the photon mean free path l. Furthermore, the minimal measured polarization value is only -1.11%, compared to the theoretical value $P_{\min} \approx -2.765\%$ computed for Rayleigh particles. However, assuming that the latter difference is explained by the finite particle size in Lyot's experiment and multiplying the theoretical polarization by a factor of 0.4, and also assuming that the actual kl was close to a realistic value of 132, it is possible to reproduce almost perfectly the angular profile of the measured polarization up to phase angles of about 15° (solid curve in Fig. 7) [47]. At larger phase angles the assumption of phase-angle-independent single-scattering and ladder contributions to the total reflection matrix is no longer valid and causes a significant deviation of the theoretical curve from the measurements.

The sharp asymmetric POE also may have been observed in more recent laboratory measurements by Geake and Geake [49] for fine alumina powders, although the smallest phase angle in their experiments was 2° and may be too large for deriving definitive conclusions. These measurements may suggest that the narrow asymmetric POE weakens and ultimately disappears with increasing particle size, which would be consistent with the similar disappearance of the spatial anisotropy of the backscattering intensity peak in the case of a linearly polarized incident beam observed by van Albada *et al.* [28].

Sharp asymmetric peaks of negative polarization at phase angles considerably smaller than 1° have been observed for Saturn's rings and the Galilean satellites of Jupiter [11,50; see also the chapter by Rosenbush *et al.*]. Because these peaks are distinctly different from the symmetric, nearly parabolic negative polarization branch commonly observed for many solar system bodies [51], they have been interpreted in terms of the asymmetric POE produced by coherent backscattering of sunlight by a surface layer composed primarily of microscopic ice particles [11,50]. The fact that the same objects exhibit comparably narrow intensity peaks centered at exactly the opposition [52] provides strong support for this interpretation.

4. Scattering by a plane-parallel layer composed of arbitrary particles in random orientation

In this section we consider multiple scattering by a plane-parallel layer consisting of discrete, randomly positioned scatterers in random orientation. As in the preceding section, the incident wave is assumed to propagate normally to the boundary of the medium. A comprehensive derivation of the equations describing the coherent and incoherent (diffuse) parts of radiation reflected by such a medium at the backscattering direction was given in [53]. It is based on the exact theory of electromagnetic scattering by systems (clusters) of particles. The equations derived allow a detailed study of the dependence of the photometric and polarization opposition effect characteristics on the microphysical properties of the scatterers (size parameter, real and imaginary parts of the refractive index, shape, number density, etc.). Here we present a brief summary of these results.

The reflection matrix is decomposed according to Eq. (2). The sum of the first-order-scattering and ladder components of the reflection matrix is given by

$$S^{\mathrm{l}}_{pn\mu\nu} + S^{\mathrm{L}}_{pn\mu\nu} = \frac{-n_0}{2k^3 \cos\vartheta} \sum_l d^l_{M_0 N_0}(\vartheta) \int_0^{kZ_0} \exp\left(2z \frac{\mathrm{Im}(\varepsilon)}{\cos\vartheta}\right) \alpha_l^{(z)(pn)(\mu\nu)} dz \tag{18}$$

where the matrix elements $S^{\mathrm{l}}_{pn\mu\nu}$ and $S^{\mathrm{L}}_{pn\mu\nu}$ are written in terms of the circular-polarization (CP) representation of the Stokes vector [54], n_0 is the particle number density, $p, n, \mu, \nu = \pm 1$, $d^l_{M_0 N_0}(\vartheta)$ are Wigner d functions [55] ($M_0 = \nu - n$, $N_0 = \mu - p$), ϑ is the scattering angle (the angle between the incidence and scattering directions), Z_0 is the geometrical thickness of the layer, and ε is the complex effective refractive index of the medium. The expansion coefficients $\alpha_L^{(z)(pn)(\mu\nu)}$ are determined from the system of equations

$$\alpha_L^{(z)(pn)(\mu\nu)} = \chi_L^{(pn)(\mu\nu)} \exp(-\tau_z) + \frac{2\pi n_0}{k^3} \sum_{qq_1} \chi_L^{(pq)(\mu q_1)} \times \int \sum_l \alpha_L^{(y)(qn)(q_1\nu)} \exp(-\tau_\rho) d^L_{M_0 N}(\omega) d^l_{M_0 N}(\omega) d\rho \sin\omega d\omega \tag{19}$$

where ρ, ω are polar coordinates of the integration point with respect to the point z, the angle ω is measured off the backscattering direction, $\tau_x = 2\,\mathrm{Im}(\varepsilon)x$, $N = q_1 - q$ ($q, q_1 = \pm 1$), and $y = z - \rho\cos\omega$. The $\chi_L^{(pn)(\mu\nu)}$ are the coefficients in the expansion of the individual-particle scattering matrix in the Wigner d functions, and the first term on the right-hand side corresponds to the first-order-scattering contribution.

Equation (19) specifies the coefficients $\alpha_L^{(z)(pn)(\mu\nu)}$ as functions of the optical depth in the layer. The exponential factor in Eq. (18) describes the extinction of radiation while it propagates from a point z to the exit point in the direction ϑ. Note that Eqs. (18) and (19) are one of many alternative representations of the vector radiative transfer equation.

The cyclical component of the reflection matrix can be written as

$$S^{\mathrm{C}}_{pn\nu\mu} = \frac{-n_0}{2k^3 \cos\vartheta} \int_0^{kZ_0} G^{(z)(pn)(\mu\nu)} \exp\left(-2z\,\mathrm{Im}(\varepsilon)\left(1 - \frac{1}{\cos\vartheta}\right)\right) dz \tag{20}$$

where the matrix $G^{(z)(pn)(\mu\nu)}$ is defined by

$$G^{(z)(pn)(\mu\nu)} = \frac{2\pi n_0}{k^3} \sum_{qq_1 LM} (-1)^L \zeta_{LM}^{*(q_1\mu)(qp)} \int \sum_{lm} \beta_{lm}^{(y)(qn)(q_1\nu)} \times d^L_{MN}(\omega) d^l_{mN}(\omega) \exp(-\tilde{\tau}_\rho) F_{m-M}(\rho, \vartheta, \omega) d\rho \sin\omega d\omega \tag{21}$$

Here $\zeta_{LM}^{(pn)(\mu\nu)}$ are the coefficients in the expansion of the phase matrix elements in the Wigner d functions,

$$F_m(\rho,\vartheta,\omega)=\sum_l i^{-l}(2l+1)j_l\left(2\rho\cos\frac{\vartheta}{2}\right)d^l_{m0}\left(\frac{\vartheta}{2}\right)d^l_{m0}(\omega)$$
$$=i^{-m}J_m(\rho\sin\vartheta\sin\omega)\exp(-i\rho\cos\omega(1+\cos\vartheta)) \tag{22}$$

$$\tilde{\tau}_\rho=\rho\,\mathrm{Im}(\varepsilon)\left(2-\cos\omega\left(1-\frac{1}{\cos\vartheta}\right)\right)+i\rho\cos\omega[\mathrm{Re}(\varepsilon)-1]\left(1+\frac{1}{\cos\vartheta}\right) \tag{23}$$

$j_l(x)$ are spherical Bessel functions, and $J_m(x)$ are Bessel functions. The coefficients $\beta^{(z)(pn)(\mu\nu)}_{LM}$ are determined from the system of equations

$$\beta^{(z)(pn)(\mu\nu)}_{LM}=\zeta^{(pn)(\mu\nu)}_{LM}+\frac{2\pi n_0}{k^3}\sum_{qq_1}\chi^{(pq)(\mu q_1)}_L\int\sum_{lm}\beta^{(y)(qn)(q_1\nu)}_{lm}$$
$$\times d^L_{MN}(\omega)d^l_{mN}(\omega)\exp(-\tilde{\tau}_\rho)F_{m-M}(\rho,\vartheta,\omega)d\rho\sin\omega d\omega \tag{24}$$

The matrix $S^{\mathrm{C}}_{pn\nu\mu}$ describes the interference of the conjugate pairs of waves propagating along the same trajectories (Fig. 1) including looped paths [53,56].

The above equations are rigorous provided that any scatterer is located in the far-field zone of all other scatterers. They are valid for arbitrary particles in random orientation. Unlike the formula for the incoherent part, Eq. (19), the kernel of the formula for the coherent part, Eq. (24), is a rapidly oscillating function. The numerical solution of such equations is a very complex problem. Therefore, below we consider several particular cases of the general equations.

4.1 Exact backscattering direction ($\vartheta=\pi$)

In this case $F_m(\rho,\pi,\omega)=\delta_{0m}$ and $\zeta^{*(\nu\mu)(np)}_{LM}=(-1)^p\chi^{(pn)(\mu\nu)}_L\delta_{M,-N_0}$. Therefore, Eqs. (18), (19), and (20)–(24) yield [53]

$$S^{\mathrm{L}}_{pn\mu\nu}=S^{\mathrm{C}}_{pn\nu\mu} \tag{25}$$

Rewriting Eq.(25) in terms of the standard Stokes representation of polarization gives Eqs. (4)–(7) [56].

It should be noted that the relations between the coherent and incoherent parts of the reflected radiation similar to Eqs. (4)–(7) and (25) can be derived for an arbitrary system of scatterers provided that they are in the far-field zones of each other. It can be done using Eqs. (19), (20), (35), and (36) of [53].

4.2 Double-scattering approximation for a semi-infinite layer

The results of the double-scattering approximation are useful for semi-quantitative or qualitative analyses of the effect of particle microphysical properties on the characteristics of the opposition effects. Instead of Eqs. (18) and (19), we now have the following [56]:

$$S^{\mathrm{l}}_{pn\mu\nu} + S^{\mathrm{L}}_{pn\mu\nu} = \frac{n_0}{4k^3\,\mathrm{Im}(\varepsilon)(1-\cos\vartheta)}\sum_l d^l_{M_0N_0}(\vartheta)A_l^{(pn)(\mu\nu)} \tag{26}$$

where

$$A_L^{(pn)(\mu\nu)} = \chi_L^{(pn)(\mu\nu)} + \frac{\pi n_0}{k^3 Im(\varepsilon)}\sum_{qq_1}\chi_L^{(pq)(\mu q_1)}\sum_l \chi_l^{(qn)(q_1\nu)}$$
$$\times\left[\int_0^{\pi/2} d^L_{M_0N}(\omega)d^l_{M_0N}(\omega)\frac{\cos\vartheta\sin\omega d\omega}{\cos\vartheta-\cos\omega}\right.$$
$$\left.+\int_0^{\pi/2} d^L_{M_0N}(\pi-\omega)d^l_{M_0N}(\pi-\omega)\frac{\sin\omega d\omega}{1+\cos\omega}\right] \tag{27}$$

The first term on the right-hand side of Eq. (27) gives the first-order-scattering contribution, whereas the second one describes the second-order-scattering contribution. The coherent part for a semi-infinite layer is given by [56]

$$S^{\mathrm{C}}_{pn\nu\mu} = \frac{\pi n_0^2}{2k^6\,\mathrm{Im}(\varepsilon)(1-\cos\vartheta)}\sum_{LMlmqq_1}(-1)^L\zeta_{LM}^{*(q_1\mu)(qp)}\zeta_{lm}^{(qn)(q_1\nu)}i^{-|m-M|}B^{(qq_1)}_{LMlm} \tag{28}$$

where

$$B^{(qq_1)}_{LMlm} = \int_0^{\pi/2} d^L_{MN}(\omega)d^l_{mN}(\omega)\frac{c^{|m-M|}\sin\omega d\omega}{\sqrt{c^2+f^2}(f+\sqrt{c^2+f^2})^{|m-M|}}$$
$$+\int_0^{\pi/2} d^L_{MN}(\pi-\omega)d^l_{mN}(\pi-\omega)\frac{c^{|m-M|}\sin\omega d\omega}{\sqrt{c^2+f^{*2}}(f^*+\sqrt{c^2+f^{*2}})^{|m-M|}} \tag{29}$$

$$c = \sin\vartheta\sin\omega \tag{30}$$

$$f = 2\,\mathrm{Im}(\varepsilon)+\cos\omega\left\{\mathrm{Im}(\varepsilon)\left(1-\frac{1}{\cos\vartheta}\right)+i[\mathrm{Re}(\varepsilon)-1]\left(1+\frac{1}{\cos\vartheta}\right)+i(1+\cos\vartheta)\right\} \tag{31}$$

The expansion coefficients $\chi_L^{(pn)(\mu\nu)}$ and $\zeta_{LM}^{(pn)(\mu\nu)}$ are determined by the microphysical properties of the scatterers. In the case of spherical particles,

$$\chi_{L_1}^{(pn)(\mu\nu)} = \frac{1}{4}\sum_{Ll}(2L+1)(2l+1)\langle a_L^{(pn)}a_l^{*(\mu\nu)}\rangle C^{L_1M_0}_{L-nl\nu}C^{L_1N_0}_{L-pl\mu} \tag{32}$$

$$\zeta_{L_1M_1}^{(pn)(\mu\nu)} = \frac{1}{4}\sum_{Ll}(2L+1)(2l+1)\langle a_L^{(pn)}a_l^{*(\mu\nu)}\rangle C^{L_1M_1}_{L-nlm_1}C^{L_1N_0}_{L-pl\mu}(-1)^l d^l_{m_1-\nu}(\vartheta) \tag{33}$$

where $m_1 = M_1 + n$, $a_L^{(pn)} = a_L + pnb_L$, a_L and b_L are the Lorenz–Mie coefficients [18], the angle brackets denote averaging over particle microphysical characteristics, and the Cs are Clebsh-Gordan coefficients [55].

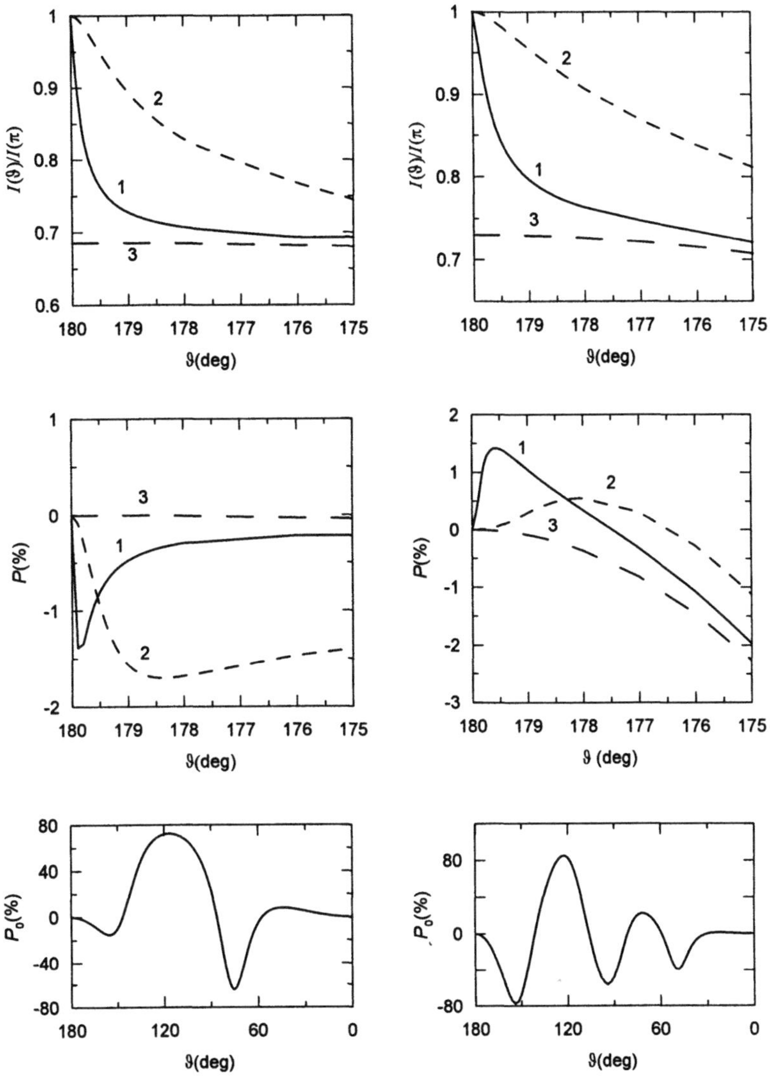

Figure 8: The coherent component (1, 2) and the sum of the diffuse and first-order-scattering components (3) of the scattered radiation for ka = 3, m = 1.35 (left-hand column) and $ka = 4.5$, m = 1.33 (right-hand column). Curves 1 correspond to ξ =0.001; curves 2 correspond to ξ =0.01. P_0 is the linear polarization for an individual isolated particle.

Figure 8 depicts the degree of linear polarization, P, and the normalized intensity, $I(\vartheta)/I(\pi)$, for a semi-infinite layer composed of identical spherical particles. The results are shown for different values of the filling factor $\xi = \frac{4}{3}\pi a^3 n_0$, where a is the particle radius. The effective refractive index of the medium is given by $\varepsilon = 1 + in_0 C_{\text{ext}}/2k$, where C_{ext} is the extinction cross section per particle.

One can see that the polarization opposition effect is more sensitive to particle properties than the photometric one (the latter always manifests itself as a sharp peak of intensity centered at the exact backscattering direction). The polarization behavior can be explained as follows. It has been demonstrated that in the case of positive polarization for an isolated scatterer, the interference of second-order waves results in negative polarization at the backscattering direction [4]. However uniformly positive polarization is a specific property of Rayleigh scatterers. The angular dependence of polarization for wavelength-sized particles becomes much more complicated [18]. Therefore, the interference of the second-order waves can result in positive backscattering polarization as well as in negative polarization.

The left-hand column of Fig. 8 corresponds to spherical particles with a size parameter $x = ka = 3$ and a refractive index $m = 1.35$. The phase function for these particles is asymmetric, most of radiation being scattered at angles $\vartheta < 60°$. Therefore, the polarization of the coherent part of the reflected radiation is determined by the behavior of P_0 in the scattering-angle ranges $\vartheta < 60°$ and $\vartheta > 120°$, where the average value of P_0 is positive (see the bottom panel of the left-hand column of Fig. 8). Hence, the interference of second-order scattering paths yields negative polarization near the opposition direction.

The right-hand column of Fig. 8 corresponds to particles with a size parameter $x = 4.5$ and a refractive index $m = 1.33$. The phase function for these particles is even more asymmetric. The average value of P_0 in the scattering-angle ranges $\vartheta < 60°$ and $\vartheta > 120°$ is negative. In this case the interference of the second-order scattering paths results in positive polarization at backscattering angles.

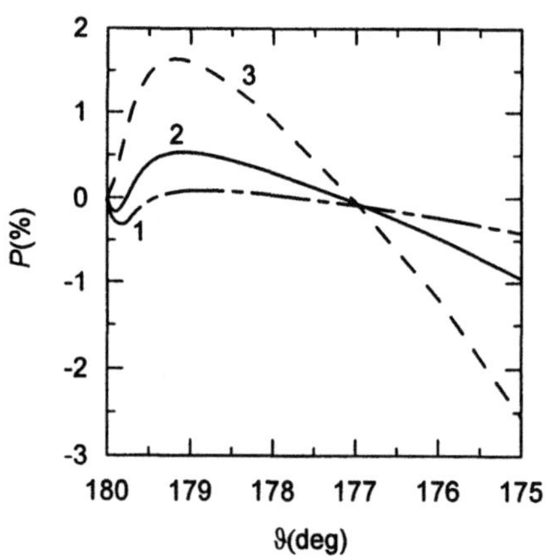

Figure 9: Degree of linear polarization for a semi-infinite particulate layer with $\xi = 0.002$ and different parameters of the particle size distribution. Curve 1: $x_{eff} = 2$, $v_{eff} = 0.1$; curve 2: $x_{eff} = 2$, $v_{eff} = 0.25$; curve 3: $x_{eff} = 3$, $v_{eff} = 0.25$.

Figures 9 and 10 show the polarization profiles for a gamma-distribution of particle size parameters $f(x) \propto x^{(1-3v_{eff})/v_{eff}} \exp[-x/(x_{eff}v_{eff})]$, where x_{eff} is the effective size parameter and v_{eff} is the effective variance, and refractive indices 1.5 and 1.33 + 0.05i, respectively. One can see that varying polarization behavior is possible depending on particle parameters. In particular, the degree of linear polarization at small phase angles can be negative as well as positive, and it is possible to obtain a bimodal angular dependence. The bimodal polarization behavior in Fig. 10 (curve 1) is formed by a superposition of the sharp polarization peak formed by constructive interference of waves scattered by different scatterers and the smooth polarization minimum caused by single scattering.

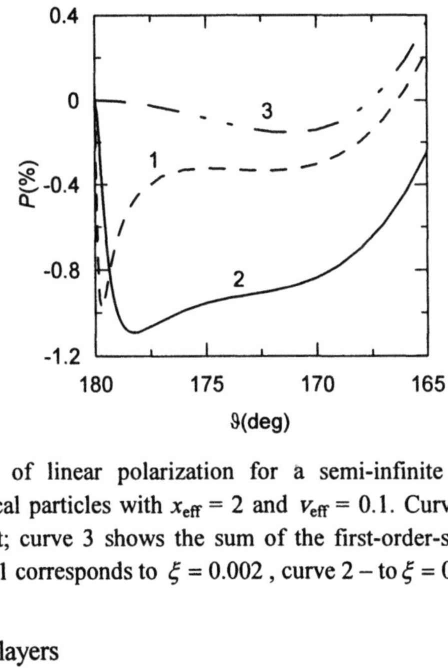

Figure 10: Degree of linear polarization for a semi-infinite layer composed of polydisperse spherical particles with $x_{eff} = 2$ and $v_{eff} = 0.1$. Curves 1 and 2 show the coherent component; curve 3 shows the sum of the first-order-scattering and diffuse components. Curve 1 corresponds to $\xi = 0.002$, curve 2 – to $\xi = 0.01$.

4.3 Optically thin layers

Figures 8–10 show a strong influence of particle microphysics and filling factor on polarization, in particular on the angular position of the polarization extremum α_m. The angular position of the extremum can be an important characteristic in analyses of experimental data. In the case of an optically thin layer, the integral over ρ in Eq. (21) is a sharp function, which has a maximum at $\omega = \pi/2$ and tends to zero for other values of ω. For such a layer composed of moderately sized particles, the scattering-angle dependence of the coefficients $\zeta_{LM}^{(np)(\nu\mu)}$ at $\vartheta \to \pi$ is weak. Moreover, we can now neglect all orders of scattering higher then the second. This allows us to derive an analytical expression for α_m:

$$\frac{\sin\alpha_m}{2\,\mathrm{Im}(\varepsilon)} = kl\sin\alpha_m = \sqrt{\sqrt{2}(2+\sqrt{2})} \approx 2 \tag{34}$$

where l is the photon transport mean free path. This expression agrees well with that for a semi-infinite layer composed of nonabsorbing point-like scatterers (Section 3). In both cases α_m is defined only by the value of l.

Equation (34) can be used in analyses of remote-sensing observations, for example, for estimating surface properties of particles forming Saturn's rings. The majority of the ring particles are believed to have sizes ranging from a few centimeters to a few meters and surface morphology similar to that of snow or frost [57]. A recent analysis of numerous earth-based observations shows a bimodal phase-angle dependence of linear polarization of light reflected by Saturn's rings and Galilean satellites of Jupiter [50]. The value of α_m for the first (sharp) polarization minimum is about $0.5°$, whereas the one for the second (smooth) minimum is about $3° - 4°$. One potential explanation of this bimodal behavior is that the smooth polarization minimum is produced by cooperative scattering within wavelength-sized aggregates consisting of a modest number of small components, whereas the sharp polarization peak is the result of constructive interference of waves scattered by different aggregates. The filling factors of aggregates then follows from

$$\xi = \frac{2\pi\langle x\rangle^3 \alpha_m}{3k^2 C_{\text{sca}}(1-\langle\cos\vartheta\rangle)} \tag{35}$$

where C_{sca} is the scattering cross section per aggregate, $\langle\cos\vartheta\rangle$ is the asymmetry parameter, and $\langle x\rangle$ is the average size parameter of the aggregates. Numerical calculations using Eqs. (35) for refractive indices $1.2 \div 1.33$ and $5 \le \langle x\rangle \le 20$ give $\xi \cong 0.02 \div 0.3$. The smooth negative polarization minimum can be produced by aggregates consisting of 10^2 to 10^3 monomers with size parameters $\cong 0.1$, which implies $\langle x\rangle \cong 10$ and $\xi \cong 0.1$. These estimates assume that the size of the aggregates relative to the wavelength and their packing density are small enough so that the effects of shadowing can be ignored. Of course, another possible explanation is that the surfaces of these objects are highly heterogeneous and that the sharp negative polarization minimum is produced by the areas covered with an optically thick layer of microscopic (Rayleigh) grains (Section 3), whereas the smooth polarization minimum is produced by the areas covered by particles of a different type.

5. Acknowledgments

M. Mishchenko acknowledges support from the NASA Radiation Sciences Program managed by Donald Anderson and from the NASA Earth Observing System program. V. Tishkovets and P. Litvinov appreciate funding received under INTAS Grant No. 1999-00652.

References

1. Yu. G. Shkuratov, Kinem. Fiz. Nebes. Tel **4**, 33 (1988).
2. K. Muinonen, Ph.D. thesis, Univ. Helsinki (1990).
3. B. Hapke, Icarus **88**, 407 (1990).
4. Yu. G. Shkuratov, Astron. Vestnik **25**, 152 (1991).
5. M. Nieto-Vesperinas and J. C. Dainty (Eds.), "Scattering in Volumes and Surfaces" (North-Holland, Amsterdam, 1990).
6. P. Sheng (Ed.), "Scattering and Localization of Classical Waves in Random Media" (World Scientific, Singapore, 1990).
7. Yu. N. Barabanenkov, Yu. A. Kravtsov, V. D. Ozrin, A. I. Saichev A. I., Progr. Opt. **29**, 65 (1991).
8. A. Lagendijk, and B. A. van Tiggelen, Phys. Rep. **270**, 143 (1996).
9. M. I. Mishchenko and J. M. Dlugach, Mon. Not. R. Astron. Soc. **254**, 15P (1992).
10. M. I. Mishchenko and J. M. Dlugach, Planet. Space Sci. **41**, 173 (1993).
11. M. I. Mishchenko, Astrophys. J. **411**, 351 (1993).
12. S. J. Ostro, D. B. Campbell, R. A. Simpson, et al., J. Geophys. Res. **97**, 18227 (1992).
13. A. Campbell, R. E. Arvidson, and M. K. Shepard, J. Geophys. Res. **98**, 17099 (1993).
14. B. J. Butler, D. O. Muhleman, and M. A. Slade, J. Geophys. Res. **98**, 15003 (1993).
15. M. I. Mishchenko, Earth Moon Planets **58**, 127 (1992).
16. M. I. Mishchenko, J. Quant. Spectrosc. Radiat. Transfer **56**, 673 (1996).
17. M. I. Mishchenko, Astrophys. Space Sci. **194**, 327 (1992).
18. M. I. Mishchenko, L. D. Travis, and A. A. Lacis, "Scattering, Absorption, and Emission of Light by Small Particles" (Cambridge University Press, Cambridge, 2002).
19. V. P. Tishkovets, Opt. Spectrosc. **85**, 212 (1998).
20. V. P. Tishkovets and P. V. Litvinov, Solar Syst. Res. **33**, 186 (1999).
21. V. P. Tishkovets, Yu. G. Shkuratov, and P. V. Litvinov, J. Quant. Spectrosc. Radiat. Transfer **61**, 767 (1999).
22. K. Muinonen, Proc. 1989 URSI Electromagn. Theory Symp., p. 428 (1989); F. M. Ismagilov and Yu. A. Kravtsov, Waves in Random Media **3**, 17 (1993).
23. M. I. Mishchenko, Opt. Lett. **21**, 623 (1996).
24. M. P. van Albada, M. B. van der Mark, and A. Lagendijk, Phys. Rev. Lett. **58**, 361 (1987).
25. M. B. van der Mark, M. P. van Albada, and A. Lagendijk, Phys. Rev. B **37**, 3575 (1988).
26. E. Akkermans, P. E. Wolf, R. Maynard, and G. Maret, J. Phys. (Paris) **49**, 77 (1988).
27. K. J. Peters, Phys. Rev. B **46**, 801 (1992).
28. M. P. van Albada, M. B. van der Mark, and A. Lagendijk, J. Phys. D **21**, S28 (1988).
29. M. P. van Albada and A. Lagendijk, Phys. Rev. B **36**, 2353 (1987).
30. C. MacKintosh and S. John, Phys. Rev. B **37**, 1884 (1988).
31. M. I. Mishchenko, J. Opt. Soc. Am. A **9**, 978 (1992).
32. V. D. Ozrin, Waves in Random Media **2**, 141 (1992).
33. M. P. van Albada and A. Lagendijk, Phys. Rev. B. **36**, 2353 (1987).
34. S. Martinez and R. Maynard, Phys. Rev. B **50**, 3714 (1994).
35. T. Iwai, H. Furukawa, and T. Asakura, Opt. Rev. **2**, 413 (1995).
36. M. I. Mishchenko, J. W. Hovenier, and L. D. Travis (Eds.), "Light Scattering by Nonspherical Particles: Theory, Measurements, and Applications" (Academic Press, San Diego, 2000).
37. S. Chandrasekhar, "Radiative Transfer" (Clarendon Press, London, 1950).

38. J. E. Hansen and L. D. Travis, Space Sci. Rev. **16**, 527 (1974).
39. J. F. de Haan, P. W. Bosma, and J. W. Hovenier, Astron. Astrophys. **183**, 371 (1987).
40. M. I. Mishchenko, Phys. Rev. B **44**, 12597 (1991).
41. D. S. Saxon, Phys. Rev. **100**, 1771 (1955).
42. M. P. van Albada, M. B. van der Mark, and A. Lagendijk, J. Phys. D **21**, S28 (1988).
43. P. E. Wolf, G. Maret, E. Ackermans, and R. Maynard, J. Phys. (Paris) **49**, 63 (1988).
44. S. Ostro, in D. Morrison (Ed.), "Satellites of Jupiter," p. 213 (University of Arizona Press, Tucson, AZ, 1982).
45. S. J. Ostro, Rev. Mod. Phys. **65**, 1235 (1993).
46. E. Amic, J. M. Luck, and Th. M. Nieuwenhuizen, J. Phys. I **7**, 445 (1997).
47. M. I. Mishchenko, J.-M. Luck, and T. M. Nieuwenhuizen, J. Opt. Soc. Am. A **17**, 888 (2000).
48. B. Lyot, NASA Tech. Transl. F-187 (NASA, Washington, D. C., 1964).
49. J. E. Geake and M. Geake, Mon. Not. R. Astron. Soc. **245**, 46 (1990).
50. V. K. Rosenbush, V. V. Avramchuk, A. E. Rosenbush, and M. I. Mishchenko, Astrophys. J. **487**, 402 (1997).
51. A. Dollfus, M. Wolff, J. E. Geake, *et al.*, in R. P. Binzel, T. Gehrels, and M. S. Matthews (Eds.), "Asteroids II," p. 594 (University of Arizona Press, Tucson, AZ, 1989).
52. P. Helfenstein, N. Currier, B. E. Clark, J. Veverka, M. Bell, *et al.*, Icarus **135**, 41 (1998).
53. V. P. Tishkovets. J. Quant. Spectrosc. Radiat. Transfer **72**, 123 (2002).
54. A. Z. Dolginov, Yu. N. Gnedin, and N. A. Silant'ev. "Propagation and Polarization of Radiation in Cosmic Media" (Gordon and Breach, Basel, 1995).
55. D. A. Varshalovich, A. N. Moskalev, and V. K. Khersonskii. "Quantum Theory of Angular Momentum" (World Scientific, Singapore, 1988).
56. V. P. Tishkovets, P. V. Litvinov, and M. V. Lyubchenko. J. Quant. Spectrosc. Radiat. Transfer **72**, 803 (2002).
57. S. J. Weidenschilling, C. R. Chapman, D. R. Davis, and R. Greenberg, in R. Greenberg, A. Brahic (Eds.), "Planetary Rings," p. 367 (University of Arizona Press, Tucson, AZ, 1984).

NUMERICAL TECHNIQUES FOR BACKSCATTERING BY RANDOM MEDIA

K. MUINONEN,[1] G. VIDEEN,[2] E. ZUBKO,[3] YU. SHKURATOV[3]
[1] *University of Helsinki, Observatory, Kopernikuksentie 1, 00130 Helsinki, Finland & Astronomical Observatory of Torino, via Osservatorio 20, 10025 Pino Torinese (TO), Italy,*
[2] *Army Research Laboratory, 2800 Powder Mill Road, Adelphi, MD 20783, USA,*
[3] *Astronomical Observatory, Kharkov National University, Sumskaya Str. 35, Kharkov 61022, Ukraine*

Abstract. Coherent backscattering of unpolarized incident light by random media of discrete scatterers leads to an opposition effect in scattered intensity (enhanced backscattering) and a surge of negative degree of linear polarization (negative polarization). We describe three computational techniques suitable for the multiple scattering problem at hand: second-order and sixth-order ray-tracing techniques for dark media, and a coherent-backscattering radiative-transfer technique for both dark and bright media. All three techniques are radiative-transfer-like approximations to the rigorous electromagnetic scattering theory. We discuss example results for miscellaneous random media of scatterers.

1. Introduction

Photometric observations for large numbers of atmosphereless solar system objects show an *opposition effect*, a nonlinear increase of brightness at small solar phase angles, the angle between the Sun and the observer as seen from the object. Polarimetric observations show *negative polarization*, a curious degree of linear polarization $(I_\perp - I_\parallel)/(I_\perp + I_\parallel)$ for unpolarized incident sunlight: at small phase angles, the disk-integrated brightness component $I_\parallel$ with the electric vector parallel to the scattering plane defined by the Sun, the object, and the observer predominates over the perpen-

G. Videen and M. Kocifaj (eds.), Optics of Cosmic Dust, 261–282.

dicular component $I_{\perp}$. In our terminology, we do not distinguish between the opposition effect and "the opposition spike" (narrow opposition effect); and the negative polarization surge and "the polarization opposition effect" (sharp negative polarization surge). For a review of the observations for atmosphereless solar system objects, see the chapter by Rosenbush *et al.*

The coherent backscattering mechanism (CBM) has recently been shown to contribute to both the opposition effects and the negative polarization surges ([1, 2, 3, 4, 5]; see the chapters by Mishchenko *et al.*, and Shkuratov and Ovcharenko), challenging the traditional shadowing-mechanism interpretations. CBM is a multiple-scattering mechanism for scattering orders higher than the second (inclusive); whereas, shadowing is a first-order multiple-scattering mechanism. There are several reviews on the mechanisms for the backscattering phenomena of atmosphereless solar system objects (e.g., [6, 7, 8]), and we refer the reader to these reviews for more historical details.

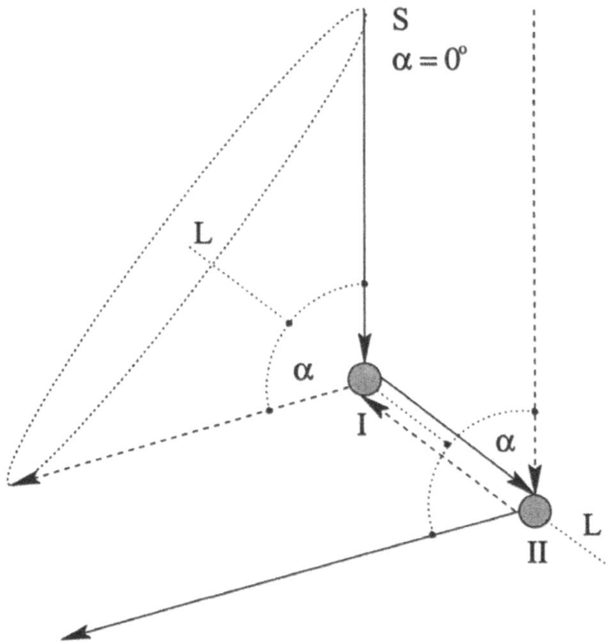

Figure 1. Coherent backscattering mechanism for the opposition effect. See text.

The CBM for the opposition effect is described in Fig. 1 for second-order scattering, readily applicable to any order of multiple scattering. An electromagnetic plane wave is incident upon the system from the top and is represented by the solid and dashed lines. This plane wave interacts with two scatterers I and II, which are of the order of the wavelength λ to hundreds of wavelengths apart before traveling to the observer to the left. The two scattered wave components deriving from the two opposite propagation directions between the scatterers interfere constructively in the conical directions defined by a rotation of the light source direction S about the axis L joining the two end scatterers. Figure 1 illustrates a scattering direction on the cone precisely opposite to the light source direction. Consequently, the exact backward direction (phase angle $\alpha = 0°$) is on the constructive-interference cone for arbitrary locations of the two scatterers. In other directions, interference varies from constructive to destructive. Three-dimensional averaging over scatterer locations results in an opposition effect. The angular width of the effect decreases for increasing orders of interactions, because of the increasing average distance between the end scatterers. The scattering processes themselves can be caused by any disorder or irregularity in the medium.

The CBM for the negative degree of linear polarization is explained for second-order scattering in Fig. 2. The incident radiation is unpolarized by definition, requiring the derivation and proper averaging of the Stokes vectors of the scattered electromagnetic fields ($\boldsymbol{E}_{\mathrm{s}}$) for two linear polarization states (perpendicular to one another) of an incident plane wave ($\boldsymbol{E}_{\mathrm{i}}$). In Fig. 2, incident polarizations parallel and perpendicular to the scattering plane (here yz-plane) are treated in the two leftmost and two rightmost panels, respectively.

Thus, in Fig. 2, an incident electromagnetic plane wave interacts with two scatterers I and II at a distance d from one another aligned either on the x-axis or the y-axis, while the observer is in the yz-plane. For the present geometries, the constructive interference cones of Fig. 1 reduce to the yz and xz-planes, depending on the alignment of the scatterers. Since first-order scattering is typically positively polarized (e.g., Rayleigh scattering and Fresnel reflections) the scatterers sufficiently far away from each other ($kd = 2\pi d/\lambda \gg 1$, where k and λ are the wave number and wavelength) interact predominantly with the electric field vector perpendicular to the plane defined by the source and the scatterers (two upper panels), while interaction with the electric field vector parallel to that plane is suppressed (two lower panels). The observer in the yz-plane will measure negative polarization from the geometry in the upper left panel of Fig. 2, and positive polarization from the geometry in the upper right panel. However, the positive polarization suffers from the phase difference $kd \sin\alpha$, whereas the phase difference for the negative polarization is zero for all phase angles. Averaging over scatterer locations will result in negative polarization near the backward direction. Scattering orders higher than the second experience similar preferential interaction geometries, and contribute to negative polarization. As above for the opposition effect, the contributions from increasing orders of scattering manifest themselves at decreasing phase angles.

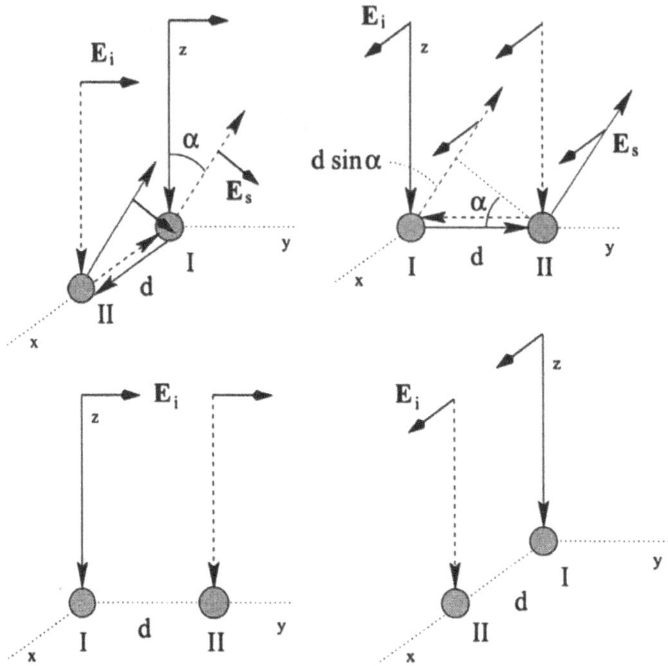

Figure 2. Coherent backscattering mechanism for the negative degree of linear polarization for unpolarized incident light. See text.

In Section 2, we describe second-order techniques for the computation of coherent backscattering by random media of discrete scatterers. Example results resemble both observations for solar system bodies and experimental laboratory measurements for relevant samples. Section 3 describes an up-to-sixth-order ray-tracing technique for coherent backscattering, with comparisons between computations and laboratory measurements for dark random media of submicron-sized soot particles. Section 4 presents a multiple scattering technique generalized from radiative transfer and capable of reaching very high orders of scattering (~ 1000) and thus applicable to both dark and bright random media of scatterers. Results from this technique suggest tentative agreement with existing reference results for conservative media of Rayleigh scatterers. Discussion closes the chapter in Section 5.

2. Second-Order Computational Techniques

2.1. RAY-TRACING FACET MODEL

Because the physical processes are transparent, a good place to start an examination of the polarization state of scattered light in the near-backward direction is through ray-tracing ([9, 10], cf. [11, 12]). In such an analysis, rays are traced from the source, through the system, to the observer. We consider a scattering system composed of surface facets that are much larger than the wavelength of the incident radiation. The effects of diffraction at individual locations are not explicitly considered and the primary physical mechanism is Fresnel reflections and refractions at the individual facet surfaces. An analysis of this type was initially undertaken by Wolff who considered both single and double reflections [13]. Unfortunately, Wolff did not include the interference of rays traversing reciprocal paths and introduced a shadowing factor which serves to enhance the contribution of the parallel polarization component. In this model, we focus on the second-order ray-tracing contribution; i.e., only the rays which interact with two facets before reaching the detector; and incorporate the interference of reciprocal rays, which serves to enhance the parallel polarization component.

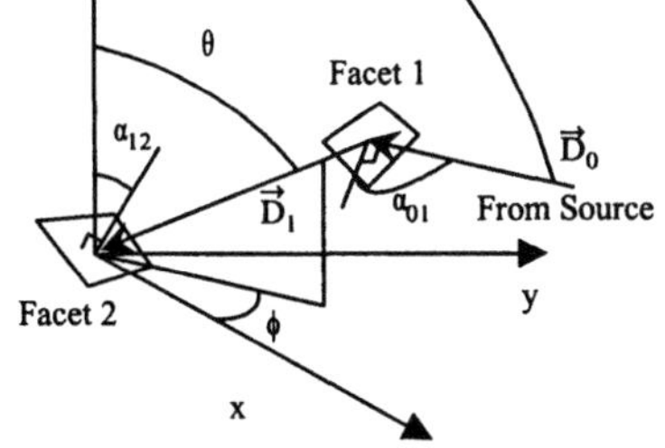

Figure 3. Geometry of two facets.

A schematic showing the orientation of two facets of the scattering system is shown in Figure 3. A light ray traveling in a plane parallel to the xz-plane, oriented at angle α from the z axis, strikes Facet 1 located at the point (r, θ, ϕ). This ray travels in the direction of vector $\boldsymbol{D}_0$. After reflecting off Facet 1, the ray, traveling in the direction of vector $\boldsymbol{D_1}$, continues to Facet 2, located at the origin where it is reflected and now travels in a direction parallel to the positive z axis, given by $\boldsymbol{D_2}$, to the observer. The relevant unit ray vectors are given by

$$\begin{aligned} \boldsymbol{D_0} &= (\sin\alpha, 0, -\cos\alpha) \\ \boldsymbol{D_1} &= (-\sin\theta\cos\phi, -\sin\theta\sin\phi, -\cos\theta) \\ \boldsymbol{D_2} &= (0,0,1). \end{aligned} \tag{1}$$

The incident angles at which the rays strike the facets are measured from the facet normals are shown in Figure 3 and are given by

$$\begin{aligned} \alpha_{01} &= \frac{1}{2}[\pi - \arccos(-\sin\theta\cos\phi\sin\alpha + \cos\theta\cos\alpha)] \\ \alpha_{12} &= \theta/2 \end{aligned} \tag{2}$$

The orientations of the planes of incidence are different for each facet. The orientation of the plane of incidence of Facet 1 with respect to Facet 2 is designated θ_{12} and is given by

$$\cos\theta_{12} = \frac{\sin\theta\cos\alpha + \cos\theta\cos\phi\sin\alpha}{\sqrt{\sin^2\theta\sin^2\phi + (\sin\theta\cos\phi\cos\alpha + \cos\theta\sin\alpha)^2}} \tag{3}$$

The last orientation of the plane of incidence from Facet 2 to the system is designated θ_{20} and is equal to ϕ. We must consider the polarization states of the detected light separately. We assume that the light from the source is unpolarized. This light undergoes a Fresnel reflection from Facet 1 before striking Facet 2. To consider the effect of the second reflection, we must rotate the coordinate system, since the plane of incidence of the two facets is not the same. We must again rotate the coordinate system to consider the light going to the observer. The electric fields of light scattered to the observer can be expressed as

$$\begin{pmatrix} E_s \\ E_p \end{pmatrix} \sim \begin{pmatrix} \cos\phi & \sin\phi \\ -\sin\phi & \cos\phi \end{pmatrix} \begin{pmatrix} r_s(\alpha_{12}) & 0 \\ 0 & r_p(\alpha_{12}) \end{pmatrix} \tag{4}$$

$$\begin{pmatrix} \cos\theta_{12} & \sin\theta_{12} \\ -\sin\theta_{12} & \cos\theta_{12} \end{pmatrix} \begin{pmatrix} r_s(\alpha_{01}) E_o \\ r_p(\alpha_{01}) E_o \end{pmatrix} \tag{5}$$

where E_o is the magnitude of the incident electric field on the system and r_s and r_p are the Fresnel reflection coefficients for s and p polarization states:

$$r_s(\theta) = \frac{\cos\theta - n\sqrt{1 - 1/n^2\sin^2\theta}}{\cos\theta + n\sqrt{1 - 1/n^2\sin^2\theta}} \tag{6}$$

$$r_p(\theta) = \frac{n\cos\theta - \sqrt{1 - 1/n^2 \sin^2\theta}}{n\cos\theta + \sqrt{1 - 1/n^2 \sin^2\theta}} \quad (7)$$

We proceed by expressing the intensities and then incorporating the coherent backscattering factors in these expressions. Two things more need to be known to derive an expression for the observed intensities. First, it is necessary to know something about the scattering system, specifically the distribution of slopes of the facets $P_{z'}$. Second, we need to know how far a ray can be expected to travel between facets 1 and 2 which we express as P_x. The total intensity is proportional to the sum of all the different orientations of the facets:

$$\begin{pmatrix} I_s \\ I_p \end{pmatrix} \sim \int\int\int \begin{pmatrix} E_s^* E_s \\ E_p^* E_p \end{pmatrix} P_{z'} P_x \cos\alpha_{12} \cos\alpha_{01} \sin\alpha_{12} d\alpha_{12} dx d\phi. \quad (8)$$

At this point, we would like to make some assumptions about the scattering system, so that we can obtain some results. First, we assume that the scattering system has no preferential orientation; i.e., the probability distribution of slopes is unity. The second quantity to be considered is the interaction distance $x = |\boldsymbol{D}_1|$; i.e., the distance that the ray travels between facets 1 and 2. For our illustrative purposes, we assume an exponential distribution:

$$I_2 = I_1 \exp(-\beta x); \quad (9)$$

i.e., the intensity at Facet 2, I_2, equals the intensity at Facet 1, I_1, multiplied by a decaying exponential, the argument of which is proportional to the distance between the two facets and inversely proportional to a characteristic length of the system $d = 1/\beta$. Note that the probability distribution is highly dependent on the type of scattering system we are analyzing. For instance, this type of facet analysis is typical when considering a surface. However, for a surface, the probability distributions on the mean free path are strongly dependent on the ray direction, and hence, the facet slope. These particular independent distributions would be highly uncharacteristic of a scattering surface. These distributions are more characteristic of unaligned particles in a volume, and the distributions more closely resemble those of a cloud of facets having no preferential orientation. One advantage of the probability distributions we have chosen is that they allow one of the integrals to be solved analytically: Eq. (8) can be written as

$$\begin{pmatrix} I_s \\ I_p \end{pmatrix} \sim \int\int\int \begin{pmatrix} E_s^* E_s \\ E_p^* E_p \end{pmatrix} \exp(-\beta x) \cos\alpha_{12} \cos\alpha_{01} \sin\alpha_{12} d\alpha_{12} dx d\phi. \quad (10)$$

At this point, there is no interference mechanism that can include backscattering enhancement. We can incorporate this effect explicitly by considering the interference between each set of reciprocal rays. The electric fields used in Eq. (10) now contain two reciprocal components:

$$E_s = E_s^{(1)} + E_s^{(2)}; E_p = E_p^{(1)} + E_p^{(2)}; \tag{11}$$

where $E_s^{(1)}$ is the s-component complex amplitude of the ray from the source that strikes Facet 1, then Facet 2 before going to the observer, and $E_s^{(2)}$ is the s-component complex amplitude of the ray from the source that strikes Facet 2, then Facet 1 before going to the observer, and likewise for the p-component. We should make particular note that the second ray cannot, in general, be directed to the observer using strict geometrical optics, but is directed to a point a certain angular extent from the z axis away from the observer. In fact, the second ray (coming from the source to Facet 2) in general is not even reflected to Facet 1. From a strictly ray-tracing point of view, this is a serious problem. However, we must remember that in general, we are considering observation positions very near the backward-scattering direction on the order of one degree. Although the facets are assumed to be large for a ray-tracing solution to be valid, they would have to be extremely large for the central lobe of the diffracted light to be less than a few degrees.

The electric field components now vary from each other by a phase difference $\exp(i\Phi)$ and Eq. (11) can be written as

$$E_s^* E_s = 2E_s^{(1)*} E_s^{(1)} (1 + \cos\Phi); E_p^* E_p = 2E_p^{(1)*} E_p^{(1)} (1 + \cos\Phi) \tag{12}$$

where

$$\Phi = k\,|\mathbf{D_1}|\,(-\cos\theta + \sin\alpha\sin\theta\cos\phi + \cos\alpha\cos\theta) = \gamma\,|\mathbf{D_1}|\,. \tag{13}$$

Integrating Eq. (10) over the separation distance between the facets now yields

$$\begin{pmatrix} I_s \\ I_p \end{pmatrix} \sim 2\int\int \begin{pmatrix} E_s^{(1)*} E_s^{(1)} \\ E_p^{(1)*} E_p^{(1)} \end{pmatrix} \left(\frac{2\beta^2 + \gamma^2}{\beta^2 + \gamma^2}\right) \cos\alpha_{12}\cos\alpha_{01}\sin\alpha_{12} d\alpha_{12} d\phi \tag{14}$$

Equation (11) provides second-order ray-tracing results of the s and p polarization intensities scattered by a cloud of facets. The only physical mechanisms present in the derivation are Fresnel reflections and interference.

Figure 4 shows sample calculations of the second-order ray-tracing results using Eq. (14) for a cloud of facets ($m = 1.55 + i0.0$) for scattering systems having three different spatial dimensions kd. The left graph shows the light-scattering total intensities for systems having three average spatial dimensions, kd = 10, 100, 1000. The plots are normalized to the value of the exact backscatter intensity without taking into account interference. The effect of the interference is to exactly double the intensity in the exact backscatter direction. This is the theoretical upper limit to enhanced backscatter and occurs only because we have taken into account only the second-order effects. Taking into account other orders would reduce this value. The width of the peak is approximately inversely proportional to the size parameter kd.

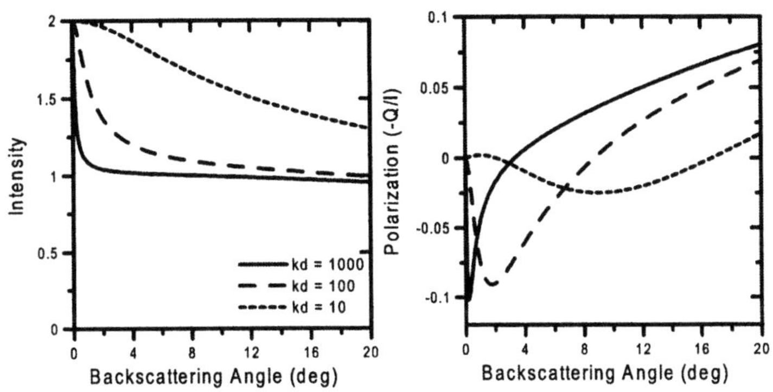

Figure 4. Backscattering intensity (left) and polarization (right) calculated for a cloud of facets.

The right graph of Figure 4 shows the polarization state of the scattered light. There are a few things to note about this graph. Because of symmetry, the polarization is zero in the exact backscatter direction. The minima location in the polarization is located approximately where the half width half maxima occurs on the corresponding intensity curve. For relatively large separation distances, $kd = 100, 1000$, the shape of the polarization curve is highly asymmetric with a minima located on the order of a degree. For systems having smaller spatial dimensions, $kd = 10$, the polarization curve becomes more symmetric, and parabolically shaped.

2.2. RAY-TRACING FROM DIPOLES

One nice feature of the second-order ray-tracing model derived in the previous section is its versatility. It can easily be modified to meet a specific requirement. For instance, if we wish to make a comparison to the scattering results from dipoles in the previous chapter, we only need to make a few minor adjustments. We first incorporate the scattering amplitudes from dipoles into the ray-tracing equation (4),

$$\begin{pmatrix} E_s \\ E_p \end{pmatrix} \sim \begin{pmatrix} \cos\phi & \sin\phi \\ -\sin\phi & \cos\phi \end{pmatrix} \begin{pmatrix} 1 & 0 \\ 0 & \cos(\pi-\theta) \end{pmatrix} \begin{pmatrix} \cos\theta_{12} & \sin\theta_{12} \\ -\sin\theta_{12} & \cos\theta_{12} \end{pmatrix} \begin{pmatrix} E_o \\ \cos(\pi-\alpha_{01})\,E_o \end{pmatrix}. \tag{15}$$

If we assume the same Beer's law probability distribution function, the intensities can be expressed as

$$\begin{pmatrix} I_s \\ I_p \end{pmatrix} \sim 2\int\int \begin{pmatrix} E_s^{(1)*}E_s^{(1)} \\ E_p^{(1)*}E_p^{(1)} \end{pmatrix} \left(\frac{2\beta^2+\gamma^2}{\beta^2+\gamma^2}\right) \sin\theta d\theta d\phi \tag{16}$$

Figure 5 illustrates the results obtained from Eq. (16). The polarization curves are similar to what is shown in Figure 5, but with some interesting differences. Most importantly, when plotted against a dimensionless parameter $kd\alpha$, the shape of the polarization curve is almost independent of the separation distance d. These results are very similar to what was seen in the previous chapter (note, $d = 2l$). The minima occurs at approximately $kd\alpha = \pi$, which is within approximately 10% of the results of the theory presented in the previous chapter, which included all orders of scattering.

2.3. SOME FURTHER APPROXIMATIONS

Since we show that the negative polarization is a result of the constructive interference mechanism, we can choose a probability function to maximize the effect [11]. This can be done by considering facets located on the x and y axes. Facets on the x axis enhance the perpendicular state, contributing to positive polarization, but are affected most by destructive interference. Facets on the y axis enhance the parallel state, contributing to negative polarization, and are least affected by destructive interference. Because the projected areas of the facets are relatively small for large incident angles, we make a further approximation that the Fresnel reflection coefficients are approximately constant and equal

$$r_s(\theta) \sim r_p(\theta) \sim r \sim \frac{n-1}{n+1}. \tag{17}$$

After integration the differences between the second-order intensities are

$$I_s^{(2)} - I_p^{(2)} \sim -\left(\frac{n-1}{n+1}\right)^4 \frac{(kd\alpha)^2}{2\left[1+(kd\alpha)^2\right]}. \tag{18}$$

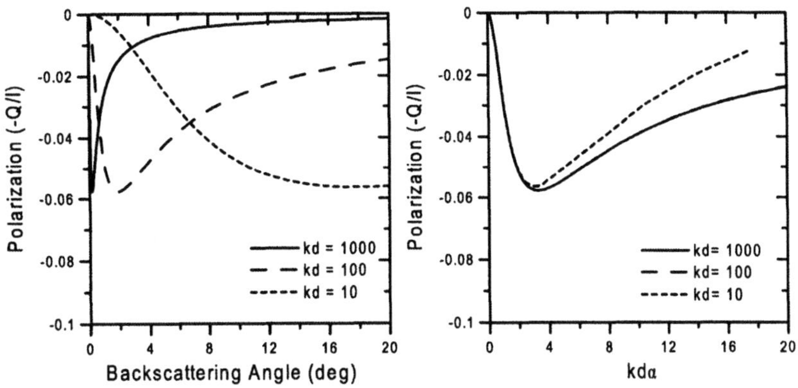

Figure 5. The polarization state of scattered light resulting from second-order ray-tracing from a cloud of dipoles. Three different spatial distributions of facets are chosen (left). When plotted as a function of dimensionless parameter $q = kd\alpha$ (right), the curves are only distinguishable when the mean-free-path approaches the wavelength. Calculations for the smallest size parameter $kd = 10$ are well beyond the range of validity of the model where α is assumed to be relatively small, yet the basic shape of the curve remains the same (cf. the chapter by Mishchenko *et al.*).

The first order reflections can be approximated as

$$I_s^{(1)} - I_p^{(1)} \sim \left(\frac{n-1}{n+1}\right)^2 \frac{\alpha^2}{2n} \tag{19}$$

and the degree of polarization as

$$P \sim \frac{\alpha^2}{2n} - \left(\frac{n-1}{n+1}\right)^2 \frac{(kd\alpha)^2}{2\left[1+(kd\alpha)^2\right]} \tag{20}$$

With this analytical equation, we can express the minimum location, the polarization minimum, and the inversion angle as

$$\begin{aligned} \alpha_{\min} &\sim \frac{1}{kd}\sqrt{kd\sqrt{n}\left(\frac{n-1}{n+1}\right) - 1}, \\ P_{\min} &\sim -\frac{1}{2}\left(\frac{n-1}{n+1} - \frac{1}{kd\sqrt{n}}\right)^2, \\ \alpha_{\min} &\sim \sqrt{n\left(\frac{n-1}{n+1}\right)^2 - \frac{1}{(kd)^2}}. \end{aligned} \tag{21}$$

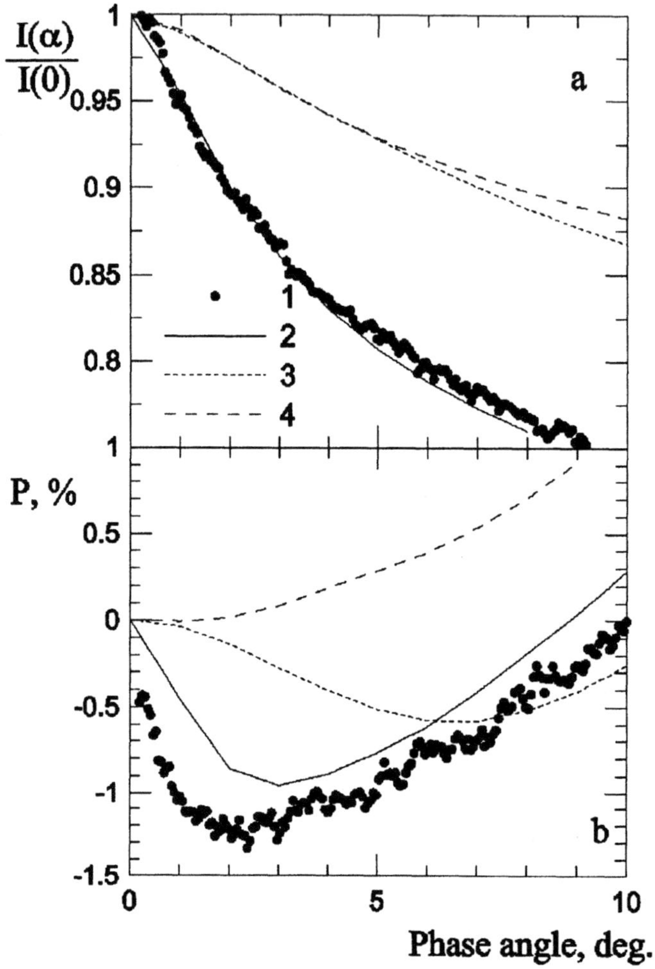

Figure 6. Measured and calculated phase curves of brightness and polarization degree for smoked soot. Experimental data are shown with points (1). Curves (2) and (3), accounting for six scattering orders, correspond to semi-infinite media of random-structure particles with $x = 2$ and $m = 1.5 + i0.2$ (albedo of single scattering is 0.5), at $v = 0.07$ and $v = 0.3$, respectively. Curve 4 corresponds to the case where the coherent backscatter mechanism is turned off. Albedo at $\alpha = 2°$ is 2.5% for soot, and 3.1%, 3.6% and 2.4% for media presented by curves (2), (3), and (4), respectively.

3. Ray Tracing in Dark Random Media

As shown previously [14, 15], our computer modeling includes the following: (1) generation of a hypothetical random medium with certain statistical properties; and (2) a ray-tracing procedure. To generate a 3D medium, we use a box (parallelepiped) consisting of scatterers with finite size. The upper boundary of the box corresponds to the average exposed top surface of the medium. The lower boundary and sides of the parallelepiped are considered to be cyclically conjugate [14, 16, 17, 18]. This means that if a ray transits out of the parallelepiped through a side (or the lower boundary), it returns to the box through the opposite side (or the upper boundary, respectively). If the medium is statistically homogeneous, it can be characterized by a single parameter: volume density. To calculate the photometric and polarimetric properties of the model medium, we apply a technique called backward tracing of rays [14, 15]. As we use ray tracing in media consisting of particles with finite sizes, we automatically take into account the shadowing effect. Such an approach is valid if the geometric-optics approximation is justified for a given medium. When the particles are small enough, ray tracing can be considered only as a technique to estimate the probability of electromanetic wave propagation in the medium by the Monte-Carlo method. In other words, using the present technique, one accounts for light attenuation in the medium. Because of this feature, we were able to make calculations only up to the sixth order of scattering, as ray tracing in such media requires long computing times. Details of the model can be found in [14, 15]. Our first calculations show that our computer model can be efficiently exploited for dark surfaces. For comparison, we used our measurements for smoked carbon soot (see the chapter by Shkuratov and Ovcharenko).

Simulating the optical properties of soot surfaces requires that we know the characteristics of soot particles. The typical particle is considered to be a submicron-sized aggregate with an irregular shape. To calculate optical properties of carbon soot particles, we use the DDA method. Particles with random shapes are generated using a method described in [19, 20]. We consider an approximately spherical volume extracted from a rather large portion of a cubic lattice. Random sites of the lattice are taken as the locations of seed domains (nuclei) of solid matter. Then, the vicinity of each nucleus is examined and the nearest neighbors of the nucleus are also marked as nuclei of solid matter. The rest of the sites can formally be considered as "nuclei of void spaces" or empty sites. We can regulate the volume ratio of occupied and void sites. Below we use random structures with equal volumes of occupied and void sites. We calculate, by means of DDA, the vector amplitudes of many sample particles of random form. The choice of single-particle parameters is optimized by matching the observed albedo and phase curves of brightness and polarization (see Fig. 6). In particular, the calculated surface albedo, when $m = 1.5 + i0.2$, $x = 2$, and volume density $v = 0.07$, equals 3.1% at $\alpha = 2°$, which is rather close to the measured value of soot albedo, 2.5% [14]. The values of $\mathrm{Re}(m) = 1.5$ and $\mathrm{Im}(m) = 0.2$ are rather typical for diesel soot in the visible spectral range (e.g., [21]). Our calculations show [19, 20] that particles with the described random structure in general exhibit a comparatively regular behavior of photometric and polarimetric indicatrices unlike the case of equivalent spherical particles. Our calculations also show that, as the carbon soot is extremely dark, accounting for scattering orders higher than five is not needed.

To model the experimental data for soot, we use a rather diluted medium with volume density $v = 0.07$. It is interesting to study the influence of these characteristics in more detail. Fig. 6 shows three curves corresponding to $v = 0.3$ for the same medium. As one can see the compressed medium has a negative polarization surge wider than that of the uncompressed one. This effect is confirmed by laboratory measurements ([14]; see the chapter by Shkuratov and Ovcharenko).

As has been noted, we include in this model both the coherent backscattering and shadowing. Thus, we can study which of these mechanisms is the most important contributor to negative polarization. We repeat our modeling without the coherent backscattering mechanism. Fig. 6 shows (curve 4) that shadowing and incoherent multiple scattering are not able to produce the negative polarization.

As we show in Fig. 6, the model of randomly structured particles gives a rather good fit to the negative polarization branch of carbon soot. The calculated polarimetric curve has an inversion angle near $\alpha = 8°$, and is somewhat shallower than the experimental one. This is probably related to an overestimation of the role of shadowing by our ray-tracing technique. For fine-structured surfaces, shadowing is weakened by diffraction.

4. Coherent-Backscattering Radiative-Transfer Technique

In the present technique, the computation of coherent backscattering by absorbing and scattering media is carried out beside a Monte Carlo radiative transfer solution for the Stokes vector of scattered intensity [22, 23]. Currently, the approach is available for finite/semi-infinite plane-parallel media and spherical media of scatterers, as well as for semi-infinite plane-parallel media embedded in a semi-infinite optically isotropic and homogeneous host medium.

For all media involved, the key requirement is the *a priori* selection of reflection/transmission angles for updating Stokes vectors during Monte-Carlo radiative transfer. Fixed angles allow the computation of electromagnetic phase differences and thus the coherent backscattering effect. The numerical technique relies on the reciprocity principle for electromagnetic scattering (see the chapter by Mishchenko *et al.*), allowing an *a posteriori* renormalization of the coherent-backscattering contributions at each multiple-scattering event. Renormalization is carried out with the help of amplitude scattering matrices. The vector radiative transfer portion of the technique has been carefully verified using reference results for a Rayleigh-scattering planetary atmosphere [24].

Currently, arbitrary Lorenz-Mie type scattering matrices (Mueller matrices) can be incorporated (using cubic splines), covering the special case of Rayleigh scattering (also treated without splines). In the generation of new propagation directions, a generalized Kepler's equation is efficiently solved using Newton's method. Close-packing effects can be approximately accounted for by the structure factor computed in the Percus-Yevick approximation. Here we constrain our computations to volume densities less than 10% to avoid invoking the structure factor formalism. Note that, as a rule, ensemble-averaged (including orientationally averaged) scattering matrices cannot be utilized, because they are typically unpolarizing: at each scattering event, the Stokes vector of scattered light has to correspond to electromagnetic field components with a well defined phase difference. The numerical technique is described in detail in [22, 23] and, below, some key parts of the technique are given.

4.1. GENERATION OF PROPAGATION DIRECTIONS

Of general interest in Monte-Carlo multiple scattering and radiative transfer computations is the generation of a new propagation direction in each scattering process. Here we outline an efficient technique for Lorenz-Mie type block-diagonal scattering phase matrices, with good prospects for future generalization to more complicated matrices.

We denote the Stokes vectors of the incoming and outgoing rays by $\boldsymbol{I}_1 = (I_1, Q_1, U_1, V_1)^T$ and $\boldsymbol{I}_2 = (I_2, Q_2, U_2, V_2)^T$, respectively, and assume that the reference system of the incoming ray is arbitrary. The outgoing ray in the reference system of the scattering plane is given by

$$\boldsymbol{I}_2 \;\propto\; \boldsymbol{P}(\theta, \phi) \cdot \boldsymbol{R}(\phi) \cdot \boldsymbol{I}_1, \tag{22}$$

where $\boldsymbol{R}$ is a 4×4 Mueller matrix for the rotation to the scattering plane, $\boldsymbol{P}$ is the 4×4 Mueller matrix for scattering (scattering phase matrix), and θ and ϕ refer to the reference system of $\boldsymbol{I}_1$).

The Monte Carlo generation of the new scattering direction is based on the angular intensity distribution $I_2(\theta, \phi)$; i.e.,

$$\begin{aligned} I_2(\theta, \phi) \propto \; & I_1 P_{11}(\theta, \phi) + \\ & Q_1[P_{12}(\theta, \phi) \cos 2\phi + P_{13}(\theta, \phi) \sin 2\phi] + \\ & U_1[-P_{12}(\theta, \phi) \sin 2\phi + P_{13}(\theta, \phi) \cos 2\phi] + \\ & V_1 P_{14}(\theta, \phi). \end{aligned} \tag{23}$$

For Lorenz-Mie type scattering, the scattering phase matrix is independent of the azimuthal scattering angle ϕ and $P_{13} = P_{14} = 0$, so we obtain

$$\begin{aligned} I_2(\theta, \phi) \propto \; & I_1 P_{11}(\theta) + \\ & Q_1 P_{12}(\theta) \cos 2\phi - U_1 P_{12}(\theta) \sin 2\phi. \end{aligned} \tag{24}$$

The azimuthal angle ϕ shows up only in the rotation sines and cosines, and $I_2(\theta, \phi)$ is independent of the Stokes parameter V_1.

We are faced with the generation of θ and ϕ for a given incoming ray. In order to obtain the probability density function $p(\theta, \phi)$, we normalize $I_2(\theta, \phi)$ and obtain

$$\begin{aligned} p(\theta, \phi) = \; & \frac{1}{4\pi} P_{11}(\theta) + \\ & \frac{1}{4\pi} P_{12}(\theta) \left(\frac{Q_1}{I_1} \cos 2\phi - \frac{U_1}{I_1} \sin 2\phi \right). \end{aligned} \tag{25}$$

The marginal probability density for θ is

$$p(\theta) = \frac{1}{2} P_{11}(\theta) \tag{26}$$

and thus independent of I_1. For ϕ, it is

$$p(\phi) = \frac{1}{2\pi} + \frac{1}{4\pi} \int_0^\pi d\theta \sin\theta P_{12}(\theta) \left(\frac{Q_1}{I_1} \cos 2\phi - \frac{U_1}{I_1} \sin 2\phi \right). \tag{27}$$

The practical generation of θ and ϕ can be carried out using, first, $p(\theta)$ to obtain the polar scattering angle θ_r and, second, the conditional probability

$$\begin{aligned} p(\phi \mid \theta_r) &= \frac{p(\theta_r, \phi)}{p(\theta_r)} \\ &= \frac{1}{2\pi} \left[1 + \frac{P_{12}(\theta_r)}{P_{11}(\theta_r)} \left(\frac{Q_1}{I_1} \cos 2\phi - \frac{U_1}{I_1} \sin 2\phi \right) \right]. \end{aligned} \tag{28}$$

to obtain the azimuthal scattering angle ϕ_r.

The generation of θ_r can best be carried out via interpolation in a precomputed table of θ_r-values as a function of random number within $]0,1[$. For the subsequent generation of ϕ_r, denote

$$
\begin{aligned}
-\frac{P_{12}(\theta_r)}{P_{11}(\theta_r)}\frac{Q_1}{I_1} &= A\cos\gamma, \\
-\frac{P_{12}(\theta_r)}{P_{11}(\theta_r)}\frac{U_1}{I_1} &= A\sin\gamma,
\end{aligned} \tag{29}
$$

where

$$
A = \sqrt{\frac{P_{12}^2(\theta_r)}{P_{11}^2(\theta_r)}\frac{Q_1^2+U_1^2}{I_1^2}} \tag{30}
$$

and the angle γ is unambiguously determined by Eq. (29). Then, the probability density function is

$$
p(\phi \mid \theta_r) = \frac{1}{2\pi}\left[1 - A\cos(2\phi+\gamma)\right], \tag{31}
$$

and the cumulative distribution function is

$$
P(\phi \mid \theta_r) = \frac{1}{2\pi}\left[\phi - \frac{1}{2}A\sin(2\phi+\gamma)\right]. \tag{32}
$$

For generating ϕ_r, we thus obtain (with $r \in]0,1[$, a uniform random deviate)

$$
(2\phi_r+\gamma) - A\sin(2\phi_r+\gamma) = 4\pi r+\gamma, \tag{33}
$$

that is, Kepler's equation $E - e\sin E = M$ with "eccentric anomaly" $E = 2\phi_r+\gamma$, "eccentricity" $e = A$, and "mean anomaly" $M = 4\pi r+\gamma$. Kepler's equation can be solved for ϕ_r efficiently using Newton's technique [25]: typically, only a few iterations are necessary to obtain ϕ_r with sufficient accuracy. Above, we have refered to Eq. (33) as "the generalized Kepler's equation," because the angle to be solved has a multiplicative factor of 2 and there is an additional angular shift of γ. Note also that the parabolic case of $A = 1$ is theoretically possible but, in order for it to affect the numerical computations, the absolute ratio $|P_{12}(\theta)/P_{11}(\theta)|$ should be equal to unity over a finite interval in θ.

4.2. RECIPROCITY PRINCIPLE

The reciprocity principle of electromagnetic scattering [26] is a key principle in the Mueller-matrix technique under consideration. In coherent backscattering, it has been extensively used in the computation of enhancement factors in the exact backscattering direction [27]. Whereas those computations offer reference results for future comparisons, the numerical techniques did not allow the computation of angular dependences of coherent backscattering peaks or negative polarization surges. The current numerical technique removes that constraint and allows us to obtain the full angular profiles of the backscattering peaks and the polarization surges.

According to the reciprocity principle, the amplitude scattering matrices of reciprocally propagating waves are interrelated by

$$\boldsymbol{S}(-\boldsymbol{e}_i, -\boldsymbol{e}_s) = \boldsymbol{Q}\boldsymbol{S}^T(\boldsymbol{e}_s, \boldsymbol{e}_i)\boldsymbol{Q}, \tag{34}$$

where

$$\boldsymbol{S} = \begin{pmatrix} S_{\vartheta\vartheta} & S_{\vartheta\varphi} \\ S_{\varphi\vartheta} & S_{\varphi\varphi} \end{pmatrix}, \quad \boldsymbol{Q} = \begin{pmatrix} 1 & 0 \\ 0 & -1 \end{pmatrix}. \tag{35}$$

The coherent backscattering contributions are here computed by applying the reciprocity principle at the exact backscattering direction for two reciprocally propagating wave components. Denote the components of the incoming field by $(u_i, v_i)^T$, the backscattering, directly propagating wave component by $(u_1, v_1)^T$, and the backscattering, reciprocally propagating wave component by $(u_2, v_2)^T$. Because of the reciprocity in Eq. (34), $(u_2, v_2)^T$ must be strictly related to $(u_1, v_1)^T$. In the current Mueller-matrix technique, we can find the complex renormalization coefficient to be applied to the wave components incident upon the final scatterer in the reciprocal direction, to guarantee reciprocity in the exact backscattering direction; the renormalization coefficient ζ is the ratio

$$\zeta = \frac{u_i u_1 - v_i v_1}{u_i u_2 - v_i v_2}. \tag{36}$$

Subsequently, the properly renormalized reciprocal wave component and the direct wave component can be used in the computation of the angular profiles in both intensity and polarization close to the backscattering direction. The absolute electromagnetic phase is computed for neither directly nor reciprocally propagating components; instead, the relative electromagnetic phase difference is properly renormalized at the exact backscattering direction, leaving the absolute phase as an arbitrary number having no influence on the results.

4.3. RESULTS

The normalized brightness and degree of linear polarization are computed for normally incident unpolarized light on an absorbing and scattering medium of Rayleigh scatterers both in free space and embedded in a host medium. For both cases, the vacuum wavelength was $2\pi/10$ μm and the mean-free-path parameter was 2.0 μm. The single-scattering albedo was 0.9 for the former and 1.0 for the latter case. The refractive index of the host medium was $1.55+i10^{-3}$, and 10^4 rays were used in the Monte Carlo computation. The coherent backscattering peaks turned out to be similar whereas the negative polarization patterns differed considerably. For Rayleigh scatterers embedded in the host medium, wide negative polarization patterns resulted. To summarize tentatively, the host medium can play a significant role in the formation of the polarization pattern, extending the inversion angle to larger phase angles. Note that in the 1980s, coherent backscattering was experimentally confirmed for scatterers in a host medium, typically polystyrene or latex spheres in liquid water.

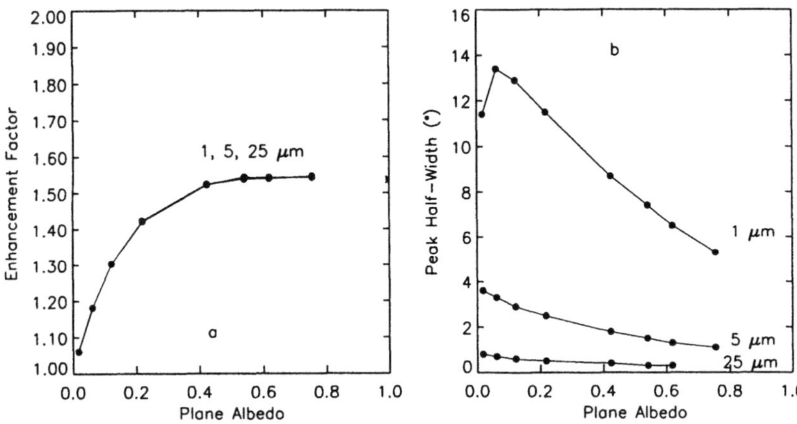

Figure 7. Coherent backscattering by semi-infinite media of Rayleigh scatterers: (a) the enhancement factor and (b) peak half width at half maximum vs. plane albedo. Asterisk denotes the benchmark enhancement factor for a plane albedo (and single-scattering albedo) of 1.0. Mean-free-path parameters are indicated for each curve—see text for the single-scattering albedos corresponding to each thick dot.

Figures 7 and 8 illustrate systematic computations for semi-infinite plane-parallel media of absorbing Rayleigh scatterers for the case of normally incident unpolarized light. The wavelength of incident radiation was $2\pi/10$ μm, and the single-scattering albedo and mean-free-path parameter are varied within 0.1, 0.3, 0.5, 0.7, 0.9, 0.95, 0.97, 0.99, and 1, 5, 25 μm, respectively. 5×10^4 rays are used in the Monte-Carlo computation in all cases, except for the albedo 0.99 case, where results are currently only partially available using 10^4 rays (explaining the statistical fluctuations in the figures).

Figure 7 shows the enhancement factor as well as the half-width at half-maximum of the coherent backscattering peak as a function of hemispherical plane albedo resulting from the computations. For increasing albedos, the enhancement factor tends to converge toward the benchmark value of 1.5368 for conservative Rayleigh scattering [28, 29, 30]. It is noteworthy that this enhancement is already reached for plane albedos of about 50%. Peak widths are smaller for larger mean free paths and reveal an almost linear trend toward narrower values.

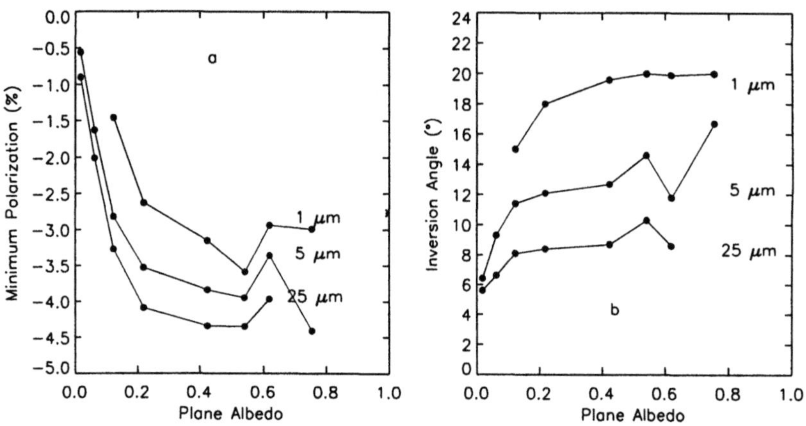

Figure 8. (a) Minimum polarization and (b) inversion angle. Asterisk denotes the benchmark minimum polarization for a plane albedo of 1.0. See Fig. 7.

Figure 8 shows the minimum polarization and inversion angle as a function of the plane albedo. It is evident that the minimum polarization depends not only on the single-scattering albedo, but also on the mean-free-path parameter with deeper patterns for larger mean free paths. For conservative scattering, the benchmark value of the pure coherent backscattering part is -2.765% [28, 29, 30]. Indeed, the present computations reveal a saturation of minimum polarization for plane albedos of about 50%, and it is realistic to predict agreement with the benchmark value for conservative scattering. To reiterate, negative polarization from coherent backscattering reaches its absolute maximum for absorbing scatterers. The inversion angle increases with increasing plane albedo (and decreasing mean free path), and a possible saturation is not yet evident in the numerical data.

5. Discussion

Exact analytical computations like those in [31, 32] are of importance for further verification of the present approximations that omit the longitudinal wave component of the electromagnetic fields. For example, rigorous electromagnetic computations of enhanced backscattering and negative polarization from small particles close to an interface between two optically isotropic and homogeneous media [31] show wide backscattering peaks and polarization surges closely resembling those from the approximate computations in Figs. 4, 5, and 8. We stress that the coherent backscattering mechanism can affect the brightness and polarization phase curves at moderate to large phase angles. It is possible that some of the single-particle measurements by Volten *et al.* [33] can be explained by coherent backscattering (see also [34]).

A single multiple-scattering technique that incorporates both coherent backscattering and shadowing mechanisms has been discussed above for dark random media. In general, there are major challenges left in combining these two mechanisms, as they are relevant in considerably differing regimes of particle sizes. Such random discrete media are highly relevant in, e.g., atmosphereless solar system bodies.

In addition to applying the current sixth-order ray-tracing technique (Sect. 3) to the interpretation of the experimental results in Fig. 6, it seems realistic to predict that similar interpretations can be carried out using the coherent-backscattering radiative-transfer technique (Sect. 4). Computations with that technique show unequivocally that absorbing random media of discrete scatterers can give rise to both sharp and shallow opposition effects and negative polarization surges, depending on the ratio of the mean-free-path length and the wavelength of incident radiation.

The various coherent backscattering techniques may be useful in the interpretation of lunar imaging data soon to be offered by the AMIE camera of the ESA SMART-1 space mission (launch in February 1, 2003). In the future, it is our goal to offer reference results for coherent backscattering by random media of absorbing Rayleigh scatterers, complementing the reference results for conservative Rayleigh scatterers [28].

6. Acknowledgments

KM is grateful to the Academy of Finland for financial support. This research is supported, in part, by INTAS grant No. 2000-0792.

References

1. Yu. G. Shkuratov, in "Astron. Circular No. 1400," (Sternberg State Astron. Inst., Moscow, 1985), p. 3.
2. K. Muinonen, in "Proc. 1989 URSI Electromagnetic Theory Symposium" (Stockholm, Sweden, 1989), p. 428.
3. K. Muinonen, "Light scattering by inhomogeneous media: Backward enhancement and reversal of polarization" (Ph.D. thesis, University of Helsinki, 1990).

4. B. Hapke, Icarus **88**, 407 (1990).
5. M. I. Mishchenko and J. M. Dlugach, Planet. Space Sci. **41**, 173 (1993).
6. K. Muinonen, in "Asteroids, Comets and Meteors 1993," Ed. by A. Milani, M. Di Martino and A. Cellino (Kluwer Academic Publishers, Dordrecht, 1994) p. 271.
7. Yu. G. Shkuratov, K. Muinonen, E. Bowell, K. Lumme, J. I. Peltoniemi, M. A. Kreslavsky, D. G. Stankevich, V. P. Tishkovets, N. V. Opanasenko and L. Y. Melkumova, Earth, Moon, and Planets **65**, 201.
8. K. Muinonen, J. Piironen, Yu. G. Shkuratov, A. Ovcharenko and B. Clark, in "Asteroids III," Ed. by W. Bottke, R. P. Binzel, A. Cellino and P. Paolicchi (University of Arizona Press, Tucson, Arizona, 2002), in press.
9. G. Videen, in "The 6th International Conference on Electromagnetic and Light Scattering by Non-spherical Particles," Ed. by B. Gustafson, L. Kolokolova and G. Videen (Army Research Laboratory, Adelphi, Maryland, 2002), p. 223.
10. G. Videen, Appl. Opt., in press (2002).
11. Yu. G. Shkuratov, Astron. Vestnik **23**, 176 (1989).
12. K. Muinonen and K. Lumme, in "IAU Colloquium 126, Origin and Evolution of Dust in the Solar System," Ed. by A.-C. Levasseur-Regourd and H. Hasekawa (Kluwer Academic Publishers, Printed in Japan, 1991), p. 159.
13. M. Wolff, Appl. Opt. **14**, 1395 (1975).
14. Yu. Shkuratov, A. Ovcharenko, E. Zubko, D. Stankevich, O. Miloslavskaya, K. Muinonen, J. Piironen, R. Nelson, W. Smythe, V. Rosenbush and P. Helfenstein, Icarus, in press (2002).
15. Eu. Zubko, A. Ovcharenko and Yu. G. Shkuratov, Opt. Spectrosc. **92**, 443 (2002).
16. D. G. Stankevich, Yu. G. Shkuratov and K. Muinonen, J. Quant. Spectr. Rad. Transfer **63**, 445 (1999).
17. D. G. Stankevich, Yu. G. Shkuratov, K. O. Muinonen and O. V. Miloslavskaya, Opt. Spectrosc. **88**, 682 (2000).
18. K. Muinonen, D. Stankevich, Yu. G. Shkuratov, S. Kaasalainen and J. Piironen, J. Quant. Spectr. Rad. Transfer **70**, 787 (2001).
19. E. S. Zubko, M. A. Kreslavsky and Yu. G. Shkuratov, Sol. Sys. Res. **33**, 296 (1999).
20. E. Zubko, Yu. Shkuratov and K. Muinonen, Opt. Spectrosc. **91**, 273 (2001).
21. C. W. Bruce, T. F. Stromberg, K. P. Gurton and J. B. Mozer, Appl. Opt. **30**, 1537 (1991).
22. K. Muinonen, in "The 6th International Conference on Electromagnetic and Light Scattering by Non-spherical Particles," Ed. by B. Gustafson, L. Kolokolova and G. Videen (Army Research Laboratory, Adelphi, Maryland, 2002), p. 223.
23. K. Muinonen, J. Quant. Spectr. Rad. Transfer, in preparation (2002).
24. K. L. Coulson, J. V. Dave and Z. Sekera, "Tables Related to Radiation Emerging from a Planetary Atmosphere with Rayleigh Scattering," (University of California Press, Berkeley & Los Angeles, 1960).
25. J. M. A. Danby, "Fundamentals of Celestial Mechanics," (Willmann-Bell, Inc., Richmond, 1992).
26. D. S. Saxon, Phys. Rev. **100**, 1771 (1955).
27. M. I. Mishchenko, J. Quant. Spectr. Rad. Transfer **56**, 673 (1996).
28. M. I. Mishchenko, J.-M. Luck and T. M. Nieuwenhuizen T. M., J. Opt. Soc. Am. A **17**, 888 (2000).
29. E. Amic, J. M. Luck and Th. M. Nieuwenhuizen, J. Phys. I **7**, 445 (1997).
30. V. D. Ozrin, Waves Random Media **2**, 141 (1992).
31. K. O. Muinonen, A. H. Sihvola, I. V. Lindell and K. A. Lumme, J. Opt. Soc. Am. A **8**, 477 (1991).
32. M. I. Mishchenko, Optics Letters **21**, 623 (1996).
33. H. Volten, O. Muñoz, J. F. de Haan, W. Vassen, J. Hovenier, K. Muinonen and T. Nousiainen, J. Geophys. Res. **106**, 17375 (2001).
34. T. Nousiainen, K. Muinonen and P. Räisänen, J. Geophys. Res., in press (2002).

THE EFFECT OF RADIATION ON THE MOTION OF METEOROIDS

I.P. WILLIAMS
Astronomy Unit,
Queen Mary, University of London
Mile End Rd, London E1 4NS

Abstract.
It has been known for over a century that radiation pressure exerts a force on small particles in the direction of the radiation flow. It has also been recognised for a similar time that when this radiation is absorbed and re-emitted, this can generate a drag with a component also in the transverse direction. The principle that governs these forces is easy to state, namely that the radiation must be considered in the rest-frame of the emitting surface. Generating actual expressions has proved harder in anything but the simplest situations. We review the development of the subject. In terms of orbital evolution of interplanetary matter, these effects can be important, but again deriving analytical expressions for the rate of change of the orbital elements is difficult. We will also review the derivation and derived solutions of these equations. Finally, we will consider their application to the orbital evolution of small grains, paying particular attention to their effect on dust ejected from asteroids and comets, the normal formation mechanism for meteoroid streams. These streams are responsible for the observed meteor showers, caused when the Earth passes through such a stream.

1. Historical Introduction

At the beginning of the last century, scientists became interested in the behavior of light (and electro-magnetic radiation in general). This interest was initiated through the experimental result in 1887 by Michelson & Morley [1] that the speed of light appeared to be constant in all frames of reference. (The actual experiment measured the speed of light in two orthogonal directions, one parallel to the direction of the motion of the Earth about

G. Videen and M. Kocifaj (eds.), Optics of Cosmic Dust, 283–300.

the Sun and the other perpendicular to this and found no difference.) The best known consequence of these works was the formulation of the Special Theory of Relativity by Einstein [2] in 1905. In this same paper, Einstein demonstrated that light would exert a pressure, (called by him the pressure of light) on any body and derived and expression for this pressure assuming that all the light was reflected as from a plane mirror placed orthogonal to the direction of the incoming light. It is clear that with this assumption, when applied to bodies orbiting about the Sun, that any force generated by the light is in the radial direction. A similar problem had been considered slightly earlier (in 1903) by Poynting [3]. He discussed the effects of solar radiation on the temperature of solar system bodies as well as the pressure that small bodies would experience. In contrast to the simple reflection assumption of Einstein, Poynting assumed that all the light was absorbed by the body and then re-emitted isotropically in the rest-frame of the body. He realized that with this assumption there would be a transverse component to the resulting force on a body moving in the solar system as well as the more obvious radial force. The incoming radiation is in a frame based on the Sun and so generates only a radial force, but the re-emitted radiation based in a frame moving with the body will have angular momentum with respect to the solar frame. The angular momentum which the radiation now possesses must have come from the body and so the body has lost angular momentum This loss can be interpreted as being equivalent to the existence of a transverse drag force.

The statement of principle embodied in the above discussion, namely that any momentum associated with the radiation must be evaluated in the reference frame within which it was emitted, is easy to state. Deriving the correct expressions for these momentum changes, or forces, is rather a different matter. For example, in two papers published at virtually the same time, Plummer [4], [5] made the same basic assumptions as Poynting but obtained different numerical coefficients for the terms in the derived expressions, not only from Poynting, but also between his two derivations. Somewhat later, in 1913, Larmor [6], again with the same assumptions as Poynting, obtained yet an other different set of coefficients. Finally, in 1937 Robertson [7] reconsidered the whole problem within the context of relativity, obtaining yet an other set of values for the numerical coefficients. Comparisons between these papers were made by Lyttleton [8] and are not reproduced in full here. However, as an illustration, we give the following. There is a term containing $\dot{r}/r^2$, where r is the heliocentric distance in the expression for the radial force. (See equations (6) and (7) in section (2) .) There is a numerical coefficient that multiplies this term, being 2 in equation (6). However, the value of this coefficient obtained by the various authors are as follows. Poynting, 2/3; Plummer in 1905, 1/3; Plummer in

1906, 4/3; Larmor, 1; and Robertson, 2.

The derivation by Roberstson [7] is difficult to follow as it uses the metric of special relativity and associated notation. With the assumptions that he made, in particular the total absorption and isotropic re-emission of the radiation from a spherical non rotating body, Robertson's derivation and results are generally accepted as being correct. Identical results have been obtained under the same assumptions by more recent authors (for example Lyttleton [8] or Burns, Lamy & Sotter [9]) using a less difficult mathematical formulation.

With the assumptions made by Robertson and many other workers, it is clear that the emitted radiation must generate both a radial and a transverse component of force on a body moving in a general way in the solar system. The absorbed radiation on the other hand can only produce a radial force. The total force, being the sum of these two forces, obviously has both a radial and a transverse components. It is also clear that the transverse component will be zero if the body is stationary whereas a radial component will still exist.

One source of confusion in many discussions of the topic is the terminology used. Virtually everybody talks of *direct radiation pressure effect* and the *Poynting-Robertson effect* but unfortunately what is included under each category can vary. Some authors, following Einstein's terminology, call the radial force due to the absorbed radiation *Radiation Pressure*, others understand *Radiation Pressure* to mean the radial component of the total force. Most commonly it is taken to mean that part of the radial force that does not depend on velocity and we shall follow this convention. Similarly, to some people the *Poynting-Robertson drag* means the total force due to emitted radiation, while to others, it is the force component in the transverse direction only. Again most commonly it is taken as those elements of the total force that depend on the velocity of the body, and we shall also adopt this. Hence, with this convention, the *Poynting-Robertson drag* is zero on a stationary body while *radiation pressure* effects still exist.

It is also important to recognize that with the assumptions made, in particular the isotropic re-emmission of radiation, there is no force component orthogonal to the orbital plane of the body. Hence, two of the five orbital elements that define the heliocentric Keplerian orbit in the solar system, the inclination, I, and the longitude of the ascending node, Ω, will not change under the influence of radiation. In general, the remaining three orbital elements, the semi major axis, a, the eccentricity, e, and the argument of perihelion, ω, will all change with time. Unfortunately, as we shall see, it is not possible to integrate the equations of motion analytically to obtain closed expressions for the change in these orbital parameters with time. However, if the equations are averaged over one complete orbit of the body around the Sun, then the secular rate of change of the three orbital elements can be obtained and this was done by Robertson [7]. In 1950, Wyatt & Whipple [10] also obtained expressions for the secular changes. The expressions for the changes in a and e are identical to those obtained by Robertson. Though they claim that the expression for the change in ω is also the same as that obtained by Robertson, this is not the case. As we shall see later, both expressions are incorrect. Wyatt & Whipple applied their expressions to the evolution of meteoroid streams, showing that the Taurid stream was related to comet Encke. They show that the very different orbital elements found currently are the consequence of orbital evolution caused by the effects of radiation. In 1976, Kresák [11], considered the effects of the *radiation pressure* term on meteoroid stream evolution, deriving limits for the minimum size of meteoroids that could survive within the solar system. Since the main purpose of this presentation is to discuss the effects of radiation on the dynamics of small bodies, in particular meteoroids in a stream, we shall return to these contributions later.

In all of the foregoing discussion, the basic assumption of Robertson, namely that the particle is spherical, non spinning and a perfect absorber, has been incorporated. Unfortunately, no real particle satisfies any of these restrictions. Lyttleton [8] considered the situation where the particle partially reflects and partially absorbs the incoming radiation, deriving corrections to the coefficients of the various terms in the Robertson derivation. It is fairly clear that including reflection will increases the effect of the *radiation pressure* term since none of the reflected light will be emitted from the *back* hemisphere and, since the total outgoing radiation has to equal the incoming, less radiation is thus re-emitted from the *back* hemisphere. This is exactly what Lyttleton found. The effect on the non-radial component is less clear. In the limit where the particle is a sphere that reflects specularly, that is, where the tangential component at any point on the surface is preserved and the normal component reversed, we only need to consider the momentum exchange due to the normal component since the same amount of momentum is brought in as is carried out by the tangential component. The effect of the normal component is the same as for the situation where all the radiation is re-radiated from the front hemisphere. Hence, in this situation, the transverse component of the drag is the same as that given by the Poynting-Robertson drag. However, it is different for any body that is not symmetrical in shape about the radial direction and one can imagine possibilities where the transverse drag is either increased or decreased depending on the selected shape. In addition to affecting the rest-frame velocity of the emitting surface, the shape of the particle also can affect the surface area to mass ratio, which is a relevant parameter when considering the acceleration. Gustafson [12] considered this problem, paying particular attention to *flakes*, where, depending on the orientation, the effective area to mass can be either much increased or much decreased.

In all the discussion so far, a further assumption has been made, namely that the radius of the body is significantly greater than the wavelength of the relevant radiation. If this is not the case, then the radiation can flow past the body without serious interaction with it. This problem has usually been dealt with by introducing a factor Q_{PR}, called the efficiency factor, into the equations. Generally, for incident radiation with a wavelength λ, Q_{PR} is roughly constant provided $2\pi b > \lambda$, where b is the radius of the (assumed spherical) particle, and is proportional to b if the inequality is reversed (see for example the work of Greenberg [13]). For solar radiation, the vast majority of the energy is carried in radiation with wavelength at a few hundred nanometers, that is, in the visible band. Hence for grains larger than a few microns we can assume that Q_{PR} is unity and effectively use the above equations unmodified.

In a series of papers in the early 1990's, Klačka [14], [15], [16], [17], [18] reconsidered the whole problem from the basic principles, but applying them to more realistic particles than the ideal one we have considered. The fundamental principle underlining any consideration always be that it is clear which frame is being used. Klačka considered the situation where not all the incoming radiation is absorbed as well as non-spherical rotating particles concluding that in many situations additional terms could be present in the equations of motion for real particles. One obvious conclusion is that if the particle is spinning and of irregular shape, a component of force orthogonal to the plane of the orbit can exist. The importance of these forces have been illustrated for example by Pittich & Klačka [19] and implies that all five of the orbital parameters can change rather than the three we discussed earlier.

The main point of this discussion is to focus on the effects of radiation on the dynamics of meteoroid streams in so far as they can be observed by us. This allows us to make some simplifications to the general problem. We shall consider the problem using the same basic assumptions regarding the properties of the particle as made by Robertson [7]. In the last section we shall discuss briefly how the results might change with different assumptions about the particle.

2. The Basic Equations

2.1. THE INCOMING RADIATION

Dealing with the effects of the solar radiation impinging on the body is fairly straightforward. If the rate of energy outflow from the Sun is L and the body is at a heliocentric distance r and presents an effective cross section area A to the radiation, then the rate of flow of momentum into the body, or force acting on it, is

$$F = \frac{LA}{4c\pi r^2}, \tag{1}$$

where c is the speed of light. The resulting acceleration of a particle of mass m is given by

$$\frac{LA}{4cm\pi r^2}. \tag{2}$$

If the body is assumed to be a sphere of radius b and density σ, then the acceleration can be written as

$$\frac{3L}{16\pi b\sigma c r^2} = \alpha c r^{-2}. \tag{3}$$

However, this is in the rest frame of the Sun, in the particle frame this becomes modified by a factor of $(1 - \dot{r}/c)$. For the situation where b is comparable to the wavelength of the incident light, then the expression has to be multiplied by Q_{PR}, the efficiency factor already discussed. Hence, the radial acceleration caused by the incoming radiation can be written as

$$\frac{Q_{PR}\alpha c(1 - \dot{r}/c)}{r^2}, \tag{4}$$

or more conveniently as

$$\frac{GM\beta(1 - \dot{r}/c)}{r^2}, \tag{5}$$

where

$$\beta = \frac{Q_{PR}\alpha c}{GM},$$

M being the mass of the Sun and G the universal gravitational constant. β is a non-dimensional quantity and is often called the ratio of radiation pressure force to the gravitational force. For visible meteors (i.e. particles where Q_{PR} can be taken as unity), β has a value $5.75 \times 10^{-5} b^{-1} \sigma^{-1}$, with both b and σ measured in *cgs* units.

2.2. THE OUTGOING RADIATION

For any given specific set of circumstances, dealing with the outgoing radiation is also easy. Problems arise because so many different sets of circumstances can exist. For example, as discussed already, the particle may reflect some radiation; it may be spinning hence presenting difficulties with the rest frame; it may be non-spherical and so on. In the following discussion, we confine ourselves strictly to the Robertson situation, namely where all the outgoing radiation is emitted isotropically relative to the particle. This situation is reasonable for a non-reflecting non-spinning spherical smooth particle that is a perfect conductor of heat. With this assumption, momentum of amount

$$\frac{GM\beta\mathbf{v}}{cr^2}$$

is lost from the particle per unit time. Here, $\mathbf{v}$ is the velocity of the particle relative to the inertial frame. This expression illustrates the difficulty of dealing with rotating bodies, since, in that situation, $\mathbf{v}$ must be interpreted as the velocity of a small radiating element and then integrated over the whole radiating surface.

The total acceleration is given by adding the incoming and outgoing effects, which in this case gives the radial component of the acceleration as

$$\frac{GM\beta(1-2\dot{r}/c)}{r^2}, \tag{6}$$

while the transverse component of the acceleration is

$$\frac{GM\beta\dot{\theta}}{cr}. \tag{7}$$

For some other situations, the outgoing radiation can be treated along similar lines. For example if the particle is a perfect reflector, then radiation is emitted only in the radial direction and the force is equal but opposite to that due to the incoming radiation. The difficulties arise when a combination of forms are present for the outgoing radiation, for example some reflected, some scattered and some absorbed and re-emitted. It is not our intention to discuss all the difficult or realistic cases here. References have already been given to the original works. The aim is to cover the basic principles and describe the simplest case, putting more effort into the discussion of the subsequent dynamics rather than on obtaining detailed models of small particles.

3. The motion of small spherical particles

With the above assumptions regarding the characteristics of the small particle and assuming that it is moving under the influence of the Sun only (gravity and radiation) its equations of motion using polar co-ordinates (r,θ) becomes

$$\ddot{r}-r\dot{\theta}^2=\frac{-GM(1-\beta)}{r^2}-\frac{2GM\beta\dot{r}}{r^2c}, \tag{8}$$

in the radial direction, while the transverse equation is

$$\frac{1}{r}\frac{d(r^2\dot{\theta})}{dt}=-\frac{GM\beta\dot{\theta}}{rc}. \tag{9}$$

Equation (9) can be integrated to give

$$h = r^2\dot{\theta} = h_o - GM\beta(\theta/c) = h_o - \alpha\theta \tag{10}$$

Hence the specific angular momentum (that is, the angular momentum per unit mass) about an axis perpendicular to the orbital plane is a continuously decreasing function of time since θ increases with time.
It is also possible to get an estimate of the survival time of a particle. If the number of completed orbits is N, then θ is $2\pi N$ so that when

$$N = \frac{h_o}{2\pi\alpha}, \tag{11}$$

h is reduced to zero. The time for this to take place is obtained simply by multiplying N and the orbital period P. Since the orbit is very close to an ellipse, then $h^2 \sim a(1-e^2)$ and $P^2 \sim a^3$ and so the survival time will be proportional to a^2/α.
Equation (8) can not be integrated to give a closed analytical solution, but it can be rewritten by substituting for $\dot{\theta}$ from equation (10), giving

$$\ddot{r} - \frac{{h_o}^2}{r^3} = \frac{-GM(1-\beta)}{r^2} + F(r,\dot{r}) \tag{12}$$

where

$$F(r,\dot{r}) = -\frac{2GM\beta\dot{r}}{r^2c} + \frac{(\alpha\theta)^2}{r^3} - \frac{2\alpha\theta h_o}{r^3}. \tag{13}$$

Using the substitution $u = 1/r$, equation (12) becomes

$$\frac{d^2u}{d\theta^2} + \frac{\alpha}{h}\frac{du}{d\theta} + u = \frac{GM(1-\beta)}{h^2}. \tag{14}$$

If we assume that h is effectively a constant over one orbit, then the orbit can be regarded as an precessing ellipse. Though h is not a constant, but is given by equation (10), we can still regard the solution as an instantaneous ellipse with the elements slowly changing with time as a consequence of the additional small terms. We shall discuss the solutions in more detail in the following sections.

3.1. ABRUPT CHANGE TO THE ORBITAL ELEMENTS

The osculating ellipse that is defined by equation (12) and ignoring $F(r,\dot{r})$ is identical to that of a standard Keplerian ellipse, but with GM replaced by $GM(1-\beta)$. In the literature, this is often referred to as the solution when the direct radiation pressure term alone is included and the effect is described as *weakening gravity*.

One conclusion is immediately obvious, if β is equal to or greater than unity then the resulting radial force is outwards and the particle will escape from the solar system. Taking the expression for β given at the end of section (2.1), namely $5.75 \times 10^{-5} b^{-1} \sigma^{-1}$, and assuming that σ is not significantly different from unity, we conclude that particles smaller than about 0.05 *microns* will be lost from the solar system. However, it is doubtful if Q_{PR} can be taken as unity for such small particles and so this conclusion needs to be investigated further for specific bodies. However, Harwit [20] in 1963 and Zook [21] in 1975 pointed out that it is not necessary for the radial force to be outwards in order that particles escape, they will become unbound if their orbital energy becomes positive. This aspect was discussed in some detail by Kresák [11] in 1976.

Consider a small particle released from a parent body that is moving on an orbit with semi-major axis a_o and eccentricity e_o with zero relative velocity at heliocentric distance r. This particle has kinetic energy per unit mass given by

$$\mathcal{T} = GM[\frac{1}{r} - \frac{1}{2a_o}] \qquad (15)$$

since the kinetic energy per unit mass will be the same as for the parent. The potential energy is given by

$$\mathcal{V} = -\frac{GM(1-\beta)}{r} \qquad (16)$$

since the force felt by the particle is reduced through the effects of radiation. Hence the orbital energy of the particle is

$$\mathcal{E} = \mathcal{T} + \mathcal{V} = GM[\frac{\beta}{r} - \frac{1}{2a_0}] \qquad (17)$$

and this will be positive provided

$$\beta \geq r/2a_o. \qquad (18)$$

Since the ejection must occur somewhere between perihelion and aphelion, then

$$a_o(1+e_o) \geq r \geq a_o(1-e_o). \qquad (19)$$

Thus, if

$$\beta \geq (1+e_o)/2$$

the particle will be lost irrespective of its ejection point.

If

$$(1+e_o)/2 \geq \beta \geq (1-e_o)/2,$$

then there is a range of heliocentric distance for the ejection point from which the particle is lost. Interestingly, it is easier to loose the particle if ejection is at perihelion than if it is ejected at aphelion.

As an illustration, consider comet 109P/Swift-Tuttle, the parent of the Perseid Meteoroid stream, where $e_o = 0.9636$. Meteoroids with β greater than 0.018 can be lost from this stream. With the same assumption regarding the density σ as before this translates to a particle radius b of 30 *microns* as the cut-off size, an increase by a factor of 50 over the size required to make $\beta = 1$. Both radio and visible meteors are greater in size than 30 *microns* and so rather obviously we conclude that the Perseid meteor stream can exist! More important, we conclude that meteoroids under 30 *microns* should not be present in the stream.

Since we have calculated the new energy and the angular momentum is unchanged at ejection, it is easy to calculate the semi major axis and eccentricity, a_1 and e_1 of the particle. We obtain

$$a_1 = \frac{a_o r(1-\beta)}{(r - 2a_o\beta)}, \tag{20}$$

$$e_1{}^2 = 1 - \frac{(1-e_o{}^2)(r-2a_o\beta)}{r(1-\beta)^2}. \tag{21}$$

We point out that the changes given above are changes to the orbit that occur immediately the particle becomes exposed to solar radiation and this change was called by Burns, Lamy & Soter [9] as *abrupt orbital changes.* It should however be realized that it is a change to the orbital elements that define the orbit that the particle is moving on. The particle itself does not do anything abruptly, it continues with the same velocity in the same location. However the force field is modified so that it moves on an orbit that is different from the original.

3.2. THE POYNTING-ROBERSTSON EFFECT

We now consider the effects of the perturbation terms in equation (12) that are given by equation (13). This perturbation, since small, can be regarded as producing small changes in the energy and angular momentum of the particle which produce small changes in the orbital elements but that the osculating orbit can still be described as an ellipse. For such orbits, we have

$$\mathcal{E} = -\frac{GM(1-\beta)}{2a} \tag{22}$$

and

$$h^2 = GM(1-\beta)p = GM(1-\beta)a(1-e^2) \tag{23}$$

where p is the semi-parameter of the orbit, that is $p = a(1-e^2)$
From the first of these equations it is easy to obtain

$$\frac{da}{dt} = -\frac{a}{\mathcal{E}}\frac{d\mathcal{E}}{dt} \tag{24}$$

and

$$\frac{dp}{dt} = \frac{2p}{h}\frac{dh}{dt}. \tag{25}$$

Since

$$\frac{dp}{dt} = (1-e^2)\frac{da}{dt} - 2ea\frac{de}{dt} \tag{26}$$

$\frac{de}{dt}$ can be found in terms of $\frac{d\mathcal{E}}{dt}$ and $\frac{dh}{dt}$.

However, we now hit the fundamental difficulty in all celestial mechanics problems, namely that θ, which is the true anomaly, and r which is related to θ by the normal equation for an ellipse, namely

$$\frac{p}{r} = 1 + e\cos\theta, \tag{27}$$

cannot be expressed as an analytical function of time. This means that we cannot obtain an analytical equation for the change in energy by integrating $F(r,\dot{r})$ from equations (12) and (13). Progress can only be made by averaging the changes over one complete orbit and hence obtaining the secular changes. This was carried out by Robertson [7] and verified by Wyatt & Whipple [10]. This produces the following two equations that have been extensively used in the literature (for example Hughes, Williams & Fox [22] in 1981 when investigating the Quadrantid shower or Arter & Williams [23] in a study of the April Lyrids .)

$$a\frac{da}{dt} = -\frac{\alpha(2+3e^2)}{(1-e^2)^{3/2}} \tag{28}$$

and

$$\frac{de}{dt} = -\frac{5\alpha e}{2(1-e^2)^{1/2}a^2}, \tag{29}$$

where α is as previously defined after equation (5).

Unfortunately these equations can not be integrated in general either (that is, a and e can not be found as analytical functions of t). However some conclusions can be reached.

(i) The right hand side of both equations are always negative, hence both a and e always decrease with time, that is the orbit becomes smaller and more circular.

(ii) A solution is possible if the initial orbit is circular, that is $e = 0$. Now, equation (29) gives $de/dt = 0$ and equation (28) then becomes

$$a\frac{da}{dt} = -2\alpha \tag{30}$$

so that the time for significant change is given by $a^2/4\alpha$.

Note however that this is a degenerate case. It is not possible to move from an orbit with radius a to an orbit with radius $a - \Delta a$ along a circular path, hence e must initially increase. Klacka & Kaufmannova [24] discussed this in some detail and concluded that averaging over one orbit is not possible when e is very small and deduced that for near circular orbits there is a long-term increase in e.

(iii) In fact even when e is not zero, we can see that the time for a significant change is still of the order of $a^2/4\alpha$, consistent with equation (11).

(iv) By combining the two equations, we obtain

$$\frac{da}{de} = \frac{2a(2+3e^2)}{5e(1-e^2)} \tag{31}$$

which has an integral

$$\frac{a(1-e^2)}{e^{4/5}} = Constant. \tag{32}$$

While this does not give a complete solution, it does show how one parameter varies in relation to the other. Thus for example, if e changes from 0.8 to 0.4, then a must reduce by a factor of approximately 4.

In general, there will also be an advancement of perihelion brought about by these forces. The secular change was given by Robertson [7] as

$$\frac{d\omega}{dt} = \frac{3(GM)^{1.5}(1-\beta)^{1.5}}{c^2a^{2.5}(1-e^2)}. \tag{33}$$

Curiously, Wyatt & Whipple [10] claimed to have obtained the same result as Robertson, but in fact gave

$$\frac{d\omega}{dt} = \frac{3(GM)^{1.5}(1-\beta)}{c^2a^{2.5}(1-e^2)}. \tag{34}$$

Lyttleton [8] claimed that both of these were incorrect and produced instead

$$\frac{d\omega}{dt} = \frac{3(GM)^{1.5}(1-\beta)^{-0.5}}{c^2a^{2.5}(1-e^2)}. \tag{35}$$

Later in the paper the exponent for $(1 - \beta)$ is changed in sign, perhaps suggesting that the former is a typographic error. Lyttleton also pointed out that a further correction existed, which gives as the claimed correct rate of change,

$$\frac{d\omega}{dt} = \frac{3(GM)^{1.5}[1 - 2\beta/3 - \beta^2/3]}{c^2 a^{2.5}(1 - e^2)(1 - \beta)^{0.5}}. \tag{36}$$

Klačka (private communications) obtained from first principles

$$\frac{d\omega}{dt} = \frac{3(GM)^{1.5}[1 + \beta/3 - \beta^2/2]}{c^2 a^{2.5}(1 - e^2)(1 - \beta)^{0.5}}. \tag{37}$$

This expression is very similar to Lyttleton's final expression, and, since β is small, application will yeild almost identical results. However there are fundamental differences of physics between the two expressions.
In all cases, when β is zero, the result reduces to the standard result in special relativity, giving for example for Mercury a change of 43 arc seconds per century. For most meteor streams it is much lower and is in general neglected.

4. Applications of the effects of radiation to Meteor showers

Meteor showers are the detectable end product of the interaction between a stream of meteoroids and the atmosphere of the Earth. The relation between meteor size and brightness has been discussed by a number of authors and this has been summarized for example by Babadzhanov *et al.* [25]. A typical meteor of magnitude +2 will have a mass of order 0.03 g, while one of magnitude -2 will be of mass about 2 g. Assuming a mean density σ of 1 gcm^{-3}, this converts to a size range of 2 to 8 millimeters. A much fainter meteor, detectable by radio alone, at a magnitude of +8 would have a radius of about 0.2 millimeters. Hence, using the expression for β given in section (2.1), namely

$$\beta = \frac{5.75 \times 10^{-5}}{b\sigma}$$

then β has a value in the range of 7×10^{-5} to 3×10^{-3}. Note that these values also fit in well with the discussion in section (3.1), about the value of β that will cause the meteoroid to be lost, that is, meteoroids smaller than these will be lost off highly eccentric orbits.

There is also a dependence on velocity in the expression for meteor magnitude, faster meteoroids producing brighter meteors for the same meteoroid size. Hence there will be some variation in the values for β in different showers. For example, the Leonid meteoroids have a very high velocity, so that the Leonids produce more bright meteors than the average stream. It also means that smaller meteoroids can produce meteors that are sufficiently bright to be observed, thus increasing the number of visible meteors in a Leonid shower. This does not really matter within this discussion, the aim is to specify the range of values for β that is applicable in meteor showers, noting in particular that values of β close to unity are unlikely.

A further important point regarding meteor showers is that the meteoroids are observed exactly at one of the nodes of their orbit. Hence if the nodal distance is not the same as the heliocentric distance of the Earth, no meteors are observed. Second, the Earth must be at the node at exactly the same time as an observed meteor. Both of these can be affected by the inclusion of radiation effects.

The change to the two orbital parameters a and e has already been discussed in section (3.2), with the time-scale for changes given in equation (30). Taking a as typically 10 au and β for the smaller meteoroids as 3×10^{-3} as given above, then a 10% change in the semi major axis will take 10^6 years. In general this is rather too long to be significant within the lifetime of a meteor shower. However, much smaller changes in a and e can affect the observability of meteor showers, for they can change either the arrival time of the meteoroids by changing the period or change the nodal distance. We consider these effects in turn.

4.1. CHANGES IN THE ORBITAL PERIOD

If a set of meteoroids are released from a parent comet onto an orbit that allows them to be visible on Earth in a given year, then they will become visible again after one orbital period. This is the simple explanation for the appearance of Leonid storms every 33 years. Hence, if Leonids are seen for a number of successive years, say 1998, 1999, 2000 and 2001, then either the parent comet must have been active for at least three years, or, more likely, the orbital periods of the meteoroids span a range about that of the comet. Since radiation effects can change the semi-major axis, then it also changes the period.

This change in the orbital period because of the effects of radiation was investigated in some detail by Ma *et. al.* [26]. If a set of meteoroids of different sizes were released from a comet at perihelion onto identical orbits, then a difference in their orbital period given by

$$P_1 - P_2 = P(\beta_1 - \beta_2)[\frac{2+e}{1-e}], \tag{38}$$

where e is the eccentricity of the comet orbit of period P, and suffix 1 and 2 refer to two meteoroids will result. For the Leonids, P is about 33.5 years and e is about 0.9. Then for meteoroids with values of β at the two ends of the range that we have obtained, namely, 7×10^{-5} and 3×10^{-3}, we have

$$P_1 - P_2 = 2.85 \; years$$

between the smaller meteoroids and the comet and 25 *days* between the larger ones and the comet. Hence, this effect can be important in determining whether or not meteors are observed in a given year. The situation is more complicated because meteoroids will be ejected from the comet with some velocity and so the initial orbit will be different from that of the comet. Large meteors ejected onto an orbit that initially had a period 25 days shorter than the comet will, because of radiation effects, arrive at the same time as the comet. A further complication is that meteoroids may have completed more than one orbit before being observed. For example, micro meteorites with β at 0.0035 may have been released onto an orbit with a period twice that of the comet. Hence those emitted in 1933 say could be visible in 1998. Thus given only that a meteor shower was seen, no unique solution exists. Unfortunately, all models for meteoroid ejection suggests that the ejection velocity is also size dependent with the smaller meteors being ejected with a higher velocity and, as was shown by Ma *et. al.* this velocity effect tends to dominate.

Ma *et. al.* [26], reported the observation of micro meteorites in 1998 some 18 hours after the main outburst and showed that this was possible with β at 0.05 for meteoroids ejected in 1933.

4.2. CHANGES TO THE NODAL DISTANCE

Since the argument of perihelion ω is defined as the angular distance of perihelion from the ascending node, it follows that the true anomaly, f of the nodes are $-\omega$ or $\pi - \omega$. The nodal distances r_N can then be derived from the standard equation for an ellipse

$(1 - e\cos\omega)r_N = p$ and $(1 + e\cos\omega)r_N = p$, where $p = a(1 - e^2)$. Hence, we obtain

$$\frac{(1 - e\cos\omega)}{r_N}\frac{dr_N}{dt} = \frac{1 - e^2}{2ea}\cos\omega\frac{da}{dt} - \frac{(e^2\cos\omega + \cos\omega - 2e)}{2ep}\frac{dp}{dt} - e\sin\omega\,\frac{d\omega}{dt} \tag{39}$$

for the first node with a similar equation for the other node.
We know from section (3.2) that $\frac{d\omega}{dt}$ is small an so, using $p = a(1 - e^2)$, and equations (28) and (29), we obtain after some algebra

$$\frac{dr_N}{dt} = -\frac{\alpha {r_N}^2(4 + e\cos\omega)}{2a^3(1 - e^2)^{1.5}}, \tag{40}$$

$$\frac{dr_N}{da} = \frac{{r_N}^2(4 + e\cos\omega)}{2a^2(2 + 3e^2)} \tag{41}$$

and

$$\frac{dr_N}{de} = \frac{r_N(4 + e\cos\omega)}{5e(1 - e\cos\omega)}. \tag{42}$$

This last equation integrates to give

$$\frac{r_N(1 - e\cos\omega)}{e^{4/5}} = Constant \tag{43}$$

an equation that could also have been obtained from equation (32) and substituting $(1 - e\cos\omega)r_N = p$

Identical equatione, except that the sign of ω terms are reversed can be obtained for the other node.

For any meteoroid stream r_N must be close to 1 *au*. Taking a typical comet with a at 10 *au* and $e = 0.9$, we have

$$\frac{dr_N}{da} = 3 \times 10^{-3}$$

The rate of change of r_N is thus considerably slower than the change in a which was estimated as about 10^6 *years* above. Thus, rather surprisingly the Poynting-Robertson drag, despite changing the orbital parameters, leaves the nodal distance almost constant for a long time.

5. Discussion

In the above we have, as many other authors have, assumed that the meteoroid is spherical, absorbs all the radiation and re-emits this isotropically in its own rest frame. In order to do this it would need to be a good conductor of heat so that all the surface is at the same temperature, not spinning or the rest frame for the emitted radiation may not be the center of mass frame and not have any reflective part. It is very unlikely that any real meteoroids will have all of these properties, hence in a strict sense, none of the above analysis is valid. In the historical introduction we cited a number of authors we have attempted to correct these deficiencies. Here, we do not wish to add to their efforts, but instead discuss which will change the results in a quantitative way only and which will also change the results qualitatively.

The inclusion of some reflection reduces the amount of re-emitted radiation, transverse component. According to Lyttleton, this has the effect of increasing slightly the size of particles that will be lost due to the *abrupt change* in the orbital elements and also increases the time taken for the orbit to be rounded by the Poynting-Robertson drag. The result may be different for non-spherical particles but, nothing qualitatively different takes place.

However, if the radiation is not emitted isotropically relative to the particle, the subsequent motion can be qualitatively different since it is now possible for a component of acceleration orthogonal to the orbital plane to exist. Such a component obviously changes the orbital plane so that changes in inclination, I, and longitude of the ascending node, Ω, will take place.

The Quadrantid stream (see Hughes *et al.* [22]) shows some evidence for differential nodal retrogression, the nodal position for radio meteors and visible meteors differing by about 0.6^o. Since the effects of radiation are mass dependent, then in principle this could be the cause of this segregation with the small radio meteors having evolved further than the larger visible ones. Unfortunately it is difficult to imagine a situation where the small meteoroids are systematically perturbed. This would require some alignment of the meteoroids so that the asymmetry in the emitted radiation is always of the same amount in the same direction. The Quadrantid stream, as was pointed out by Wu & Williams [27] has close encounters with Jupiter and so the observed phenomena may be a consequence of the more standard effects of radiation, namely that the smaller meteoroids have been more significantly affected so that they are on orbits that take them fractionally nearer to Jupiter than the visible ones.

In conclusion, the basic principle in considering the effects of radiation on the dynamics of small bodies is very simple. It is that the incoming radiation is in the rest frame based on the Sun while the outgoing radiation is relative to a rest-frame based on the emitting surface. The difficulties and the disagreements arise when an attempt is made to apply this principle to real situations, in particular through claiming that a particular effect is small, when averaged over an orbit, when in fact this might not be the case. Fortunately meteoroid streams are fairly short-lived and so variations between different formulations may not have time to manifest themselves and gravitational perturbations from planets tend to dominate.

6. Acknowledgments

I would like to record my thanks to Jozef Klačka for a number of communications and comments made on various aspects of this work. I also acknowledge the support of NATO in providing travel support which stimulated me to write this chapter.

References

1. A.A. Michelson & E.W. Morley, 1887, On the Relative Motion of the Earth and Luminiferous Ether, *Am. Jl. Sci., Series 3*, **34**, 333-345
2. A. Einstein, 1905, On the Electrodynamics of Moving Bodies, *Annalen der Physik*, **17**
3. J.M. Poynting, 1903, Radiation in the solar system, its effect on temperature and its pressure on small bodies, *Proc. Roy. Soc.*, **72**, 265-267
4. H.C. Plummer, 1905, On possible effects of Radiation on the motion of comets with special reference to Encke's comet, *Mon. Not. R. astr. Soc.*, **65**, 229-237
5. H.C. Plummer, 1906, On the effects of radiation on the motion of comets, *Mon. Not. R. astr. Soc.*, **67**, 63-67
6. J. Larmour, 1913, On the Dynamics of Radiation, *Proc. Int. Math. Cong. Cambridge*, 213
7. H.P. Robertson, 1937, Dynamical effects of Radiation in the Solar System, *Mon. Not. R. astr. Soc.*, **97**, 423-438
8. R.A. Lyttleton, 1976, Effects of solar radiation on the orbits of small particles, *Ap. Sp. Sci.*, **44**, 119-140
9. J.A. Burns, P.L. Lamy & S. Soter, 1979, Radiation forces on small particles in the Solar System, *Icarus*, **40**, 1-48
10. S.P. Wyatt & F.L. Whipple, 1950, The Poynting-Robertson effect on meteor orbits, *Ap. J.*, **111**, 134-141
11. L. Kresák, 1976, Orbital evolution of the dust streams released from comets, *Bull. Astron. Inst. Czech.*, **27**, 35-46
12. B.Å.S. Gustafson, 1989, Comet ejection and dynamics of nonspherical dust particles and meteoroids, *Ap. J.*, **337**, 945-949
13. J.M. Greenberg, 1972 in *The origin of the Solar System*, Ed Reeves H., CNRS
14. J. Klačka, 1992a, Poynting-Robertson effect I. Equations of motion, *Earth, Moon, & Planets*, **59**, 41-59
15. J. Klačka, 1992b, Poynting-Robertson effect II. Perturbation Equations, *Earth, Moon, & Planets*, **59**, 211-2185
16. J. Klačka, 1993a, Interplanetary dust particles and solar electromagnetic radiation, *Earth, Moon, & Planets*, **63**, 9-13
17. J. Klačka, 1993b, Radiation forces and nonspherical dust particles, *Earth, Moon, & Planets*, **62**, 145-148
18. J. Klačka, 1994, Interplentary dust particles and solar radiation, *Earth, Moon, & Planets*, **64**, 125-132
19. E.M. Pittich & J. Klačka, 1996, On the importance of the Poynting-Robertson effect on meteoroids, *Earth, Moon & Planets*, **72**, 339-342
20. M. Harwit, 1963, Origins of the Zodiacal dust cloud, *Jl. Geophys. Res.*, **86**, 2171-2180
21. H.A. Zook, 1975, Hyperbolic cosmic dust - its origin and astrophysical significance, *Plan. Sp. Sci.*, **23**, 1391-1397
22. D.W. Hughes, I.P. Williams & K. Fox, 1981, The mass segregation and nodal retrogression of the Quadrantid meteor stream, *Mon. Not. R. Astr. Soc.*, **195**, 625-637
23. T.R. Arter & I.P. Williams, 1997, The mean orbit of the April Lyrids, *Mon. Not. R. Astr. Soc.*, **289**, 721-728
24. J. Klačka & J. Kaufmannova, 1992, Poynting-Robertson effect: Circular orbit, *Earth, Moon, & Planets*, **59**, 97-102
25. P.B. Babadzhanov, Z. Wu, I.P. Williams & D.W. Hughes, 1991, The Leonids, Comet Biela and Biela's associated meteoroid stream, *Mon. Not. R. Astr. Soc.*, **253**, 69-74
26. Y. Ma, Y. He & I.P. Williams, 2001, A micrometeor component of the 1998 Leonid shower, *Mon. Not. R. Astr. Soc.*, **325**, 457-462
27. Z. Wu & I.P. Williams, 1992, On the Quadrantid meteoroid stream complex, *Mon. Not. R. Astr. Soc.*, **259**, 617-628

COVARIANT EQUATION OF MOTION FOR A PARTICLE IN AN ELECTROMAGNETIC FIELD

J. KLACKA
Astronomical Institute, Faculty of Mathematics, Physics and Informatics, Comenius University, Mlynská dolina
842 48 Bratislava, Slovak Republic

Abstract. A relativistically covariant form of the equation of motion for an arbitrarily shaped dust particle under the action of electromagnetic radiation is derived. The particle is neutral in charge. The equation of motion is expressed in terms of the particle's optical properties, standardly used for stationary particles. The derived covariant equation of motion yields, as special cases, the well-known examples presented in the first half of the twentieth century by Einstein and Robertson.

1. Introduction

The first case closely connected with the relativistic equation of motion for a particle under the action of electromagnetic radiation was presented by Einstein [1], who calculated the change of energy and light pressure for a plane mirror. Later on, Robertson [8] presented the equation of motion for a perfectly absorbing spherical particle – this result has been applied to astronomical situations for several decades as the well-known Poynting-Robertson (P-R) effect. Robertson's case was relativistically generalized by Klačka [2]. However, Klačka [2] has shown that an even more generalized P-R effect holds only for a very special case of interaction between the particle and the incident electromagnetic radiation: the total momentum of the outgoing (scattered) radiation per unit time is in the direction of the incident radiation (either orientation is possible) in the proper frame of reference, i. e. the rest frame of the particle (see Eq. (120) or Eq. (122) in [2]). It is worth mentioning that Einstein's example is more general than the P-R effect with respect to the condition of interaction.

G. Videen and M. Kocifaj (eds.), Optics of Cosmic Dust, 301–312.

It is convenient to have at our disposal the general equation of motion for all types of particles, not only for the special cases treated by the above cited authors. Because of complex morphology and optical properties, real particles scatter radiation in a more complicated manner. This chapter details the derivation of the relativistically covariant equation of motion for an arbitrarily shaped (cosmic dust) particle. The covariant formulation proves that the derived equation of motion is physically correct [3], [5].

2. Proper reference frame of the particle – stationary particle

The term "stationary particle" denotes a particle that does not move in a given inertial frame of reference. Primed quantities denote quantities measured in the proper reference frame of the particle, the rest frame of the particle.

The flux density of photons scattered into an elementary solid angle $d\Omega' = \sin\vartheta' \, d\vartheta' \, d\varphi'$ is proportional to $p'(\vartheta', \varphi') \, d\Omega'$, where $p'(\vartheta', \varphi')$ is the phase function. The phase function depends on particle orientation with respect to the direction of incident radiation and on the particle characteristics; angles ϑ', φ' correspond to the direction of travel of the scattered radiation, ϑ' is the polar angle and equals zero for the case of rays traveling in the orientation identical with the unit vector $\vec{e}'_1$ of the incident radiation. The phase function fulfills the condition

$$\int_{4\pi} p'(\vartheta', \varphi') \, d\Omega' = 1 \, . \tag{1}$$

The momentum of the incident beam of photons, which is lost in the process of its interaction with the particle is proportional to the extinction cross-section C'_{ext}. The part proportional to absorption C'_{abs} is completely lost and the part proportional to scattering $C'_{ext} - C'_{abs} = C'_{sca}$ is reemitted.

The momentum per unit time of the scattered photons into an elementary solid angle $d\Omega'$ is

$$d\vec{p}'_{sca} = \frac{1}{c} \, S' \, C'_{sca} \, p'(\vartheta', \varphi') \, \vec{K}' \, d\Omega' \, , \tag{2}$$

where the unit vector in the direction of scattering is

$$\vec{K}' = \cos\vartheta' \, \vec{e}'_1 \; + \; \sin\vartheta' \, \cos\varphi' \, \vec{e}'_2 \; + \; \sin\vartheta' \, \sin\varphi' \, \vec{e}'_3 \, . \tag{3}$$

and S' is the flux density of radiation energy. The system of unit vectors used on the RHS of the last equation forms an orthogonal basis. The total momentum per unit time of the scattered photons is

$$\vec{p}'_{sca} = \frac{1}{c} \, S' \, C'_{sca} \int_{4\pi} p'(\vartheta', \varphi') \, \vec{K}' \, d\Omega' \, . \tag{4}$$

The momentum per unit time obtained by the particle due to the interaction with radiation is

$$\frac{d\,\vec{p}'}{d\,t'} = \frac{1}{c}\,S'\,\left\{C'_{ext}\,\vec{e}'_1\,-\,C'_{sca}\int_{4\pi} p'(\vartheta',\varphi')\,\vec{K}'\,d\Omega'\right\}\,. \tag{5}$$

We expect that energy is conserved: the energy per unit time of the incoming radiation E'_i equals the energy per unit time of the outgoing radiation after interaction with the particle E'_o. We will use the fact that time $t' = \tau$, where τ is proper time.

For brevity, we use "effective efficiencies" Q'_x instead of effective cross-sections C'_x: $C'_x = Q'_x\,A'$, where A' is the geometrical cross-section of a sphere of volume equal to the volume of the particle. Equation (5) can be rewritten to the form

$$\begin{aligned}\frac{d\,\vec{p}'}{d\,\tau} \;=\; & \frac{1}{c}\,S'\,A'\;\{[Q'_{ext}\;-\;<\cos\vartheta'>\;Q'_{sca}]\;\vec{e}'_1\;+\\ & [\;-\;<\sin\vartheta'\;\cos\varphi'>\;Q'_{sca}]\;\vec{e}'_2\;+\\ & [\;-\;<\sin\vartheta'\;\sin\varphi'>\;Q'_{sca}]\;\vec{e}'_3\}\;,\end{aligned} \tag{6}$$

or, in a short form

$$\frac{d\,\vec{p}'}{d\,\tau} = \frac{S'\,A'}{c}\sum_{j=1}^{3} Q'_j\,\vec{e}'_j\,. \tag{7}$$

2.1. SUMMARY OF THE IMPORTANT EQUATIONS

Using the text concerning energy below Eq. (5) and the last Eq. (7), we may describe the total process of interaction in the form of the following equations (energies and momenta per unit time):

$$\begin{aligned}E'_o &= E'_i = A'\,S'\;,\\ \vec{p}'_o &= (1\;-\;Q'_1)\;\vec{p}'_i\;-\;(Q'_2\;\vec{e}'_2\;+\;Q'_3\;\vec{e}'_3)\;E'_o/c\;,\\ \vec{p}'_i &= (E'_i/c)\;\vec{e}'_1\;,\end{aligned} \tag{8}$$

The index "i" represents the incoming radiation, beam of photons, the index "o" represents the outgoing radiation.

The changes of energy and momentum of the particle due to the interaction with electromagnetic radiation are

$$\begin{aligned}\frac{d\,E'}{d\,\tau} &= E'_i\;-\;E'_o = 0\;,\\ \frac{d\,\vec{p}'}{d\,\tau} &= \vec{p}'_i\;-\;\vec{p}'_o\;.\end{aligned} \tag{9}$$

The condition for energy in Eq. (8) is equivalent to Eq. (121) in [2]. The condition for momentum $\vec{p}'_o$ in Eq. (8) is a generalization of Eq. (122) in [2] which can be obtained from Eq. (8) by setting $Q'_2 = Q'_3 = 0$.

3. Stationary frame of reference

"Stationary frame of reference" refers to a frame of reference in which the particle moves with a velocity vector $\vec{v} = \vec{v}(t)$. The physical quantities measured in the stationary frame of reference are denoted by unprimed symbols.

Our aim is to derive the equation of motion for the particle in the stationary frame of reference. We use the fact that we know this equation in the proper frame of reference. We use the Lorentz transformation for the purpose of making the transformation from the proper frame of reference to the stationary frame of reference.

If we have a four-vector $A^\mu = (A^0, \vec{A})$, where A^0 is its time component and $\vec{A}$ is its spatial component, the generalized special Lorentz transformation yields

$$\begin{aligned} A^{0'} &= \gamma \, (A^0 - \vec{v} \cdot \vec{A}/c) \, , \\ \vec{A}' &= \vec{A} + [(\gamma - 1) \, \vec{v} \cdot \vec{A}/\vec{v}^2 - \gamma \, A^0/c] \, \vec{v} \, . \end{aligned} \tag{10}$$

The inverse generalized special Lorentz transformation is

$$\begin{aligned} A^0 &= \gamma \, (A^{0'} + \vec{v} \cdot \vec{A}'/c) \, , \\ \vec{A} &= \vec{A}' + [(\gamma - 1) \, \vec{v} \cdot \vec{A}'/\vec{v}^2 + \gamma \, A^{0'}/c] \, \vec{v} \, . \end{aligned} \tag{11}$$

The γ factor is given by the well-known relation

$$\gamma = 1/\sqrt{1 - \vec{v}^2/c^2} \, . \tag{12}$$

We introduce four-momentum:

$$p^\mu = (p^0, \vec{p}) \equiv (E/c, \vec{p}) \, . \tag{13}$$

3.1. INCIDENT RADIATION

Applying Eqs. (11) and (13) to quantity $E'_i/c, \vec{p}'_i$ (four-momentum per unit time – proper time is a scalar quantity) and taking into account also Eqs. (8), we can write

$$\begin{aligned} E_i &= E'_i \, \gamma \, (1 + \vec{v} \cdot \vec{e}'_1/c) \, , \\ \vec{p}_i &= \frac{E'_i}{c} \left\{ \vec{e}'_1 + \left[(\gamma - 1) \, \vec{v} \cdot \vec{e}'_1/\vec{v}^2 + \gamma/c \right] \vec{v} \right\} \, . \end{aligned} \tag{14}$$

Since $p^\mu = (h\,\nu, h\,\nu\,\vec{e}_1)$ for photons, the Lorentz transformation yields

$$\begin{aligned} \nu' &= \nu\, w\,, \\ \vec{e}_1' &= \frac{1}{w}\left\{\vec{e}_1 + \left[(\gamma - 1)\ \vec{v}\cdot\vec{e}_1/\vec{v}^2 - \gamma/c\right]\vec{v}\right\}, \end{aligned} \tag{15}$$

where abbreviation

$$w \equiv \gamma\,(1 - \vec{v}\cdot\vec{e}_1/c) \tag{16}$$

is used.

Inserting the second of Eqs. (15) into Eqs. (14), one obtains

$$\begin{aligned} E_i &= (1/w)\ E_i'\,, \\ \vec{p}_i &= (1/w)\ (E_i'/c)\ \vec{e}_1\,. \end{aligned} \tag{17}$$

We have four-vector $p_i^\mu = (E_i/c, \vec{p}_i) = (1, \vec{e}_1)E_i/c = (1/w, \vec{e}_1/w)w\ E_i/c \equiv b_1^\mu\ w\ E_i/c$.

3.2. OUTGOING RADIATION

Outgoing radiation is analogous to incident radiation. It is only a little more algebraically complicated, since radiation may spread out also in directions given by unit vectors $\vec{e}_2$ and $\vec{e}_3$. The relations between $\vec{e}_2'$ and $\vec{e}_2$, $\vec{e}_3'$ and $\vec{e}_3$, are analogous to that presented by the second of Eqs. (15):

$$\vec{e}_j' = \frac{1}{w_j}\left\{\vec{e}_j + \left[(\gamma - 1)\ \vec{v}\cdot\vec{e}_j/\vec{v}^2 - \gamma/c\right]\vec{v}\right\},\quad j = 1,2,3\,, \tag{18}$$

where

$$w_j \equiv \gamma\,(1 - \vec{v}\cdot\vec{e}_j/c)\,,\quad j = 1,2,3\,. \tag{19}$$

Using Eqs. (11) and (13) on quantity $E_o'/c, \vec{p}_o'$ (four-momentum per unit time – proper time is a scalar quantity), we can write

$$\begin{aligned} E_o &= \gamma\,(E_o' + \vec{v}\cdot\vec{p}_o')\,, \\ \vec{p}_o &= \vec{p}_o' + \left[(\gamma - 1)\ \vec{v}\cdot\vec{p}_o'/\vec{v}^2 + \gamma\,\frac{E_o'}{c^2}\right]\vec{v}\,. \end{aligned} \tag{20}$$

Using also $\vec{p}_i' = E_i'\ \vec{e}_1'/c$, Eqs. (8), (18), (20) and the second of Eqs. (15),

$$\begin{aligned} E_o &= Q_1'\ w\ E_i\,\gamma + (1 - Q_1')\ E_i + \\ &\quad w\ E_i\ (Q_2' + Q_3')\,\gamma - w\ E_i\ (Q_2'/w_2 + Q_3'/w_3)\,, \\ \vec{p}_o &= (1 - Q_1')\,\frac{E_i}{c}\,\vec{e}_1 + Q_1'\,\frac{w\ E_i}{c^2}\,\gamma\,\vec{v} - \\ &\quad \sum_{j=2}^{3} Q_j'\,\frac{w\ E_i}{c^2}\,(c\,\vec{e}_j/w_j - \gamma\,\vec{v})\,. \end{aligned} \tag{21}$$

3.3. EQUATION OF MOTION

Analogous with Eqs. (9), we have for the changes of energy and momentum of the particle due to the interaction with electromagnetic radiation

$$\frac{d\,E}{d\,\tau} = E_i - E_o\,,$$
$$\frac{d\,\vec{p}}{d\,\tau} = \vec{p}_i - \vec{p}_o\,. \tag{22}$$

Substituting Eqs. (21) into Eqs. (22), using also $\vec{p}_i = (E_i/c)\,\vec{e}_1$, one easily obtains

$$\frac{d\,E}{d\,\tau} = \sum_{j=1}^{3} Q'_j\; w\; E_i\; (1/w_j - \gamma) \quad ; \quad w \equiv w_1\,,$$
$$\frac{d\,\vec{p}}{d\,\tau} = \sum_{j=1}^{3} Q'_j\; \frac{w\,E_i}{c^2}\; (c\,\vec{e}_j/w_j - \gamma\vec{v})\,. \tag{23}$$

Equations (23) may be rewritten in terms of four-vectors:

$$\frac{d\,p^{\mu}}{d\,\tau} = \sum_{j=1}^{3} Q'_j\; \frac{w\,E_i}{c^2}\; \left(c\,b_j^{\mu} - u^{\mu}\right)\,, \tag{24}$$

where p^{μ} is four-vector of the particle of mass m

$$p^{\mu} = m\;u^{\mu}\,, \tag{25}$$

four-vector of the world-velocity of the particle is

$$u^{\mu} = (\gamma\;c, \gamma\;\vec{v})\,. \tag{26}$$

We have also other four-vectors

$$b_j^{\mu} = (1/w_j, \vec{e}_j/w_j)\,,\quad j = 1, 2, 3\,. \tag{27}$$

The following can be verified easily:
i) the quantity $w\;E_i$ is a scalar quantity – see first of Eqs. (17);
ii) Eq. (24) reduces to Eq. (7) and to the first of Eqs. (9) for the case of proper inertial frame of reference of the particle;
iii) Eq. (24) yields $d\,m/d\,\tau = 0$.

We introduce

$$b_j^0 \equiv 1/w_j = \gamma\,(1 + \vec{v}\cdot\vec{e}\,'_j\,/\,c)\,,\quad j = 1, 2, 3,$$
$$\vec{b}_j \equiv \vec{e}_j\,/\,w_j = \vec{e}\,'_j + [(\gamma - 1)\,\vec{v}\cdot\vec{e}\,'_j\,/\,\vec{v}^2 + \gamma\,/\,c]\,\vec{v}\,,$$
$$w \equiv w_1 = \gamma^{-1}\,(1 + \vec{v}\cdot\vec{e}\,'_1\,/\,c)^{-1} \tag{28}$$

for the purpose of practical calculations. Physics of these relations corresponds to aberration of light.

We have derived the equation of motion for a real dust particle under the action of electromagnetic radiation. It is assumed that the equation of motion is represented by Eqs. (8) and (9) in the proper frame of reference of the particle. The final covariant form is represented by Eq. (24), or using $E_i = w\ S\ A'$ (see Eq. (38) in [2]),

$$\frac{d\,p^{\mu}}{d\,\tau} = \left(w^2\ S\ A'\ /\ c^2\right) \sum_{j=1}^{3} Q'_j \left(c\, b_j^{\mu} - u^{\mu}\right) . \tag{29}$$

Within first-order accuracy in $\vec{v}/c$, Eqs. (28)-(29) yield

$$\begin{aligned} \frac{d\,\vec{v}}{d\,t} &= \frac{S\ A'}{m\ c} \sum_{j=1}^{3} Q'_j\ [(1 - 2\,\vec{v}\cdot\vec{e}_1/c + \vec{v}\cdot\vec{e}_j/c)\ \vec{e}_j - \vec{v}/c]\ , \\ \vec{e}_j &= (1 - \vec{v}\cdot\vec{e}\,'_j/c)\ \vec{e}\,'_j + \vec{v}/c\ \ ,\ \ j = 1,2,3\ . \end{aligned} \tag{30}$$

It is worth stressing that values of Q'−coefficients depend on particle orientation with respect to the incident radiation: these values are time dependent. A slightly different derivation of the equation of motion within first-order accuracy in $\vec{v}/c$ is presented in [4]; the derivation to the same accuracy can be found in [7].

It is left to the reader to verify that Eq. (29) (or Eqs. (30) within the accuracy to the first order in $\vec{v}/c$) yields as a special cases the situations discussed in [1] and [8] – Robertson's case is obtained by substituting $Q'_1 = 1$, $Q'_2 = Q'_3 = 0$, Einstein's results require a little more calculations. The result is given in [6].

4. Another formulation of the equation of motion

Section 3.3 represents crucial results for orbital evolution of particles in the universe. This section represents a slightly different formulation of the problem.

4.1. PROPER REFERENCE FRAME OF THE PARTICLE – STATIONARY PARTICLE

The equation of motion of the particle in its proper frame of reference is taken in the form

$$\begin{aligned} \frac{d\,E'}{d\,\tau} &= 0\ , \\ \frac{d\,\vec{p}\,'}{d\,\tau} &= \frac{1}{c}\ S'\ (C'\ \vec{e}\,'_1)\ , \end{aligned} \tag{31}$$

where E' is the particle's energy, $\vec{p'}$ its momentum, τ is proper time, c is the speed of light, S' is the flux density of radiation energy (energy flow through unit area perpendicular to the ray per unit time), C' is the radiation pressure cross section (a 3×3 matrix). Unit vector $\vec{e'_1}$ is directed along the path of the incident radiation. It is assumed that a beam of photons propagates in parallel lines and its orientation corresponds to the orientation of light propagation.

Eqs. (31) describe the equation of motion of the particle in the proper frame of reference due to its interaction with electromagnetic radiation. It is assumed that the energy E' of the particle is unchanged: the energy of the incoming radiation equals to the energy of the outgoing radiation, per unit time.

4.2. COVARIANT EQUATION OF MOTION – FIRST CASE

Let the components of the radiation pressure cross section (3×3 matrix C') be given in a basis defined by orthonormal vectors $\vec{e'_1}$, $\vec{e'_2}$ and $\vec{e'_3}$. We then may write

$$C' \, \vec{e'_1} = \sum_{j=1}^{3} \left(\vec{e'_j}^{T} \, C' \, \vec{e'_1} \right) \, \vec{e'_j} \, . \tag{32}$$

Substitutions $Q'_j \, A' \equiv \vec{e'_j}^{T} \, C' \, \vec{e'_1}$, $j = 1, 2, 3$ immediately yield Eq. (7) which has already been rewritten in the covariant form represented by Eq. (29).

4.3. COVARIANT EQUATION OF MOTION – SECOND CASE

We want to derive an equation of motion for the particle in the frame of reference in which the particle moves with actual velocity $\vec{v}$. We use the fact that we know this equation in the proper frame of reference – see Eqs. (31).

Inspiration comes from the fact that space components of four-momentum are written as a product with unit vector $\vec{e'_1}$. We know that this unit vector can be generalized to a four-vector

$$\begin{aligned} b^{\mu} &= (1 \, / \, w, \vec{e}_1 \, / \, w) \\ w &\equiv \gamma \, (1 \, - \, \vec{v} \cdot \vec{e}_1 / c) \, , \quad \gamma \equiv 1 \, / \, \sqrt{1 \, - \, (\vec{v} \, / \, c)^2} \, . \end{aligned} \tag{33}$$

Moreover, we know that $w^2 \, S \, / \, c$ is a scalar quantity – invariant of the Lorentz transformation.

The idea is to write the covariant equation of motion in the form

$$\frac{d \, p^{\mu}}{d \, \tau} = \frac{w^2 \, S}{c} \, G^{\mu \, \nu} \, b_{\nu} \, . \tag{34}$$

The problem is to find components of the tensor of the second rank $G^{\mu\nu}$.

In order to find $G^{\mu\nu}$, we proceed in two steps. First, we rewrite Eq. (34) in the proper frame of reference of the particle. Comparison with Eqs. (31) yields components of $G'^{\mu\nu}$. Second, we transform from $G'^{\mu\nu}$ to $G^{\mu\nu}$.

Eq. (34) yields

$$\begin{aligned} \frac{d\,E'}{d\,\tau} &= \frac{w^2\,S}{c}\left\{G'^{\,0\,0} - \sum_{j=1}^{3} G'^{\,0\,j}\,(\vec{e}_1')_j\right\}, \\ \frac{d\,(\vec{p}')_k}{d\,\tau} &= \frac{w^2\,S}{c}\left\{G'^{\,k\,0} - \sum_{j=1}^{3} G'^{\,k\,j}\,(\vec{e}_1')_j\right\}, \end{aligned} \tag{35}$$

where the term $1/w'$ in the brackets is omitted since it equals 1 in the proper frame of reference. Comparison with Eqs. (31) yields

$$\begin{aligned} &G'^{\,0\,0} = G'^{\,0\,j} = G'^{\,j\,0} = 0\,,\quad j = 1,2,3\,, \\ &G'^{\,k\,j} = -\,C'_{k\,j}\,,\quad j = 1,2,3\,,\quad k = 1,2,3\,. \end{aligned} \tag{36}$$

In order to find $G^{\mu\nu}$, we use the Lorentz transformation

$$G^{\mu\,\nu} = \Lambda_\alpha{}^{\mu}\,\Lambda_\beta{}^{\nu}\,G'^{\,\alpha\,\beta}\,, \tag{37}$$

where summation over repeated indices is supposed (and also in all the following equations) – 0, 1, 2, 3 for Greek letters and 1, 2, 3 for Latin letters – and

$$\begin{aligned} \Lambda_\alpha{}^{\beta} &= \eta_{\alpha\,\rho}\,\eta^{\beta\,\gamma}\,\Lambda^{\rho}{}_{\gamma}\,, \\ \eta_{\alpha\,\beta} &= diag(+\,1, -\,1, -\,1, -\,1)\,. \end{aligned} \tag{38}$$

The generalized special Lorentz transformation yields

$$\begin{aligned} \Lambda_0{}^{0} &= \gamma \\ \Lambda_0{}^{i} &= \Lambda_i{}^{0} = \gamma\,(\vec{v}/c)_i \equiv \gamma\,(\vec{\beta})_i\,,\quad i = 1,2,3\,, \\ \Lambda_i{}^{j} &= \delta_{i\,j} + (\gamma - 1)\,(\vec{\beta})_i\,(\vec{\beta})_j\,/\,\vec{\beta}^2\,,\quad i = 1,2,3\,,\quad j = 1,2,3 \end{aligned} \tag{39}$$

Inserting Eqs. (36) and (39) into Eq. (37), one can find all components of the tensor $G^{\mu\nu}$:

$$\begin{aligned} G^{0\,0} &= -\,\gamma^2\,(\vec{\beta})_i\,(\vec{\beta})_j\,C'_{i\,j} \equiv -\,\gamma^2\,\vec{\beta}^T\,(C'\,\vec{\beta})\,, \\ G^{0\,i} &= -\,\gamma\,(\vec{\beta})_j\,C'_{j\,k}\,\left\{\delta_{k\,i} + (\gamma - 1)\,(\vec{\beta})_k\,(\vec{\beta})_i\,/\,\vec{\beta}^2\right\} \\ &\equiv -\,\gamma\,(\vec{\beta})_j\,C'_{j\,i} - \gamma\,(\gamma - 1)\,\left[(\vec{\beta})^T\,(C'\,\vec{\beta})\right]\,(\vec{\beta})_i\,/\,\vec{\beta}^2\,, \end{aligned}$$

$$
\begin{aligned}
G^{i\,0} &= -\,\gamma\,(\vec{\beta})_k\,\left\{C'_{i\,k} + (\gamma - 1)\,(\vec{\beta})_i\,(\vec{\beta})_j\,C'_{j\,k}\,/\,\vec{\beta}^2\right\} \\
&\equiv -\,\gamma\,(C'\,\vec{\beta})_i - \gamma\,(\gamma - 1)\,\left[(\vec{\beta})^T\,(C'\,\vec{\beta})\right]\,(\vec{\beta})_i\,/\,\vec{\beta}^2\,, \\
G^{i\,j} &= -\,C'_{k\,l}\,[\delta_{i\,k} + (\gamma - 1)\,(\vec{\beta})_i\,(\vec{\beta})_k\,/\,\vec{\beta}^2] \times \\
&\quad [\delta_{l\,j} + (\gamma - 1)\,(\vec{\beta})_l\,(\vec{\beta})_j\,/\,\vec{\beta}^2] \\
&\equiv -\,C'_{i\,j} - (\gamma - 1)\,(\vec{\beta})_j\,(C'\,\vec{\beta})_i\,/\,\vec{\beta}^2 - \\
&\quad (\gamma - 1)\,(\vec{\beta})_k\,C'_{k\,j}\,(\vec{\beta})_i\,/\,\vec{\beta}^2 - \\
&\quad [(\gamma - 1)\,/\,\vec{\beta}^2]^2\,\left[(\vec{\beta})^T\,(C'\,\vec{\beta})\right]\,(\vec{\beta})_i\,\vec{\beta}_j\,. \qquad (40)
\end{aligned}
$$

It is worthwhile to mention that $C'_{i\,j} = A'\,\delta_{i\,j}$ leads to the symmetric tensor $G^{\mu\,\nu}$.

Instead of making algebra in putting Eqs. (40) into Eq. (34), we discuss verification. We want to prove that Eq. (34) yields results consistent with the more transparent Eq. (29).

4.4. CONSISTENCY OF THE COVARIANT FORMULATIONS

We have obtained the equation of motion in the form of Eq. (34) and we have determined the components of quantities on its right-hand side. Another form of covariant equation is presented in Eq. (29). Are these equations consistent?

Eq. (29) is the covariant equation of motion in the form corresponding to

$$
\left(\frac{d\,p^\mu}{d\,\tau}\right)_I = \frac{w^2\,S\,A'}{c}\,\sum_{j=1}^{3} Q'_j\,\left(b^\mu_j - \beta^\mu\right)\;; \qquad (41)
$$

it is supposed that $b^\mu_1 = b^\mu$. In this section we discuss an equation of motion of the form

$$
\left(\frac{d\,p^\mu}{d\,\tau}\right)_{II} = \frac{w^2\,S}{c}\,G^{\mu\,\nu}\,b_\nu\,. \qquad (42)
$$

We want to show that Eqs. (41) and (42) are equivalent, i. e., that $(dp^\mu/d\tau)_I = (dp^\mu/d\tau)_{II}$.

Multiplication of Eq. (41) by four-vector $b_{k\,\mu}$ and summation over μ yields

$$
\left(\frac{d\,p^\mu}{d\,\tau}\right)_I\,b_{k\,\mu} = \frac{w^2\,S\,A'}{c}\,\sum_{j=1}^{3} Q'_j\,\left(b^\mu_j\,b_{k\,\mu} - \beta^\mu\,b_{k\,\mu}\right)\;,\; k = 1,2,3\,. \qquad (43)
$$

It easily can be verified, on the basis of Eq. (33) and analogous equations for b^μ_2 and b^μ_3 (see Eqs. (27)), that

$$
\beta^\mu\,b_{k\,\mu} = 1\;,\; k = 1,2,3\,. \qquad (44)
$$

In calculations of $b_j^{\mu}\ b_{k\ \mu}$ we use the fact that this represents a scalar product of two four-vectors. Thus, its value is independent of the frame of reference. For the proper frame of reference

$$b_j^{\mu}\ b_{k\ \mu} = 1\ -\ \vec{e}_j^{\,\prime} \cdot \vec{e}_k^{\,\prime} = 1\ -\ \delta_{j\ k}\ ,\ j = 1,2,3\ ,\ k = 1,2,3\ , \tag{45}$$

since in the optics of scattering processes it is assumed that unit vectors $\vec{e}_1^{\,\prime}$, $\vec{e}_2^{\,\prime}$ and $\vec{e}_3^{\,\prime}$ are orthogonal. Thus, we obtain (inserting results of Eqs. (44) and (45) into Eq. (43))

$$\left(\frac{d\ p^{\mu}}{d\ \tau} \right)_I\ b_{k\ \mu} = -\ \frac{w^2\ S\ A'}{c}\ Q_k'\ ,\ \ k = 1,2,3\ . \tag{46}$$

Multiplication of Eq. (42) by four-vector $b_{k\ \mu}$ and summation over μ yields

$$\left(\frac{d\ p^{\mu}}{d\ \tau} \right)_{II}\ b_{k\ \mu} = \frac{w^2\ S}{c}\ G^{\mu\ \nu}\ b_{\nu}\ b_{k\ \mu}\ ,\ \ k = 1,2,3\ . \tag{47}$$

The value value is independent of the frame of reference. For the proper frame of reference the relations of Eq. (36) yield

$$\begin{aligned} \left(\frac{d\ p^{\mu}}{d\ \tau} \right)_{II}\ b_{k\ \mu} &= -\ \frac{w^2\ S}{c}\ C'_{j\ i}\ (\vec{e}_1^{\,\prime})_i\ (\vec{e}_k^{\,\prime})_j \\ &\equiv -\ \frac{w^2\ S}{c}\ (\vec{e}_k^{\,\prime})^T\ (C'\ \vec{e}_1^{\,\prime})\ ,\ \ k = 1,2,3\ . \end{aligned} \tag{48}$$

It was previously shown in section 4.2 that right-hand sides of Eqs. (46) and (48) are identical. Thus, also left-hand sides of Eqs. (46) and (48) are identical:

$$\left(\frac{d\ p^{\mu}}{d\ \tau} \right)_I\ b_{k\ \mu} = \left(\frac{d\ p^{\mu}}{d\ \tau} \right)_{II}\ b_{k\ \mu}\ ,\ \ k = 1,2,3\ . \tag{49}$$

If we take into account that four-vectors b_k^{μ} may be taken in various ways, Eq. (49) yields

$$\left(\frac{d\ p^{\mu}}{d\ \tau} \right)_I = \left(\frac{d\ p^{\mu}}{d\ \tau} \right)_{II}\ . \tag{50}$$

Thus, Eqs. (41) and (42) are equivalent, QED.

5. Conclusion

We have shown that the equation of motion for arbitrarily shaped particle under the action of electromagnetic radiation, a parallel beam of photons, can be written in the form of Eq. (29), or equivalently, by Eq. (34). For most common applications to systems in the universe (e. g., meteoroids in the Solar System, dust particles in circumstellar disks), the equation of motion in the form of Eqs. (30) is sufficient. The consequence of the general equation of motion may be a completely different orbital evolution of dust particles than is standardly used based on the P-R effect.

To treat electrically charged particles, one has to make a superposition of the right-hand side of Eq. (29) with a force acting on a charged particle in the presence of electromagnetic fields; Lorentz force acting on a point charge q in the presence of electromagnetic fields is $q \; F^{\mu \; \nu} \; u_{\nu}$, where $F^{\mu \; \nu}$ is the well-known field strength tensor.

Acknowledgements

This paper was partially supported by VEGA grant No. 1/7067/20.

References

1. A. Einstein, 1905, Zur Elektrodynamik der bewegter Kőrper, *Annalen der Physik* **17**, 891-920.
2. J. Klačka, 1992, Poynting-Robertson effect. I. Equation of motion, *Earth, Moon, and Planets* **59**, 41-59.
3. J. Klačka, 2000, Electromagnetic radiation and motion of real particle, http://xxx.lanl.gov/abs/astro-ph/0008510
4. J. Klačka, 2000, Aberration of light and motion of real particle, http://xxx.lanl.gov/abs/astro-ph/0009108
5. J. Klačka, 2001, Motion of electrically neutral particle in the field of electromagnetic radiation, *Meteor reports* **22**, 21-28.
6. J. Klačka, 2002, On equation of motion for arbitrarily shaped particle under action of electromagnetic radiation, http://xxx.lanl.gov/abs/astro-ph/0204148
7. J. Klačka, M. Kocifaj, 2001, Motion of nonspherical dust particle under the action of electromagnetic radiation, *J. Quant. Spectrosc. Radiat. Transfer* **70/4-6**, 595-610.
8. H. P. Robertson, 1937, The dynamical effects of radiation in the Solar System, *Mon. Not. R. Astron. Soc.* **97**, 423-438.

Participants

L.G. Astafieva, Stepanov Institute of Physics, Belarus National Academy of Sciences, F.Skaryna ave. 68, Minsk 220072, Belarus, astafev@dragon.bas-net.by

Victor A. Babenko, Stepanov Institute of Physics, Belarus National Academy of Sciences, F.Skaryna ave. 68, Minsk 220072, Belarus, babenko@dragon.bas-net.by

Sergey Beletsky, Astronomical observatory of Kharkov National University, Sumskaya St.,35, Kharkov, 61022, Ukraine, sbelets@mail.ru

John Robert Brucato, Astronomical observatory of Capodimonte, via Moiariello 16, I-80131 Napoli, Italy, brucato@na.astro.it

Daria N. Dubkova, Sobolev Astronomical Institute, Saint-Petersburg State University, 5th line V.O., 46-22, Saint-Petersburg, 199004, Russia, daria@DD8103.spb.edu, dasha@astro.spbu.ru

George J. Flynn, Department of Physics, SUNY-Plattsburgh, NY 12901, USA, e-mail: flynngf@splava.cc.plattsburgh.edu

Evgeniy S. Grynko, Dep. Remote Sensing of Planets, Astronomical Observatory, Kharkov National University, Sumskaya st., 35 Kharkov, Ukraine, 61022, egrinko@hotmail.com

Bo A. S. Gustafson, University of Florida, 211 Bryant Space Science Center P.O. Box 112055, Gainesville, Florida 32611, USA, gustaf@astro.ufl.edu

Thomas Henning, Astrophysical Institute and University Observatory (AIU) University of Jena, Schillergaesschen 3, D-07745 Jena, Germany, henning@astro.uni-jena.de

Alfons Hoekstra, University of Amsterdam, Kruislaan 403, 1098 SJ Amsterdam, The Netherlands, alfons@science.uva.nl

Helmuth Horvath, Univ. of Vienna, Institute of Experimental Physics, Boltzmanngasse 5, Vienna, Austria, HOR@AP.UNIVIE.AC.AT

Igor Kapišinský, Astronomical Institute, Slovak Academy of Sciences, Dubravska cesta 9, 842 28 Bratislava, Slovak Republic, e-mail: astrorom@savba.sk

Nikolai Kiselev, Astronomical Observatory of Kharkov National University Sumskaya str. 35, Kharkov, 61022, Ukraine, kiselev@linmpi.mpg.de

Jozef Klačka, Astronomical Institute, Comenius University, 842 48, Bratislava, Slovak Republic, klacka@fmph.uniba.sk

Miroslav Kocifaj, Astronomical Institute, Slovak Academy of Sciences, Dubravska cesta 9, 842 28 Bratislava, Slovak Republic, e-mail: kocifaj@astro.savba.sk

Jacek Krelowski, Torun Center for Astronomy, N. Copernicus University, Gagarina 11, Pl-87-100 Torun, Poland, jacek@astri.uni.torun.pl

Pavel Litvinov, Institute of Radio Astronomy of NASU, Chervonopraporna St.,4, Kharkov,61002, Ukraine, litvinov@ira.kharkov.ua

Andrzej Megier, Torun Center for Astronomy, Nicolaus Copernicus University, Gagarina 11, Pl-87-100 Torun, Poland, anmeg@astri.uni.torun.pl

Vito Mennella, Osservatorio Astronomico di Capodimonte, via Moiariello, 16 80131 Napoli Italy, mennella@na.astro.it

Michael Mishchenko, NASA Goddard Institute for Space Studies, 2880 Broadway, New York, NY 10025, USA, crmim@giss.nasa.gov

Fernando Moreno, Instituto de astrofisica de Andalucia, Cno. Bajo de Huetor, 24, 18008 Granada, Spain, fernando@iaa.es

Karri Muinonen, Observatory, P.O. Box 14, Kopernikuksentie 1, FIN-00014 University of Helsinki, Finland Karri.Muinonen@helsinki.fi

Olga Muñoz, Instituto de Astrofísica de Andalucía, P.O. Box 3004, Granada 18080, Spain, olga@iaa.es

Andriy Ovcharenko, Dep. Remote Sensing of Planets, Astronomical Observatory Kharkov National University, Sumskaya st., 35, Kharkov, Ukraine, 61022, ovcharenko@astron.kharkov.ua

Anri Perelman St. Petersburg Academy of Forestry Engineering, St. Petersburg, 194018 Russia, perelman@TV3107.spb.edu

Vladimír Porubčan, Astronomical Institute, Slovak Academy of Sciences, Dubravska cesta 9, 842 28 Bratislava, Slovak Republic, e-mail: astropor@savba.sk

Vera Rosenbush, Main Astronomical Observatory, NAS of Ukraine, Golosiiv, 03680 Kyiv-127, Ukraine, rosevera@mao.kiev.ua

Martin Schnaiter, Forschungszentrum Karlsruhe, P.O. Box 3640, D-76021 Karlsruhe, Germany, martin.schnaiter@imk.fzk.de

Dmitry Semenov, Astrophysikalisches Institut und Universitaets-Sternwarte, Schillergaesschen 3, D-07745, Jena, Germany, dima@astro.uni-jena.de

Yurij Shkuratov, Astronomical Observatory of Kharkov National University, 35 Sumskaya St. Kharkov, 61022, Ukraine, shkuratov@ygs.kharkov.ua

Martin Šolc, Astronomical Institute, Charles University, V Holesovickach 2, 180 00 Prague 8, Czech Republic, e-mail: martin.solc@mff.cuni.cz

Ján Svoreň, Astronomical Institute, Slovak Academy of Sciences, 059 60 Tatranska Lomnica, The Slovak Republic, e-mail: astrsven@ta3.sk

Victor Tishkovets, Astronomical observatory of Kharkov National University, Sumskaya St.,35, Kharkov, 61022, Ukraine, tishkovets@astron.kharkov.ua

Gorden Videen, Army Research Laboratory, AMSRL-CI-EM, 2800 Powder Mill Road, Adelphi MD 20783, USA, gvideen@arl.army.mil

Hester Volten, Amolf, Kruislaan 407, 1098 SJ Amsterdam, Netherlands, volten@amolf.nl

Nikolai Voshchinnikov, Sobolev Astronomical Institute, St. Petersburg University, Bibliotechnaya pl. 2, St. Petersburg - Peterhof, 198504 Russia, nvv@astro.spbu.ru

Iwan Williams, Queen Mary: University of London, Mile End Rd, London E1 4NS, U.K., I.P.Williams@qmul.ac.uk

Gerhard Wurm, Westfälische Wilhelms-Universität, Institute für Planetologie, Wilhelm-Klemm-Strasse 10, 48149 Münster, Germany, gwurm@nwz.uni-muenster.de

Evgenij Zubko, Dep. Remote Sensing of Planets, Astronomical Observatory, Kharkov National University, Sumskaya st., 35, Kharkov, Ukraine, 61022, zubko@astron.kharkov.ua

Index

GPSR Compliance

The European Union's (EU) General Product Safety Regulation (GPSR) is a set of rules that requires consumer products to be safe and our obligations to ensure this.

If you have any concerns about our products, you can contact us on ProductSafety@springernature.com

In case Publisher is established outside the EU, the EU authorized representative is:

Springer Nature Customer Service Center GmbH
Europaplatz 3
69115 Heidelberg, Germany

Batch number: 10683547

Printed by Printforce, the Netherlands